Systematic Training
Program Design

Systematic Training Program Design

Maximizing Effectiveness and Minimizing Liability

Sallie E. Gordon
University of Idaho

P T R Prentice Hall, Englewood Cliffs, New Jersey 07632

Library of Congress Cataloging-in-Publication Data

Gordon, Sallie E.
    Systematic training program design : maximizing effectiveness and
minimizing liability / Sallie Gordon.
      p. cm.
    Includes bibliographical references and index.
    ISBN 0-13-100389-5
    1. Employees—Training of—United States. 2. Educational
technology—United States. I. Title. II. Title: Training program
design.
    HF5549.5.T7G548   1994
    658.3′124—dc20   93-31031
                                                                      CIP

Editorial/production
  and interior design: *bookworks*
Acquisitions editor: *Michael Hays*
Cover designer: *Tweet Graphics*
Cover illustration from: *Stockworks, Stock*
  Illustration © *Rob Saunders*
Manufacturing manager: *Alexis R. Heydt*

 © 1994 by P T R Prentice Hall
Prentice-Hall, Inc.
A Paramount Communications Company
Englewood Cliffs, NJ 07632

The publisher offers discounts on this book when ordered
in bulk quantities. For more information, contact:

  Corporate Sales Department
  PTR Prentice Hall
  113 Sylvan Avenue
  Englewood Cliffs, NJ 07632

  Phone: (201) 592-2863
  FAX: (201) 592-2249

Printed in the United States of America

10 9 8 7 6 5 4 3

ISBN 0-13-100389-5

Prentice-Hall International (UK) Limited, *London*
Prentice-Hall of Australia Pty. Limited, *Sydney*
Prentice-Hall Canada Inc., *Toronto*
Prentice-Hall Hispanoamericana, S.A., *Mexico*
Prentice-Hall of India Private Limited, *New Delhi*
Prentice-Hall of Japan, Inc., *Tokyo*
Simon & Schuster Asia Pte. Ltd., *Singapore*
Editora Prentice-Hall do Brasil, Ltda., *Rio de Janeiro*

Contents

PREFACE xiii

1 INTRODUCTION 1

Training and Performance Support Systems 1
    Training: The Basic Concept 1
    Current Instructional Systems 2
    Performance Support Systems 3
Enhancing Performance: Training or (What)? 6
    Life's Little Complexities 6
    Priorities for Performance Enhancement 7
    Trainers as Design Team Members 9
System Design 9
    Engineering Design 9
    An Ergonomic Approach to Instructional System
      Design 10
Overview of the Design Model 11
    Phase I: Front-End Analysis 11
    Phase II: Design and Development 14
    Phase III: System Evaluation 16
    Iterative Processes 18

About This Book   19
     Philosophy of Instructional Design   19
     Text Organization   20
References   22
Additional Resources   23

2 PROCEDURE: ORGANIZATIONAL ANALYSIS   25

Rationale for Organizational Analysis   25
Types of Information to be Acquired   26
     When Current Employees Exhibit Low Performance
       Levels   27
     When New Employees Lack Necessary Knowledge
       and Skills   29
     When It Is a New Job/Task   29
Methods for Obtaining Information   32
     Interviews   32
     Questionnaires   33
     Observation   34
     Obtaining Information About Employee Performance   35
Summary   35
References   36
Additional Resources   36

3 BACKGROUND: KNOWLEDGE AND EXPERTISE   37

Types of Knowledge   38
     Declarative vs. Procedural Knowledge   39
     Subtypes of Declarative and Procedural/Skill
       Knowledge   42
     Rasmussen's Levels of Cognitive Processing   46
     Controlled vs. Automatic Processing   47
     Gagne's Five Kinds of Learned Capabilities   48
     Knowledge, Skills, and Abilities   50
     Terminology To Be Used in This Book   51
Mental Models   51
     Definition of a Mental Model   51
     Relationship Between Mental Models and Types
       of Knowledge   53
     Application of Mental Models to Training   54
Novice and Expert Performance   55
     Anderson's Three-Stage Model of Expertise   55
     Empirical Evidence for Models   56
     Characteristics of Expert Performance   56

References   58
Additional Resources   61

4 PROCEDURES: TASK ANALYSIS AND TRAINEE ANALYSIS   62

Traditional Task Analysis   63
    Task Lists and Hierarchies   63
    Declarative Knowledge and Skills   66
Cognitive Task Analysis   67
Task Analysis for Human Factors Design   69
Methods for Performing Task Analysis   69
    General Methods for Data Collection   70
    General Methods for Data Representation   81
    Specific Task Analysis Methods   90
Choosing Task Analysis Methods   108
Summary of Task Analysis   110
Trainee Analysis   110
References   111
Additional Resources   117

5 PROCEDURES: TRAINING NEED AND RESOURCE ANALYSES   118

Determining the Need, Finding the Best Answer   118
    Design Approach #1: System Design or Redesign   119
    Example of a System Redesign Solution   120
    Design Approach #2: Performance Support Systems
      and Job Aids   120
    Example of an Electronic Performance Support System
      Solution   122
    Example of a Job Aid Solution   123
    Design Approach #3: Training Programs   125
    Example of Combined Redesign, Performance Support,
      and Training Program   125
Training Resource Analysis   126
Summary   128
References   128

6 BACKGROUND: LEARNING AND MOTIVATION   129

A Basic Information-processing Model of Memory   130
    Sensory Memory   131
    Working Memory   132

Long-Term Memory   134
Acquiring Knowledge   135
    Acquiring Declarative Knowledge   136
    Roles of Declarative and Procedural/Skill Knowledge   137
    Acquiring Procedural/Skill Knowledge   139
Transfer of Training   141
    Identical Elements Theory of Transfer   142
    Positive and Negative Transfer   143
Part-task Training and Automation   144
    Sequencing of Instructional Materials   144
    Part-Task Versus Whole-Task Training   145
    Automation of Subtasks   146
Motivation   146
    The ARCS Model of Motivation   147
    Weiner's Attribution Theory   149
    ARCS Strategies   151
Principles and Guidelines for Instructional Design   153
    Attention   154
    Acquiring Declarative Knowledge   154
    Acquiring Procedural/Skill Knowledge   156
    Cognitive Workload   158
    Motivation   158
References   158
Additional Resources   163

7 PROCEDURE: WRITING FUNCTIONAL SPECIFICATIONS   164

Background   164
    Instructional Goals and Objectives   165
    Liability Issues   166
Functional Specifications   167
    Program Goal   168
    Post-training Goals and Objectives   170
    System Performance Requirements   172
    System Constraints   173
Final Comments   174
Summary   175
References   176

8 BACKGROUND: TRAINING SYSTEMS   177

Taxonomies for Instructional Techniques   178
Review of Instructional Techniques   180

      Text   180
      Audiovisual Techniques   182
      Lecture   184
      Inquiry Learning   186
      Tutoring   188
      Programmed Instruction   189
      Computer-Assisted Instruction   197
      Intelligent-Computer Assisted Instruction (ICAI)   201
      Simulations   203
      On-the-Job (OTJ) Training   205
Combining Methods   207
Training for Leadership and Interpersonal Skills   208
      Simulations and Business Games   208
      Role Playing and Behavior Modeling   209
      Human Relations Training: Group Interaction   210
Team Training   211
References   212

9 BACKGROUND: COMPUTER-ASSISTED INSTRUCTION 216

Background: Trends in CAI   216
More Technological Advances Multimedia   218
      Multimedia in a Nutshell   218
      The Technologies Behind Multimedia   219
      Implications for Training   222
CAI Designs   223
      Drill and Practice   223
      Tutorials   224
      Information Access and Generation   225
      Simulations   229
      Instructional Games   230
      Hybrid Designs   231
      Case Study: Antibody Identification for Blood
         Transfusion   232
Authoring Tools   233
      Levels of Authoring Tools   234
      Basic Programming Languages   234
      Midlevel Hybrid Tools   235
      High-End Graphical Interface Tools   238
      Comparison of Authoring Tools   240
Comments   240
References   241
Additional Resources   243

10 PROCEDURE: DEVELOPING THE DESIGN CONCEPT 246

Design Concept Generation 247
    Step 1: Generate General Idea of Instructional Design 247
    Step 2: Adjust the Design Concept(s) 249
    Step 3: Fill in Design Concept 250
Product Development Plan 253
Obtaining Project Approval 253
System Design Specifications 256
A Case Study 256
Final Comments 258
Summary 259
References 259

11 BACKGROUND: DOCUMENT AND COMPUTER INTERFACE DESIGN 261

Document Design 261
    Instruction and Job Aid Documents 262
    Factors to Consider in Document Design 263
    Principles of Document Design 264
    Document Evaluation Tools 272
Design of Job Aids 272
Interactive Computer Interface Design 280
    Static Screen Design 282
    Use of Color 284
    Designing for Interaction 285
References 288
Additional Resources 291

12 PROCEDURE: INITIAL DEVELOPMENT AND PROTOTYPING 293

Origins Of Prototyping 293
    Traditional Engineering Design 293
    Human Factors Engineering 294
Prototyping and Rapid Prototyping for Training Programs 295
    Traditional Instructional Design Models 295
    Incorporating Prototyping into the Instructional Design
      Model 297
    Rapid Prototyping 298
    Prototyping for Instructional Design 299

Designing Instructional Materials 302
    Determining Overall Instructional Content 303
    Determining Instructional Sequence 303
    Determining Unit Content 306
Development Methods for Major Instructional Techniques 309
    Tutoring, OTJ Training 310
    Lectures and Workshops 311
    Linear Documents 312
    Hardcopy Job Aids 313
    Videotape 314
    CAI and Nonlinear Electronic Documents 318
    Simulations 323
Combining Procedures 323
Summary 326
References 326
Additional Resources 329

13 PROCEDURE: FORMATIVE EVALUATION AND USER TESTING 330

Formative Evaluation 330
User Testing 331
    Variables Measured in User Testing 332
    Collecting Data for User Testing 334
User Testing For Instructional Strategies 335
    Tutoring and OTJ Training 335
    Lectures and Workshops 335
    Linear Documents 336
    Videotape 337
    CAI and Nonlinear Documents 338
    Job Aids and Performance Support Systems 340
Final Comments 341
Summary 341
References 342

14 PROCEDURES: FULL-SCALE DEVELOPMENT AND FINAL USER TESTING 343

Full-scale Development 343
    Tutoring and OTJ Training 343
    Lectures and Workshops 344
    Linear Documents 345
    Videotape 345

CAI and Nonlinear Documents   347
Job Aids and Performance Support Systems   348
Supporting Documents   348
Final User Testing   351
References   351

15 PROCEDURE: PROGRAM EVALUATION   353

Defining Program Evaluation   353
Importance of Program Evaluation   355
Accountability   356
Liability   357
Why Program Evaluation Is Neglected   357
Goals of Program Evaluation   359
Overview   359
Internal Validity   360
External Validity   363
Methods for Conducting Program Evaluation   364
Choosing Criteria   365
Research Designs: Giving Power to Our Conclusions   371
When to Collect Posttest Data   375
Choosing a Context   377
Jobs Aids   378
Summary   379
References   379

16 MINIMIZING PROFESSIONAL LIABILITY   382

Introduction To The Liability Issue   383
Recent Trends in the Courtroom   383
Courtroom Scenarios   384
Implications for Training Program Design   388
Summary   390
Guidelines For Trainers   390
Five Basic Rules   390
Design of Warnings   394
References   396

INDEX   399

Preface

This book provides principles and guidelines for the design of training programs and basic performance support systems within the context of a systematic instructional design model. The instructional design model is derived from the field of instructional technology, but also reflects recent research and practice in the fields of cognitive psychology, education, and most important, *human factors* or *ergonomics*. In ergonomics, a system is designed to support the user's informational needs, sensory and mental capabilities, and physical characteristics. Because it incorporates ergonomic methods and principles, the book presents an instructional design model that emphasizes the following principles:

- Designers should adequately evaluate the potential or need for system design (or redesign) and performance support *before* assuming that training is the solution to a performance problem.
- Training programs and performance support systems should be systematically designed and developed using a human-centered, or ergonomic, design philosophy. This means extensive involvement with trainee/users from the beginning, adequate task and environment analysis, application of ergonomic principles to document and system interface design,

and a strong emphasis on prototyping and user testing during design and development phases.

- Training programs should incorporate active learning and authentic skill performance to the extent possible. Evaluation should assess job performance or a reasonable facsimile, not just immediate learning.

The book presents, and is structured around, a three-phase instructional design model. The phases are front-end analysis, design and development, and system evaluation. The book provides background information and specific procedural guidelines for each phase. It is written to serve as a basic introduction to instructional design as well as a reference or job aid for professionals as they develop training and/or performance support systems.

In addition to the instructional design model and its associated procedures, the book also discusses a particular type of *liability issue* for trainers. More specifically, some training programs instruct people how to perform potentially dangerous tasks or activities. An example would be the owner's manual that accompanies a chain saw, or a videotape discussing forklift safety. A person who is subsequently injured from a task covered by the training or performance support program may sue the person or company who developed and/or delivered the training program (the exception to this is, of course, when the person is an employee of that company). The person or company who developed the training program may be found liable for damages. The last chapter of the book covers this liability issue, and provides some basic guidelines for designers to protect themselves against future litigation and findings of *professional liability*.

This book is intended primarily for three types of audience: (1) people in industry and government who are not professional trainers or instructional designers, but who nevertheless are responsible for the design of training programs and basic performance support systems; (2) training and instructional design students and professionals who wish to learn of the application of cognitive psychology and ergonomics to their field; and (3) anyone who is responsible for training tasks where subsequent injury or death is possible.

TEXT ORGANIZATION

The book has been written and organized to meet the needs of diverse audiences. As noted, the text presents a systematic instructional design model reflecting a strong human factors emphasis. The chapters contain specific guidelines and procedures, but also some of the background theory upon which the guidelines rest. This background material is necessary because at the current point in our field, some basic understanding of the background information is necessary in order to apply the procedures and guidelines.

Chapter 1 provides an introduction to the philosophies of the book and an overview of the instructional design model. Chapters 2 through 15 present the specific instructional design *procedures*, as well as appropriate *background* material provided in a just-in-time fashion. For example, Chapter 3 provides background material for Chapter 4. Once the reader has read and used the text, the procedure chapters can be used for reference and guidelines.

Each *background* chapter includes an overview of the content material, general guidelines when appropriate, references cited in the chapter, and additional resources for learning more about the chapter topic. Each *procedure* chapter includes an extensive presentation of the procedural steps for that particular phase of the design model, applicable guidelines, a summary of the procedures, references cited in the chapter, and additional resources.

It is my hope that by using this text and others listed in the resource sections, readers will be able to develop sound and effective training programs, in addition to basic performance support systems or job aids. And equally important, while the professional trainers of the world may take issue with this, I hope it will also encourage people to press toward the design of more user-friendly systems that don't require training at all.

ACKNOWLEDGMENTS

I would like to thank the many students at the University of Idaho who used the first draft of this book as a text for our graduate course on training program design. Their patience and feedback as students and "users" were invaluable. I would also like to thank many colleagues and reviewers for their comments and suggestions, and in particular, Brock Allen, for his encouragement in writing a book of this type and helpful feedback on the content. I am also indebted to Doug Chambers for creating the text illustrations with great professionalism and patience.

Finally, and most important, I thank my family for tolerating my absence, and especially my husband for both professional and moral support.

Sallie E. Gordon

Systematic Training
Program Design

Introduction

TRAINING AND PERFORMANCE SUPPORT SYSTEMS

Most of us spend a large proportion of our lives accumulating knowledge and acquiring skills. Some of this activity takes place through the course of normal daily events, some of it takes place through structured instructional programs such as those offered in high school and college, and much is eventually covered by training in industry and governmental agencies.

The knowledge and skills we acquire go on to serve us as we function in an increasingly complex world. Today's environment requires a great deal of specialized knowledge; from knowing how to navigate through a complex airport to understanding how to operate a programmable VCR. Thus, a great deal of our lives is spent acquiring basic knowledge and also more specific skills to allow us to successfully engage and function within this complex environment.

Training: The Basic Concept

We acquire knowledge and skills in a variety of settings, and this knowledge acquisition is variously termed learning, education, and training. What is the difference between these concepts? Traditionally, educators have been asked to teach *general knowledge* in formal educational institutions, and training professionals develop more specific *job-oriented* or *task-specific* training programs for employees. That is, training involves teaching information or procedures that are directly relevant to the performance of a particular set of tasks, such as driving a car.

In addition to *what* is taught, there has also traditionally been a difference in the length of time for instruction. Educational institutions have emphasized teaching of more general knowledge for longer periods of time. Training, on the other hand, has mostly consisted of short-term activities such as workshops that provide knowledge and skills for immediate use. Thus, educational instruction has traditionally been general and long-term, while organizational training has typically been specific and short-term.

While still accurate, this distinction between formal education and industrial training is becoming blurred. For example, recent concern expressed by the American public (Education Commission of the States, 1985; U.S. Department of Education; 1983) has caused educators to re-evaluate traditional methods of teaching. One trend is to bring in more specific experiential learning and real-world problems to augment the general, abstract topics typically taught in isolation (e.g., Carr, Eppig & Nonether, 1986; Esler & Sciortino, 1991; O'Neil, 1992). The emphasis on making abstract knowledge more tightly tied into concrete experience has resulted in an increase of more *specific* knowledge and skills integrated into the curriculum.

Likewise, industry has found itself teaching remedial courses for general skills such as reading and basic mathematics. And many organizations are beginning to take a "learning organization" perspective (Greenwood, Wasson & Giles, 1993), where employees are expected to identify their own relevant learning goals and pursue more long-term activities to support goal attainment. As a result of these changes in need and philosophy, we have seen a blending of the line that separates institutional education from organizational training.

Current Instructional Systems

Most countries are becoming acutely aware of the need for economic competitiveness in a global economy. Industrial leaders have made it clear that one important factor for attaining economic competitiveness is adequate levels of knowledge and skills in the work force. People entering the job market are often lacking in basic skills, and better education of the country's work force has become a high priority among industrial and congressional leaders (Office of Technology Assessment, 1986). Training has always been big business, but renewed emphasis is being placed on ways of finding more effective job skill training and also basic remediation of workers. This effort is being focused on both formal education and enhanced training in industrial settings.

Traditionally, training efforts have relied strongly on several instructional favorites: lecture, textbooks or student manuals, and on-the-job training. However, many job tasks and their requisite knowledge are becoming increasingly complex as we enter the "Information Age." This increase in job complexity is causing training professionals to turn to a wide range of educational methods to make their instructional programs more effective and more efficient. Some of these methods include intelligent computer-aided instruction, on-line help systems, simulations, and various types of interactive multimedia systems. Computer-delivered "virtual realities" are on the not-too-distant horizon. One of the goals of this book is to provide some basic guidelines for determining which instructional design solution is most appropriate for a given performance problem.

Performance Support Systems

While some new training technologies may be more effective and efficient than older methods, we still frequently see below-expectation-level job performance. While there are many reasons for this problem, two have to do with the changing nature of the jobs themselves. First, many jobs have simply become more cognitively complex. Consider the maintenance trouble-shooting required on the single prop planes of yesterday as compared with today's supersonic jets. The second factor is a related one; this is simply the vast amount of information that many employees must have at their disposal to perform their various job duties. For example, most employees who are responsible for phone-in customer support are required to either know a vast amount of information off the top of their heads or at least know where to find it within a few seconds.

Training programs alone are insufficient to address the amount and complexity of information retrieved and used in many such jobs. In addition to the sheer amount of material that must be retrieved and mentally integrated at the time of job performance, there is the additional problem of retention. A manufacturing firm may be able and willing to spend the time and money to train a person to know the attributes of 300 products, but realistically, how long will they retain that information? And how long will that information be accurate and current?

These difficulties and others related to system complexity have spawned an increasing number of systems designed to support a person *as they perform their job,* rather than teaching them the information or procedures ahead of time. I will refer to this class of systems as **performance support systems,** or simply "support systems." Performance support systems are any means of helping a person do their job, other than changing the actual task environment or interface. They are auxiliary systems that may or may not reside on a computer. Performance support systems are essentially informational or advisory in nature (as opposed to any support system such as a calculator).

One of the difficulties in discussing performance support systems is the fact that the term refers to an entire class of products or systems that range from very simple to very complex. Table 1.1 lists the most commonly used types of performance support systems. Each designer has a favorite term, with some being "in," and others being older terms (e.g., job aid).

As you may note, some of the terms seem to refer to essentially the same thing, and all except job aid usually mean some type of computer-based system. **Electronic performance support systems** (alternatively termed *information integration systems*) are a popular technology in the 1990s. They are some blend of factual and procedural knowledge, and usually combine training, job aid, and extensive on-line reference information into one system. Adaptive aiding and intelligent support systems are more complex or "intelligent" systems, capable of providing a high level of expertise and directive support for complex cognitive tasks.

Finally, in their original form, decision support systems were computer-based systems that supported the user in making choices or judgments by structuring problems, delineating alternative choices, identifying consequences, etc. They have more recently been expanded to include support of a wide variety of tasks including problem formulation and analysis, information retrieval and management, reasoning, representation, judgment, situational assessment, planning, execution, and monitoring (Rouse, 1991). In short, they support almost any information-processing task requiring information access, evaluation, and judgment.

It is often the case that the best design solution for enhancing performance is to provide some type of performance support system in addition to, or in place of, training. One of the things the designer must do is to evaluate the trade-offs between the two. This issue will be addressed at appropriate points in the following chapters.

The goal of this book is to present a structured set of principles, guidelines, and procedures for the sound design of any type of training program and/or job aid. It is secondarily meant to provide principles and guidelines for the design of basic performance support systems. In this text, I will use these terms as follows:

- **Training or instruction**—any learning activities occurring significantly before actual job performance
- **Job aid**—short and relatively simple hardcopy or electronic documents to be used while performing a task
- **Performance support system**—a more complex system that provides multiple types of structured information, on paper or computer, to support tasks during and/or immediately prior to their performance

TABLE 1.1   Alternative Types of Performance Support Systems

Term	Definition
Performance Support System	**Any "information" system that directly supports the user in performance of a task (at the time of task performance).**
Job Aid	Any product, on or off the computer, that provides assistance or support to a person performing a task. Job aids are usually very short and succinct. Examples include checklists, procedure guides, technical documentation, references, etc.
Electronic Performance Support System (Gery, 1991), also termed Integrated Information System (Johnson, Norton & Utsman, 1992)	A computer-based integration of one or more of the following: information databases, on-line reference, learning experiences and simulations, assessment systems, productivity software, and expert advisory systems.
Aiding[1] (Rouse, 1991)	Any functionality that is separate from, and added to, the basic system. Examples are an airplane autopilot, automobile automatic transmission, expert system to aid decision making, etc. It is any secondary system designed to help or enhance human performance.
Adaptive Aiding (Rouse, 1991)	Aiding that changes in nature and/or degree from one task to another, and from one user to another.
Intelligent Support System	Computer-based system that has an internal representation of the domain as well as the user to provide expert assistance to the user performing essentially any type of task.
Decision Aiding, Decision Support System	Computer-based system that supports information access, evaluation, and decision making or judgment.

[1]While Rouse defines aiding as any system that supports performance (such as an automatic transmission), he generally restricts his discussions to aiding systems that are computer-based and informational or advisory in nature.

The specific design of highly complex and *intelligent* performance support systems, such as adaptive aiding or intelligent support systems, is essentially beyond the scope of this text. Readers are referred to Rouse (1991), or Wickens (1992) for a more thorough coverage of this topic. For the sake of efficiency I will normally refer simply to the design of an instructional system. Most of the design principles will apply to the design of performance support systems, but where it is appropriate and necessary, I will address the two programs separately.

ENHANCING PERFORMANCE: TRAINING OR (WHAT?)

When a person becomes aware of a performance problem and/or training need, the first thing that they should determine is whether training is the appropriate or optimal solution. In this section, I introduce different ways that performance can be enhanced, and some general considerations for choosing among them. In later chapters, I give more specific procedures for making these design decisions.

Life's Little Complexities

Historically, any system to be used by people was first designed, tested, manufactured, and then finally given over to training analysts. The complexities inherent in use of the system were viewed by the engineering design team as being things that the owners and operators could overcome through "learning." While this approach has always made sense to system designers, there is an increasing awareness among certain professionals and most of the general public that something has gone awry (e.g., Harbour, 1993). Technology has marched forward, launching itself onto the masses without adequate regard for its success. Voss (1986) has noted that 40–70% of the technology implemented in manufacturing has been less successful than originally hoped. Norman (1992) writes about this clash between people and technological advances in an excellent book, *Turn Signals Are the Facial Expressions of Automobiles.* In the preface he states:

> I watch the way people interact with technology. I am not happy with what I see. Much of modern technology seems to exist solely for its own sake, oblivious to the needs and concerns of the people around it, people who, after all, are supposed to be the reason for its existence. (Norman 1992, ix)

The theme of Norman's book is how to humanize technology, and it should be read by anyone who is interested in, or deals with, the interaction between people and technology (which is most of us).

Consider the recent "improvements" made on the home VCR, resulting in a system known as a *programmable VCR.* These systems have become so "powerful," as the designers call it, that no one can use them. An article in a popular magazine referred to the programmable VCRs as having "buttons from Hell." This reflects the typical consumer's view of the absurd difficulty of using such systems. After two years, we saw two design solutions: use of a bar code system, and a commercial training program specifically designed to teach people how to use their programmable VCR! Two more years and we have a universal remote to walk us through the task.

The VCR design failure is unfortunate because it didn't have to be this way. It *is* possible to design even a complex VCR to be relatively easy to use, and that doesn't require a special training program or a retrofit remote control (Norman, 1988).

Why does a system such as the programmable VCR fail? One reason is that engineers simply don't adequately consider the needs of the system users; either because they haven't thought of it, they don't have the time, or they don't have the skills. As Norman (1986) explains, when a person must use some system, there is an interactive "gap" that must be bridged. A system that is very simple and easy to use has a small gap that the person must cross, for example, picking up a hammer and hitting something. A system that is complex and difficult to use has a large gap and the user has trouble crossing over that gap. That is, the user has trouble *translating* his or her own *goals* into the necessary actions that must be performed upon that system. When designers build a system, they usually focus on the *internal mechanisms* of the system, making it function to fulfill its stated purpose. The design rule seems to be: The more it does, the better the system. Designers rarely focus on building a system with a front or *interface* such that the person understands the system and knows how to use it to accomplish his or her goals. The end result is a complex system that people have difficulty operating (such as the VCR, most software programs, and any other number of systems). As we saw with the VCR, trainers are then brought in to "fix" this problem, when in reality the system and/or system interface should have been designed differently from the beginning.

Priorities for Performance Enhancement

In the previous section, we saw that often a system is designed where training is inappropriately used as a crutch. This suggests that there may be certain priorities in system design, and that training should be the last resort. As an analogy, a similar priority classification can be seen in the design of physical systems where safety is a concern. When an engineer designs a system, there are several methods of eliminating or controlling hazards. There is a standard priority schedule regarding which procedures are best (Hammer, 1989). In general, designers should strive to do the following *in the order listed:*

1. **Design the Hazard Out:** Design the system so that either (a) the hazard is eliminated entirely, or (b) the hazard is limited to a level below which it can do no harm.
2. **Safeguard:** Where safety by design is not feasible, protective safeguards should be employed.
3. **Warn:** When designing the hazard out and safeguarding are not feasible, automatic warning devices should be employed.
4. **Train:** When none of the above procedures are feasible, adequate training should be used.

An example of eliminating a hazard is cleaning oil from the floor as soon as it is spilled. An example of safeguarding is providing a chain saw with a chain guard. Some hazards, such as a child swallowing a balloon, do not allow a safeguard device, so warnings should be provided. It can be seen that for each product, an engineer proceeds through the priority list, using training only as a last resort. As an example, workers at a paper plant may have to walk around dangerous machinery at certain times. Their safety officer is responsible for training proper procedures for adjusting or turning off the machinery before moving into hazardous areas behind the safeguards.

In product liability suits, experts are asked to testify in court regarding the safety of the product. During this process, the original design engineers must show that they adequately followed the priority list described above. If they failed to design out the hazard, they must justify the decision (e.g., the equipment could not perform the job). Whatever approach they finally used, *warning* for example, they or an expert on the subject must show that they did so in a way that was in accordance with current standards and practices.

We can develop an analogous prescription for enhancing performance:

1. **Ease of use in the job/system itself**
2. **Provide performance support**
3. **Train**

When possible, design the task interface such that a person can use it without help. An example is design of a room thermostat or television control panel. People should be able to walk up to these systems and use them easily and correctly without external support (see Norman, 1988 for a discussion of the difficulties in doing this design task).

When it is not possible to design the interface to be usable without support, the next approach should be a job aid or performance support system available to guide the person through the task. For example, many gas pumps have a place where a customer can use a credit card to buy gas. These systems usually have a short procedural job aid placed directly on the pump, instructing the user on the steps necessary to perform the task.

When it is not possible to design the interface to support performance, and it is not possible to provide a performance support system, then training will be necessary. For example, a sales representative must know the basic skills of interacting appropriately with customers. There is no interface to design, and a performance support system for "social skills" would probably be unacceptable to both parties. In this case, instruction or training at some point before the actual task performance is most appropriate.

Finally, there will be many circumstances where some combination of these design solutions will be most efficient. For example, a medical receptionist may need to ask for certain information from patients on the phone. The basic information to be requested can be shown on the computer in front of him/her, along with form boxes to be filled in. In addition to this performance support system, the receptionist should receive training beforehand to provide an overview, rationale, and details of the task.

Trainers as Design Team Members

Given the priority system described, we can see that one important role for both human factors engineers and training analysts is to defend the position of future users by promoting the user-centered design of systems and their associated performance support systems. When trainers are members of the system design team, one of their most critical tasks *before* actually designing a training program is to make sure that the system itself is designed to optimally allocate functions to the physical components and to the human operator (with the system supporting and adapting to the person instead of the other way around). This means pushing designers to develop a system that is as easy to use and to learn as possible, given the functional goals and constraints of the system. Thus, a critical role of the training analyst is to make sure that much of the gap we described earlier between human and system is bridged by the *system itself*, and *not* by the user and the training program. The training can then focus on primary task performance, not on the use of a system that was incorrectly designed in the first place.

SYSTEM DESIGN

Engineering Design

In the design of physical systems, engineers first evaluate the environment in which the system will operate, the functions which should and should not be provided or supported by the system, and any constraints that will affect the design of the system. Design concepts are then developed to meet these requirements, evaluated using prototyping and other methods, and modified.

Recently, the engineering design process has been expanded in many domains to include a greater consideration of the end user's needs. In this ergonomic approach, the engineer takes into account the characteristics and capabilities of the people who must interact with the system, thus evaluating the human factors aspects (e.g., Sanders & McCormick, 1987; Wickens, 1992). An *ergonomic* approach to design places a large emphasis on user needs throughout the entire design process, and incorporates extensive user testing during the design and development phases.

An Ergonomic Approach to Instructional System Design

An instructional program is a product or system just as much as any physical system such as a chair, automobile, or software program. As such, it can be developed using design principles analogous to those we see used in engineering design. More specifically, the human factors principles, as well as general engineering design methodologies, can work well for training program design and development. If a user-centered design model is used to guide system development, the analyst is more likely to end up with a design product that meets the needs of an organization and its employees. Notice that this approach is entirely consistent with the recent emphasis on **Total Quality Management** in industry today (e.g., Deming, 1986; Gitlow & Gitlow, 1987).

This text uses an instructional design model that has been changed to reflect an ergonomic approach. We begin by adopting a generic instructional design model that is similar to most engineering design models (see Andrews & Goodson, 1980, for a review of standard instructional design models). Instructional design models typically include a front-end needs analysis phase, a design and development phase, and a final system evaluation phase. However, the design process is inefficient in that most of the user-based evaluation is done at the very end of the process. The evaluation comes at a time when development costs have already been incurred. It is really too late to make substantial modifications to the program.

The design model used in this text modifies a traditional instructional design model to reflect the ergonomic principles discussed above. For one thing, this means putting a great deal of emphasis on the task performer or system user from the beginning, and using input from them to aid in the design process during all phases. It also means following the priority schedule outlined above (design, support, train), and doing a significant amount of user testing in the early stages of design and development.

OVERVIEW OF THE DESIGN MODEL

The design model presented in this text is a highly structured three-phase model including **front-end analysis, design and development,** and **system evaluation** (see Figure 1.1). The model includes traditional instructional design techniques, such as needs assessment, but also incorporates useful ergonomic design methodologies such as writing functional specifications, prototyping, and user testing (e.g., Dreyfuss, 1974; Wilson & Wilson, 1965). The procedures contained within the model are briefly described in the following sections.

Phase I: Front-End Analysis

The first phase in the design process is performance of a thorough front-end analysis (sometimes referred to simply as "analysis"), which includes all of the initial work which should be performed before the program is actually developed. This phase consists of preliminary information gathering and analysis activities, and writing of the functional specifications document (see Figure 1.1). Just as an automobile designer must identify all of the relevant needs and constraints before designing the system (such as identification of the 95th percentile in driver arm length), so the trainer should identify needs and constraints before design of an instructional or performance support system. Throughout this text, it will be emphasized that this analysis should be done *before* the designer identifies any one particular training technique. Otherwise, the odds of using an inappropriate design solution are significantly increased.

     Needs assessment.   Needs assessment for instructional design has traditionally consisted of three activities, *organizational analysis, task analysis,* and *person analysis* (e.g., Goldstein, 1986; Latham, 1988). In this text, the term person analysis has been changed to *trainee analysis,* or *trainee/user analysis,* for the sake of clarity.[1] These three activities are briefly described below, and will be presented in detail in Chapters 2 and 4.

     Historically, **organizational analysis** has had two components: evaluation of the organization and/or industry in which the trainee performs the job, and evaluation of the organization expected to provide the training. These may or may not be the same institutions. In this book, I have broken the two analyses apart and treated them in separate procedures. Analysis of the orga-

---

[1]Some analysts use the term "performer." I personally feel that this term simply comes too close to describing what people and animals do in a circus.

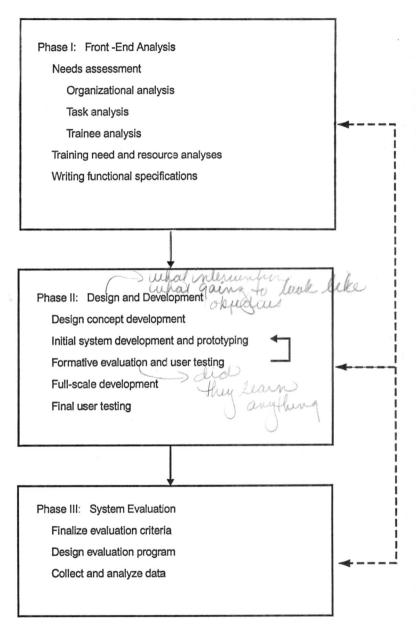

Figure 1.1   Design model for training and performance support systems.

nization which serves as a context for the job is termed **organizational analysis.** Evaluation of the *training* organization and associated resources are covered later in the step "Training need and resource analyses."

The organization providing the ultimate job setting is analyzed to determine whether training or performance support of some type is the most appropriate solution, *given the larger job context.* The analysis focuses on organizational goals, work climate and norms, etc. Sometimes a task may not be performed in an organization per se, as in the use of equipment in one's home. In this case, the organizational analysis may be omitted or replaced with a more appropriate evaluation of the context.

Task analysis is performed to determine the tasks required in a job, the subtasks performed for each task, and the knowledge and skills required for successful performance of the subtasks. This step is one of the most critical in the design process; if the knowledge and skill components are not adequately specified, the designer will have difficulty determining the necessary and sufficient contents for the instructional or performance support system. In addition, careful analysis of the task may reveal ways in which the task and/or system interface can be redesigned to eliminate the need for training or performance support in the first place.

Trainee analysis is performed to identify all relevant characteristics of the people who will be participants in the program. Several basic types of information are usually obtained. The first is demographic information such as age, gender, current and previous occupations, etc. The second, and most important type of information, is trainee knowledge and/or skills relevant to the potential program. A third type of information that is useful to the designer is what the potential trainees themselves regard as important; the analysis should identify trainees' perceptions of job-related needs, skill deficiencies, and other problems (e.g., Berryman-Fink, 1985).

Training need and resource analyses.   Once the Needs Assessment has been performed, the analyst must evaluate the various options for enhancing performance. These options include changing the nature of the job or task interface(s), providing performance support, and training, or some combination of these approaches. In this step, the analyst evaluates the "training need" and identifies the most appropriate option(s) for the given circumstances. If either training or performance support components are chosen, a **training resource analysis** is carried out. This step consists of identifying the resources available for development and delivery of the final system.

Writing functional specifications.   As noted earlier, when engineers design physical systems, they begin by describing functions the system must serve, under what conditions, and also any known constraints. This knowledge is written as a set of functional specifications. Similarly, for instructional design, the previous steps will have yielded information about the organizational goals and needs, employee goals and needs, and various constraining factors, such as availability of equipment, time, money, and instructional personnel. This information is transformed into a set of **functional design specifications**; a specific list of training goals, system requirements, and constraints that will all act to "bound" the design of the training program. This initial specification document supports the designer in generating appropriate design solutions. Later, after one or more design concepts have been generated, the functional specifications may be elaborated to include detailed and exact design specifications for program development.

Phase II: Design and Development

After the front-end analysis phase has been completed and the functional specifications written, the designer will have a firm definition of what must be accomplished by the training program or performance support system, and the factors constraining its final form. Given this knowledge, the analyst is ready to design and develop the system(s) as shown in Phase II of Figure 1.1. This phase takes place by performing several procedures briefly described below.

Generating the design concept. By considering the information generated in Phase I, the designer develops one or more design concepts. The design concept is a simple description of the overall approach to be taken by the instructional program and/or performance support system. It is at this point that the designer must evaluate the trade-offs between providing performance support or training (or both). More often than not, the design concept includes several complementary methodologies. Examples of preliminary design concepts include:

- A one- to two-hour videotape describing and demonstrating safe procedures for using a chain saw, followed by hands-on practice of each procedure with feedback from an instructor.
- A computer-based procedural checklist for task performance.

In addition to the basic idea, the design concept includes a description of one or more delivery methods as well as form and content of the actual material. In some cases, several alternative design concepts might be generated for preliminary testing and evaluation.

Initial system development and prototyping.   In this procedure, one or more mock-ups or prototypes are developed to test each basic design concept. A prototype in engineering is often a complete or near-complete version of the final product, although it might be much smaller. In instructional design, a prototype is a small and incomplete version of the final training system. However, it should have the look and feel of the final product to the extent possible.

Formative evaluation and user testing.   The goal of prototyping for training systems is to provide a product for **formative evaluation**—getting feedback from clients, managers, peer professionals, etc. It is also a product to be used for **user testing.** User testing consists of trying out the prototype with typical or representative users. It is performed to determine whether the training system as a whole, or various components of the system, are (a) easy to use and learn, (b) positively accepted by the learner, and (c) effective. The last thing a system designer wants is to find out after a system is fielded that the program is too difficult for learners to use, that trainees are not adequately learning the material, or that the program requires unrealistic retention by trainees. In other words, we want to know as early in the design cycle as possible that the program needs modification.

Another function of user testing is to determine whether certain instructional strategies are necessary and sufficient to accomplish the training goals. It will often be the case that the ideal instructional strategies for training of knowledge and skills are also the most expensive and time-consuming to carry out (such as use of real-time task performance, simulations, and microworlds). The designer may want to test various instructional options with actual learners to determine whether a less resource-intensive alternative will still result in satisfactory learning. Even if only one design alternative has been selected, user testing should be carried out to determine that the program will fulfill the needs identified in the functional specifications.

Based on the data obtained through user testing, the prototype is modified and subjected to further user testing. Iterative prototyping and user testing is a method that is very efficient for identifying a good design concept without full-scale system development. Prototyping and user testing is an iterative process that is usually repeated several times throughout the design and development stages.

I will stress throughout this book that while the training analyst may feel that he/she can reasonably evaluate the developing system without user testing, this is almost never the case. Designers are simply too close to the product to be able to see it from a user's point of view. Regardless of the type of system being developed, even if it is simply a procedural job aid or text, prototyping and user testing should be performed. Finally, notice that this iterative prototyping and user testing occurs *within* the middle design and development phase. It should *augment* the traditional analyze/design/evaluate instructional design model, rather than *replace* it (as Tripp and Bichelmeyer, 1991, have suggested).

Full-scale development.   After prototyping and user testing procedures, full-scale development of the program is begun. For physical systems such as automobiles, this means gearing up for the manufacturing process. For instructional programs, it means developing specific materials in addition to whatever has been done for the prototype. This work can range from writing lecture outlines to producing videotapes, creating simulations, and developing procedural job aids. Any computer-based interfaces are completed, intelligent tutoring modules are finalized, and so forth. This is usually the most time-consuming step of the entire design process. Iterative user testing often continues simultaneously with development to help make decisions on final program details.

Final user testing.   After the system is finished, it should be submitted to final user testing before being fielded or implemented. This is especially true for any computer-based training system, or electronic performance support system. Bugs in the programming, inconsistencies in interface design, poor organization of system elements, and other problems should all be identified by having representative trainees use the system in a very extensive manner. This process is performed in a fashion similar to the preliminary user testing except that trainees are asked to use all components of the full system, which obviously will take more time. In addition, measurement of task performance may sometimes need to be delayed for a period of time after training to more accurately reflect the conditions of fielded systems.

Phase III: System Evaluation

In our earlier example of automobile production, we discussed how designs are optimized through prototyping and testing of the prototypes. However, the ultimate goal of a car manufacturer is to sell a large number of automobiles. Therefore, the ultimate goal is to cause an increase in sales of the vehicle. It is tempting to say that the manufacturing company should just look at sales figures to evaluate whether their engineers did a good job of designing the car. Unfortunately, we cannot assume that an increase or decrease in sales is only caused by the success of the engineers in making the product safe, comfortable, user-friendly, and so forth. Obviously, the number of sales might be affected by many factors in addition to these design elements. For this reason, when manufacturers want to assess whether the engineers did an adequate job, they collect data that is closer to the questions they are asking. That is, they collect data that will bear more directly on whether their design has been successful at meeting its goals. For example, manufacturers measure the number of repairs on the vehicles, and they ask owners to fill out questionnaires at various periods of time after the date of purchase. Owners are asked to comment on various aspects of the product such as reliability, comfort, ease-of-use, etc.

A training or performance support system should be evaluated after it is fielded in the same way that engineering products are evaluated. That is, designers must determine whether a program has actually attained the ultimate goals. However, they must be cautious about using economic figures for the same reasons that the car manufacturers must be cautious. As an example, imagine that you have developed a program for car salespersons on how to interact more effectively with customers. It would be short-sighted to use number of car sales to evaluate the success of your program, because there are too many other factors that affect car sales. Better measures are those closer to the question you are really asking: "Are the salespeople doing a better job of interacting with customers?"

You can see that just as in engineering design, the last job of training analysts is to design ways of assessing the success of their product. While this is frequently not done in the training industry, it is becoming increasingly important. Many practitioners are not aware of this fact, but courts are now holding companies liable for failure to train employees adequately on the safe use and operation of equipment. This liability can extend to the people or company who developed the training program. So the final evaluation of a training program is necessary not only from a "good business" point of view, but also as a protection against litigation. A third reason for a complete evaluation of the fielded program is to determine, in a more molecular sense, which parts of the program were successful and which need revision.

All of these types of evaluation are performed through the procedures in Phase III (presented in Chapter 15). The designer must choose criteria to be used for evaluation, identify a method for collecting evaluation data, and analyze the data to form conclusions about training program effectiveness. These conclusions relate to the overall success of the program as well as more specific objectives. Based on the results, the designer may have to go back to some point in the design process and perform the process over again. In rare cases, this may require a completely new needs assessment along with all of the ensuing steps (see Figure 1.1).

Iterative Processes

Up to this point, the design steps shown in Figure 1.1 have been described as occurring in a relatively linear fashion. In reality, many steps in the design process are likely to be iterative because at certain points the designer will need to go back and work through some of the previous steps. However, some types of iteration are relatively routine, while other types of iteration signify that a procedure was probably not performed properly in the first place.

Iterations that can be expected to occur in the normal course of events are:

- Moving back and forth between the needs assessment steps in Phase I (organizational analysis, task analysis, and trainee analysis)
- Looping back through steps within Phase II (especially between prototyping and user testing)
- Going back from Phase II to Phase I to do more extensive task analysis as the system is designed
- Going back from Phase II to Phase I to do more extensive analysis of trainee knowledge as the system is designed

Other reasons for iteration are less common. For example, the designer may have to reiterate previous steps because of changes in the environment beyond his or her control: the job environment itself changes, company goals change, jobs are eliminated and added, demographics of employees change, and so forth. Especially in a technology-driven world, certain aspects of many jobs change frequently, or are performed with new and different equipment. Some training analysts have noted that this requires designers to focus on not only what is done in a job now, but also what skills will be needed by a worker in the future (Fossum, Arvey, Paradise, & Robbins; 1986; Latham, 1988).

Finally, the designer may have to iterate from Phase III back up to the other two phases. Under normal circumstances, feedback from the evaluation phase should confirm the original design. That is, if the first two phases are carried out properly, the chances of an unsuccessful program being fielded should be very small. However, sometimes modifications will be required.

In summary, the designer should progress through an analysis of the problem, determination of system requirements and constraints, concept development and user testing, and final system evaluation. It is extremely important, as I will repeat often, that the analyst generally follow the procedures in the *order specified.* Do not let the design process be *methodology driven* (e.g., Goldstein, 1986; Gagne, Briggs, & Wager, 1988; Johnston, 1987). That is, the designer should not decide at the beginning that he or she wants to use a particular delivery system, such as interactive multimedia, then proceed with the analysis phase. This procedure causes tunnel vision with resultant biased data collection and conclusions. Because of this bias, a person may fail to consider training methodologies that would be much more effective in attaining the true training goals. Carefully evaluate the problem and analyze the tasks to be taught, *then* choose one or more complementary instructional systems that will be most effective for accomplishing your unique goals.

ABOUT THIS BOOK

Philosophy of Instructional Design

Training is one of those fields where trends seem to perennially come and go. The trends are always accompanied by people who promote them and people who want to test to see whether the new methods are "better" than old ones. Most serious designers who have studied training and instructional design have concluded that it is *not technology* that really determines the effectiveness of a training program, but rather the soundness of the *instructional design process* itself (e.g., Gagne, Briggs, & Wager, 1988; Goldstein, 1986; Hannafin & Peck, 1988). As Hannafin and Peck (1988) state:

> As with text, videotape, lecture, and other educational media, the medium is neither good nor bad. A given computerized lesson may succeed or fail, depending on the expertise of the developer and the care with which it is developed. (p. 8)

The philosophy embodied here is that sound instructional design is necessary for any instructional program to be successful. And, in fact, this factor is usually more important than the particular training medium that is used. This is because, by default, if sound instructional design principles are used, the designer will choose instructional or other technologies that meet the functional requirements.

In this text, the instructional analyst will learn to evaluate the type of knowledge or skill that is needed, and determine the instructional activities, or performance support systems, that must be implemented for those specific needs. Choice of instructional medium should follow from that analysis, and not from any preconceptions that certain technologies are better than others. Therefore, nowhere in this text will you find statements suggesting that certain instructional methods are uniformly *best*. Each method has certain strengths and weaknesses, and must be evaluated within light of the unique needs of the project.

Finally, there are many skills helpful or even necessary for development of effective training and performance support systems that are not covered in this book. In particular, these are skills such as interviewing, graphical design, computer interface design, communications, writing, questionnaire development (testing and measurement), statistical data analysis, etc. The reader is encouraged to seek other sources such as texts, courses, workshops, or seminars to gain this knowledge.

Text Organization

The goal of this text is to present a structured method for training and instructional system design. The chapters are directly organized around the model presented in Figure 1.1. That is, the chapters progress from beginning to end of the system design process. However, several of the chapters present design methodologies and instructional strategies that can only be adequately understood if readers have certain background knowledge. For that reason, in addition to the **procedural** chapters, I have included a few chapters that explicitly provide what I view as essential **background** knowledge. The background knowledge is provided immediately prior to the procedural chapter where it is first needed, a *just-in-time* approach. The relationships between this background knowledge and the procedures of the design model are shown in Figure 1.2. The background chapters should be read carefully because the concepts and principles will be used repeatedly in the procedural chapter(s) that follow.

Each background chapter includes basic conceptual material needed to fully implement the procedures in the next chapter. Background chapters will also contain summary lists of principles and guidelines based on the conceptual material, where such guidelines are appropriate. Procedural chapters give a full explanation and discussion of the steps to be performed, with examples where necessary for comprehension. Most procedural chapters also have procedural summaries at the end to serve as a later reference or job aid.

Finally, in addition to the background and procedure chapters, there is a concluding chapter that reviews the issue of professional liability. There are several factors related to liability when one develops a training program or performance support system. Some of these have to do with personnel selected for training, or other processes that may promote some type of discrimination. The discussion of liability in this book only addresses a different type of litigation and professional liability. That category is when a person who was trained via a particular instructional system is injured, or injures someone else, while performing the tasks that were trained or supported.

BACKGROUND KNOWLEDGE   DESIGN PROCEDURES

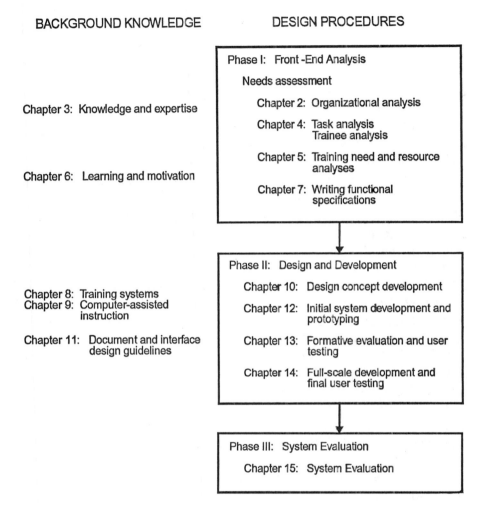

Figure 1.2   Book chapter organization relative to the design model.

Many instructional designers are not aware that they can be considered liable for their designs just as an engineer might be. If a designer does a poor job of developing and/or delivering a training program, and the trainee or another person is subsequently injured due to inadequate job performance, that designer (or more likely his or her firm) can be found professionally liable and held responsible for damages. Chapter 16 presents some of the means by which designers can protect themselves and their firm against findings of professional negligence.

REFERENCES

ANDREWS, D. H., & GOODSON, L. A. (1980). A comparative analysis of models of instructional design. *Journal of Instructional Development, 3(4),* 2–16.

BERRYMAN-FINK, C. (1985). Male and female manager's views of the communication skills and training needs of women in management. *Publication of Personnel Management, 14,* 307–313.

CARR, J., EPPIG, P., & NONETHER, P. (1986). Learning by solving real problems. *Middle School Journal, 2,* 14–16.

DEMING, W. E. (1986). *Out of the crisis.* Cambridge, MA: Massachusetts Institute of Technology, Center for Advanced Engineering Study.

DREYFUSS, H. (1974). *Designing for people.* New York: Grossman Publishers.

EDUCATION COMMISSION OF THE STATES. (1985). *Reconnecting youth: The next stage of reform.* Report of the Business Advisory Commission to the Education Commission of the States. Washington, DC: Author.

ESLER, W. K., & SCIORTINO, P. (1991). *Methods for teaching: An overview of current practices.* Raleigh, NC: Contemporary Publishing.

FOSSUM, J. A., ARVEY, R. D., PARADISE, C. A., & ROBBINS, N. E. (1986). Modeling the skills obsolescence process: A psychological/economic integration. *Acad Management Review, 11,* 362–374.

GAGNE, R. M., BRIGGS, L. J., & WAGER, W. W. (1988). *Principles of instructional design* (3rd ed.). New York: Holt, Rinehart and Winston.

GERY, G. J. (1991). *Electronic performance support systems.* Boston, MA: Weingarten.

GITLOW, H. S., & GITLOW, S. J. (1987). *The Deming guide to quality and competitive position.* Englewood Cliffs, NJ: Prentice Hall.

GOLDSTEIN, I. L. (1986). *Training in organizations: Needs assessment, development, and evaluation* (2nd ed.). Monterey, CA: Brooks/Cole.

GREENWOOD, T., WASSON, A., & GILES, R. (1993). The learning organization: Concepts, processes, and questions. *Performance & Instruction, 32,* 7–11.

HAMMER, W. (1989). *Occupational safety management and engineering* (4th ed.). Englewood Cliffs, NJ: Prentice Hall.

HANNAFIN, M. J., & PECK, K. L. (1988). *The design, development, and evaluation of instructional software.* New York: Macmillan.

HARBOUR, J. L. (1993). False hopes: The promise of technology. *Performance & Instruction, 32,* 30–33.

JOHNSON, W. B., NORTON, J. E., & UTSMAN, L. G. (1992). Integrated information for maintenance training, aiding, and on-line documentation. *Proceedings of the Human Factors Society 36th Annual Meeting* (pp. 87–91). Santa Monica, CA: Human Factors Society.

JOHNSTON, J. (1987). *Electronic learning: From audiotape to videodisc.* Hillsdale, NJ: Lawrence Erlbaum Associates.

LATHAM, G. P. (1988). Human resource training and development. *Annual Review of Psychology, 39,* 545–582.

NORMAN, D. A. (1986). Cognitive engineering. In D. A. Norman & S. W. Draper (eds.), *User centered system design: New perspectives on human-computer interaction* (pp. 31–61). Hillsdale, NJ: Lawrence Erlbaum Associates.

NORMAN, D. A. (1988). *The design of everyday things.* New York: Doubleday.

NORMAN, D. A. (1992). *Turn signals are the facial expressions of automobiles.* Reading, MA: Addison-Wesley.

OFFICE OF TECHNOLOGY ASSESSMENT. (1986). *Assessment, technology, and the American economic transition: Choices for the future* (OTA-TET-283). Washington, DC: U.S. Government Printing Office.

O'NEIL, J. (1992). Rx for better thinkers: Problem-based learning. *ASCD Update, 34,* 1–4.

ROUSE, W. B. (1991). *Design for success: A human-centered approach to designing successful products and systems.* New York: John Wiley & Sons.

SANDERS, M. S., & MCCORMICK, E. J. (1987). *Human factors in engineering and design* (6th ed.). New York: McGraw-Hill.

TRIPP, S. D., & BICHELMEYER, B. (1991). Rapid prototyping: An alternative to instructional design strategy. *Educational Technology Research and Development, 38,* 31–44.

U.S. DEPARTMENT OF EDUCATION (1983). *A nation at risk: The imperative for educational reform.* National Commission on Excellence in Education. Washington, D.C.: U. S. Government Printing Office.

VOSS, C. A. (ed.) (1986). *Managing advanced manufacturing technology.* Kempston, UK: Kempston.

WICKENS, C. D. (1992). *Engineering psychology and human performance* (2nd ed.). New York: Harper Collins.

WILSON, I. G., & WILSON, M. E. (1965). *Information, computers and system design.* New York: John Wiley & Sons.

ADDITIONAL RESOURCES

Human Factors and Ergonomics

BAILEY, R. W. (1989). *Human performance engineering: Using human factors/ergonomics to achieve computer system usability* (2nd ed.). Englewood Cliffs, NJ: Prentice Hall.

BROWN, O., & HENDRICK, H. W. (eds.) (1986). *Human factors in organizational design and management II.* Amsterdam: North Holland.

CASEY, S. (1993). *'Set phasers on stun' and other true tales of design, technology, and human error.* Santa Barbara, CA: Aegean.

HELANDER, M. (ed.) (1988). *Handbook of human-computer interaction.* The Netherlands: Elsevier.

MAYHEW, D. J. (1992). *Principles and guidelines in software user interface design.* Englewood Cliffs, NJ: Prentice Hall.

RUBINSTEIN, R., & HERSH, H. (1984). *The human factor: Designing computer systems for people.* Burlington, MA: Digital Press.

SALVENDY, G. (ed.) (1987). *Handbook of human factors.* New York: John Wiley & Sons.

Instructional System Design Models

DICK, W., & CAREY, L. (1985). *The systematic design of instruction* (2nd ed.). Glenview, IL: Scott-Foresman.

GAGNE, R. M., & BRIGGS, L. J. (1979). *Principles of instructional design* (2nd ed.). New York: Holt, Rinehart, and Winston.

REIGELUTH, C. M. (1983). *Instructional-design theories and models: An overview of their current status.* Hillsdale, NJ: Lawrence Erlbaum Associates.

ROBLYER, M. D. (1988). Fundamental problems and principles of designing effective courseware. In D. H. Jonassen (ed.), *Instructional designs for microcomputer courseware.* Hillsdale, NJ: Lawrence Erlbaum Associates.

Systems Engineering

BLANCHARD, B. S., and FABRYCKY, W. J. (1990). *Systems engineering and analysis* (2nd ed.). Englewood Cliffs, NJ: Prentice Hall.

2

Procedure:
Organizational Analysis

The first phase in the design process is front-end analysis. In essence, this phase involves needs assessment (analyzing the job, the person doing the job, and the job environment), identifying the best method for performance enhancement, and writing the results in a functional specification. The various types of analysis for the needs assessment are usually conducted somewhat simultaneously. For the purposes of clarity, each analysis activity will be described separately, with this chapter covering procedures for the organizational analysis (see Figure 1.1 for a graphical description of where organizational analysis fits into the big picture).

RATIONALE FOR ORGANIZATIONAL ANALYSIS

The first step in the needs assessment is to analyze the problem from a very broad perspective. The designer must ask, what is the root of the problem? If employees are performing poorly in a particular job, there may be a great many reasons. They may have the necessary knowledge and skills, but be unmotivated or otherwise uninclined to perform the task. They may face physical, emotional, or psychological barriers which severely hamper their job performance. And finally, their less than acceptable performance may stem from a lack of adequate knowledge and skills.

To assess all of these possibilities, the analyst starts by evaluating the entire job environment. This is termed an **organizational analysis.** Some training specialists include an evaluation of the *training organization* in the organizational analysis. We will be discussing that process as a separate step because in this particular design model, it occurs later in the design process. If the task occurs within an environment that is not an organization per se, the analyses which follow must be modified to fit the particular circumstances, or possibly omitted altogether.

TYPES OF INFORMATION TO BE ACQUIRED

The organizational analysis will vary depending on the particular circumstance that initiates the design process. Within a standard industrial setting, these circumstances generally fall into three categories:

1. There are currently workers who are failing to perform some aspect of their job at a satisfactory level, and the training program would be used to remediate these workers.

2. There are new employees who do not have the requisite knowledge and skills to perform a job, and the instructional program is to train them to do this job.

3. A new job (or job task) has been created and there are no workers who already know how to accomplish the job or task.

The initiating circumstance may be one, two, or all three of these situations.

Certain types of information should be obtained regardless of the initiating circumstance, while other types will be specific to the particular initiating circumstance. Therefore, the analyst should first determine which of the initiating circumstances are descriptive of the current situation and then obtain the appropriate types of information. In the three sections below, I list the initiating circumstance, and describe both the generic and circumstance-specific types of information that should be obtained.[1] These three sections are followed by a description of techniques for obtaining the information.

---

[1]This has the disadvantage of listing the same general background requirements in each of the three sections, but the reference advantage of having all information requirements in one place for each of the initiating circumstances.

When Current Employees Exhibit Low Performance Levels

The job environment can be conceptualized as an envelope that exists around all of the job tasks. This envelope impacts job performance in a variety of ways, both physically and psychologically, positively and negatively. For example, a janitor may be unable to perform a job because some portion of the necessary equipment is not provided. He/she may not perform well because there are no incentives. I once heard of a situation where janitors are rewarded for good performance by being moved to locations where there is more work, and the work is more difficult. The employees letting me in on this fact made it very clear that they deliberately performed at a mediocre level (not their exact words) because of these disincentives for good performance.

All of this is to show that the analyst must be very careful to look for *any* factors that might lower performance. Any factors other than lack of knowledge or skill should be identified because if they are not, costly and time-consuming training programs might be implemented with no impact on job performance. This analysis can be done in a variety of ways. One place to start is to determine the extent to which the employees *cannot* perform the job tasks at an acceptable level as opposed to finding that employees *do not* perform them at acceptable levels. If they can perform them but do not, then the analyst should attempt to identify the causal factors (or at least report that lack of knowledge and skill is not the cause).

In summary, the overall goal for this analysis is to determine characteristics of the organization that currently do, or in the future will, impact the worker. Data should be obtained for the job environment at two levels: the organization as a whole, and the employee's unit, including management and the employees themselves. For the organization as a whole, the analyst should identify the general goals of the organization, the organizational structure, and where the organization is headed. If the training system is not targeted for a specific organization, but rather simply a certain job or job class (such as tax accountant), the analyst should identify the goals and trends typical of the industry in which the job takes place. Similar analyses should be conducted to identify goals and attitudes within the employee's unit, for both managers and employees. The top two sections in Table 2.1 list some of the important types of information that should be gathered.[2]

In addition, the analyst identifies factors associated with job perfor-

---

[2]If the project involves both low performance levels and new employee training, use the analysis described in this section.

Below acceptable

**TABLE 2.1   Information to Obtain for Project Where Current
Employee Performance Is Below Acceptable Levels**

The Organization
- Goals of the organization as a whole
- Methods by which the organization hopes to accomplish those goals
- Values of the company/industry in general
- Policies relevant to the job in question
- Policies relevant to performance enhancement and training
- Trends in the company/industry as a whole
- Trends in the organizational structure
- Perceived performance-related problems within the organization

The Organizational Unit
- Management goals
- Management values
- Management written and unwritten policies
- Attitudes of managers; what are their perceptions of actual and expected task performance levels
- Trends in the target job, and jobs above or below it in the company structure (e.g., are they considering automating part of the job or augmenting the worker with high-level computer software, etc.)
- Unit policies regarding performance enhancement and/or training
- Management attitude toward training and/or performance support systems (e.g., will managers be supportive of the program and skills taught, how much would they be willing to invest in training of potential or current employees)
- Implicit and explicit expectations and attitudes of employees regarding their job, job environment, company as a whole, managers, peers, etc.

Job Performance
- Job performance level of which employees are capable
- Job performance level at which employees are currently working
- Job performance level expected by management
- Job performance level expected by employee and peers
- Job performance level employees perceive that they could conceivably attain, and circumstances that would result in those levels
- Factors that managers feel interfere with job performance
- Factors that employees feel interfere with job performance level, including physical, social, political environments, lack of knowledge and skills, etc.
- Management and employee attitudes regarding safe or effective task performance
- Stories and anecdotal information about other employees who have had difficulties in task performance; causes of difficulties

mance, what employees are capable of doing and what they are actually doing. Once previous and current levels of performance have been established, the analyst must determine what the organization and management perceive to be acceptable knowledge and performance for that job class. The analyst can then attempt to use various tests of knowledge and skills to determine whether employees do in fact have the knowledge and skills but are simply not using them.

The goal of this analysis is to try to rule out any factors other than deficiencies in knowledge and skills that might cause poor performance. One other possibility should be noted. Sometimes, performance is low because of worker beliefs and attitudes rather than ability per se. If such a situation is identified, it is possible to design training programs that are targeted at changing employee (and perhaps management) attitudes rather than knowledge and skills.

When New Employees Lack Necessary Knowledge and Skills

This initiating circumstance is characterized by current employees performing the job at acceptable levels, and the existence of new employees who will need some amount of training for the job.[3] This training might take place within the organization itself, or in an educational institution. In this case, there is less of a need to look for organizational factors that impact job performance. However, it is still necessary to evaluate the organizational climate, particularly regarding values, goals, and attitudes toward various training approaches. The analyst will also obtain information about the attitude of current employees toward the job, whether there is any aspect of the job that is problematic that might be addressed in the training program, and other similar questions. Table 2.2 lists the typical types of information that would be collected for this initiating circumstance.

When It Is a New Job/Task

In this third circumstance, there is no "deficiency" in knowledge or skill so much as a need to train for some particular job or task from scratch. In this case, in addition to the general types of question seen earlier (such as evaluation of the organizational values) information should be collected concerning the role of the new job or task within the context of the surrounding organiza-

---

[3]If the project involves both low performance levels and new employee training, use the analysis described in the previous section.

new

TABLE 2.2   Information to Obtain for Project Where Employee Performance Is Adequate, but Training Is Needed for New Employees

The Organization
- Goals of the organization as a whole
- Methods by which the organization hopes to accomplish those goals
- Values of the company/industry in general
- Policies relevant to the job in question
- Policies relevant to performance enhancement and training
- Trends in the company/industry as a whole
- Trends in the organizational structure

The Organizational Unit
- Management goals
- Management values
- Management written and unwritten policies
- Attitudes of managers; what are their perceptions of actual and expected task performance levels
- Trends in the target job, and jobs above or below it in the company structure
- Unit policies regarding performance enhancement and/or training
- Management attitude toward training and/or performance support systems (e.g., will managers be supportive of the program and skills taught, how much would they be willing to invest in training of potential or current employees)
- Implicit and explicit expectations and attitudes of employees regarding their job, job environment, company as a whole, managers, peers, etc.

Job Performance
- Job performance level at which employees are currently working
- Whether current job performance is at the level expected by management
- Factors that employees and/or managers feel interfere with job performance, including physical, social, political environments, lack of knowledge and skills, etc.
- Management and employee attitudes regarding safe or effective task performance

tion. The reason is to ensure that the trainees have a positive work environment and ultimately perform the job satisfactorily. Ask questions such as: Is the position or task being implemented with a positive attitude by all employees, some employees, or virtually no one? Is it being foisted on the company by only one or a few persons? Will there be adequate resources and support equipment? and other similar questions.

The physical and psychological climate should be evaluated in a *predictive* fashion, where the training analyst must try to *foresee* any factors that might undermine performance despite a successful training program. Again, factors that are known to impact job performance should be sought out. If any are found, they should be brought out and addressed before implementation of training, or the program may be undermined. Table 2.3 lists the major types of information that should be sought for training, or other type of support, for a new task.

TABLE 2.3   Information to Obtain for Project Where Employees Must Learn New Task

The Organization
- Goals of the organization as a whole
- Methods by which the organization hopes to accomplish those goals
- Values of the company/industry in general
- Policies relevant to the new job or task in question
- Policies relevant to performance enhancement and training
- Trends in the company/industry as a whole
- Trends in the organizational structure

The Organizational Unit
- Management goals
- Management values
- Management written and unwritten policies
- Reasons for implementation of the new job or task
- Importance or priority of new job or task
- Attitudes of managers toward the new job or task
- Trends in the target job, and jobs above or below it in the company structure
- Unit policies regarding performance enhancement and/or training
- Resources being provided to train the new task
- Management attitude toward training and/or performance support systems
- Implicit and explicit expectations and attitudes of employees regarding the new job or task (whether it is helpful, difficult, irrelevant, likely to create a work overload, etc.)

Job Performance
- Job performance level most likely to be expected by management
- Job performance level most likely to be expected by employee and peers
- Job performance level employees perceive that they could conceivably attain, and circumstances that would result in those levels
- Factors that managers predict might interfere with job performance
- Factors that employees predict might interfere with job performance level
- Management and employee attitudes regarding safe or effective task performance

METHODS FOR OBTAINING INFORMATION

The information described in the previous three sections can and should be gained by using a number of different methods. Some of the more frequently used ones are listed in Table 2.4. The approach that is probably most efficient is a combination of document analysis, interview, and questionnaire. That is, the analyst would start by obtaining and analyzing organizational documents relevant to the items listed above. Once this has been completed, a few key interviews will yield much of the relevant information. This information may then be verified through a larger number of people via questionnaires.

Interviews

Interviews are almost a necessity for performing the organizational analysis. There are few methods that yield as much useful and relevant information. Interviews can be used to obtain objective information as well as *perceptions*

TABLE 2.4　Methods for Obtaining Data for Organizational Analysis

Document Analysis
- Organizational statements of mission and long-range goals
- Organizational policies and procedures
- Budgets
- Productivity reports
- Performance appraisals
- Annual reports
- Financial reports
- Annual reviews
- Schedules
- Training materials

Interviews and Questionnaires
- Upper management
- Unit management
- Work groups
- Individual employees
- Prospective/new employees

Observation of Employees
- Observation of employees who are performing below expectation levels
- Observation of "experts"

Testing Written or performance tests of knowledge and skills

Observation/Analysis of Physical or Social Environment

of problems and difficulties relevant to the job under consideration. Face-to-face interviews, if done professionally, can increase interest in the project and elicit sensitive information.

Before conducting any interview, the analyst should contact the interviewee to explain the purpose of the interview, obtain permission, and set up the meeting. A set of questions should be developed to structure the interview. The questions can be focused around the topics listed in the tables above, with some items omitted depending on the document analysis. However, it is usually best to obtain all information types in the interview, because it is common for an individual's answers to differ from the "official" company statements.

Interviewing requires certain skills that are difficult to convey in a text such as this. Readers are encouraged to obtain appropriate training if they will be performing this task frequently. In general, there are certain rules that apply to interviewing (e.g., see Nilson, 1989; Singleton, Straits & Straits, 1993):

1. Keep the interview situation as *private* as possible.
2. Put the respondent at ease, using a *relaxed* and natural conversational style.
3. Assure the respondent that all information will be kept *confidential.* One way to do this is to assume the policy that general results will be available to management and relevant other personnel, but no data will be associated with any one particular person.
4. Treat the respondent with respect, and be *courteous, tactful,* and *non-judgmental.* Your job is to elicit interviewee's ideas without changing them.
5. *Dress* and *act* professionally but in a way that is appropriate given the position of the interviewee.
6. Ask the questions in their proper *sequence,* in the order written.
7. *Do not assume* the answer to any question, or put answers in the respondent's mouth.
8. Use positive words of *encouragement* such as "yes," "good," "I see," "that's a good observation," etc.
9. *Don't allow* respondents to fall into a *trend of complaining* or use you to air all of their gripes.

Questionnaires

While information from a few key interviews is often enough, sometimes it is important that the information be valid for others as well. In this case, a questionnaire can be developed based on the information obtained from documents and interviews. This questionnaire can serve to validate the findings obtained thus far, and can also be used to finish obtaining more detailed types of information.

A questionnaire for organizational analysis usually consists of several closed-ended questions and a fewer number of open-ended questions. Closed-ended questions are items that require the respondent to choose from a set of alternatives; examples include rating scales, yes/no items, multiple choice, etc. Open-ended questions allow respondents to answer the question in their own words. Table 2.5 shows examples of closed-ended and open-ended questions for obtaining information relevant to organizational unit goals. In writing the questionnaire, follow guidelines such as those provided in Singleton et al. (1993). It is a good idea to write the questionnaire and try it out on one or two people before sending it out to the entire sample of respondents.

Observation

Direct observation of job performance as well as the job environment will often yield information not obtained through interviews or questionnaires. For example, an analyst might observe the line workers at a paper mill and notice that certain informal "standards" exist for what is and is not safe behavior. If the employees were asked about what constitutes safe behavior, they may respond with answers consistent with formal company policies.

TABLE 2.5   Examples of Closed-ended and Open-ended Questions

Open
What are the goals of your department?
What are the methods by which you accomplish those goals?
What are the most important duties performed by your sales representatives?

Closed
Please rate the importance of the following duties by placing an X on the appropriate line:
Check product updates daily

| Not at all important | __ | __ | __ | __ | __ | __ | __ | Very Important |

Attend weekly briefings

| Not at all important | __ | __ | __ | __ | __ | __ | __ | Very Important |

Understand customer needs

| Not at all important | __ | __ | __ | __ | __ | __ | __ | Very Important |

Obtaining Information
About Employee Performance

To obtain information relevant to employee performance, more creative and energetic tactics must often be used. To begin, managers can be questioned or interviewed to determine current employee performance levels as well as what they consider to be acceptable performance. Employees can be interviewed and/or observed, and any relevant records can be assessed to determine current levels of task performance.

However, it is often also necessary to determine what employees are *capable* of doing. Obviously, this might be a difficult thing to determine; the employee is certainly going to be biased *against* showing anyone that he/she is more capable than normally exhibited on the job. However, if the analyst can establish that the worker does in fact already have the prerequisite knowledge and skills, then this can be used as convincing evidence that training is not the answer that is needed. One way of accomplishing this is to give the employee a paper-and-pencil test of knowledge and skills. This must be done with an accompanying innocuous rationale for the test.

If the analyst cannot directly measure the worker's underlying ability to perform the job, there are other methods of evaluating the organizational environment and its impact on performance. One of the most direct is to simply interview the employees. Ask them for their opinion: Do they think they are performing the various job tasks at a reasonable level? If not, why not? Depending on the circumstances, the designer might have to promise not to release the content of these interviews to the organization.

Another means of evaluating the job context is to observe employees performing the job. Take special care to notice any physical circumstances which might inhibit performance. Also, note any social or psychological events that might impact the employee, such as overly heavy workloads, sexual harassment, lack of rewards, etc. Texts in the area of industrial/organizational psychology can be useful in providing ideas about what factors will be most likely to affect worker performance.

SUMMARY

To summarize, organizational analysis consists of essentially two steps:

1. **Determine the initiating circumstance (what seems to be the problem?).**
2. **Given the initiating circumstance, obtain the appropriate informa-**

tion that will be most useful in identifying solutions to the problem. This includes information about organizational goals and policies, management values and perceptions, and employee values and perceptions relevant to job performance.

The first three tables in this chapter provide a guideline for the information to be collected; however, the analyst should also be open to other types of information because each case is unique. The data is typically collected by using a mixture of methods, including document analysis, interviews, questionnaires, and observation. Results of the organizational analysis will then be summarized in the functional specifications, as described in a later chapter.

REFERENCES

NILSON, C. (1989). *Training program workbook and kit.* Englewood Cliffs, NJ: Prentice Hall.

SINGLETON, R. A., STRAITS, , B. C., & STRAITS, M. M. (1993). *Approaches to social research.* (2nd ed.). New York: Oxford University Press.

ADDITIONAL RESOURCES

FOSSUM, J. A., ARVEY, R. D., PARADISE, C. A., & ROBBINS, N. E. (1986). Modeling the skills obsolescence process: A psychological/economic integration. *Academy of Management Review, 11,* 362–374.

GOLDSTEIN, I. L. (1986). *Training in organizations: Needs assessment, development, and evaluation* (2nd Ed). Monterey, CA: Brooks/Cole.

HUSSEY, D. E. (1985). Implementing corporate strategy: Using management education and training. *Long Range Plan, 18(5),* 28–37.

ROTHWELL, W. J., & KAZANAS, H. C. (1992). *Mastering the instructional design process: A systematic approach.* San Francisco, CA: Jossey-Bass.

TESSMER, M. (1990). Environmental analysis: A neglected stage of instructional theory. *Educational Technology Research and Development, 38,* 55–64.

WEXLEY, K. N., & LATHAM, G. P. (1991). *Developing and training human resources in organizations* (2nd ed). Glenview, IL: Scott, Foresman.

3

Background:
Knowledge
and Expertise

Training programs are products and activities whose goals are to support the acquisition of knowledge and skills, and also sometimes to affect trainee attitudes. The design of an instructional system should ideally be based on what is currently known in the areas of cognition, education, technical communication, artificial intelligence, and human factors. Unfortunately, more is known in these areas than could be fit in a dozen books this size. Therefore, I will provide just a brief overview of certain theoretical work that is useful in the development of training programs. For the sake of simplicity, I will divide the information into two general subject areas, to be termed "knowledge" and "learning."

Knowledge will refer to the mental facts, rules, procedures, skills, and strategies one relies upon in performing a task. By understanding something about the basic types of knowledge that experts and novices use in performing tasks, we can enhance the processes involved in front-end analysis, instructional system design, and program evaluation. The term **learning** will be used as a broad umbrella covering material on how people learn or acquire knowledge and skills, and the factors that support or interfere with learning and skill acquisition processes.

While it is certainly possible to simply specify some rules and guidelines for the designer to apply during front-end analysis and system design, I feel that it serves the designer best in the long run to be familiar with certain terms and research findings relevant to instructional design. This allows us to discuss specific methods for promoting specific types of knowledge, where knowledge is referred to using explicitly defined terms. In addition, knowledge of these terms and concepts will allow the designer to more easily read the current training and educational publications.

Because it is also my goal to constrain this text with respect to theoretical material, I have selected a relatively small and representative sample of work on the topics of knowledge and learning that appears to bear most directly on the design of training programs and performance support systems. A review of the work related to *knowledge* is presented in the remainder of this chapter under sections entitled types of knowledge, mental models, and novice vs. expert performance. The material on how people learn and factors affecting learning will be presented in Chapter 6.

TYPES OF KNOWLEDGE

As researchers have studied the nature of knowledge and how we acquire it, they have sometimes found it useful to categorize knowledge into qualitatively different types. As a result, there are taxonomies which have been developed in fields such as cognitive psychology, linguistics, artificial intelligence, education, training, and human factors. These taxonomies have become increasingly used as a basis or rationale for some particular type of training program design. Unfortunately, because there are so many different taxonomies, it is often difficult to remember the meaning of categories for any one particular model. In addition, sometimes a term, such as "procedural knowledge," is used in different ways.

Most models take essentially the same knowledge "pie" and simply cut it up into different numbers of pieces, or make the cuts in different places. Figure 3.1 shows how the knowledge pie is cut by various researchers and their taxonomies. In the left-hand column, I have divided knowledge into enough pieces to accommodate all of the taxonomies. In the remaining columns, I have listed five well-known taxonomies, with the key researcher's name at the top. Each column shows how that particular taxonomy groups all of the individual or separate types of knowledge shown on the far left. The sections below describe each of the five taxonomies. After I describe the first basic distinction in Figure 3.1, Anderson's declarative vs. procedural knowledge, I will explain the individual units listed on the left. Then I will continue with a description of the remaining four models.

Knowledge Categories:	Authors				
	Anderson	Rasmussen	Schneider Shiffrin	Gagne	Prien
DECLARATIVE CONCEPTUAL	Declarative	Knowledge-based processing	Controlled	Verbal information	Knowledge
DECLARATIVE RULE		Rule-based processing			
PROCEDURAL/SKILL	Procedural	Skill-based processing	Automatic	Intellectual skills	Abilities
Automated Rules					
Perceptual Skills					
Domain-general Strategies				Strategies	
Domain-specific Strategies					
Psychomotor Skills			Controlled ↓ Automatic	Motor skills	Skills
ATTITUDES	—	—	—	Attitudes	—

Figure 3.1 Various taxonomie of types of knowledge and cognitive processing.

Declarative vs. Procedural Knowledge

Probably the most basic and well-known knowledge classification system in psychology, and perhaps in education as well, is the distinction between declarative and procedural knowledge. This distinction was first made by Ryle

(1949), and was then popularized by John Anderson (1983) in his ACT* model of cognitive processing. The model has also been incorporated in a well-known educational text (E. D. Gagne, 1985), and has recently received support from researchers studying physiological bases of memory and cognition (Squire, 1987).

Declarative knowledge consists of what we know about objects, concepts, events, relationships between objects or concepts, etc. It is knowledge about things, systems, the world, and how to interact with the world. Declarative knowledge is knowledge that is verbalizable, things that we can overtly declare or state. For this reason it is sometimes described as **explicit** knowledge. Examples of declarative knowledge are shown in Table 3.1.

Declarative knowledge is often referred to as "knowledge that," while procedural knowledge is referred to as "knowledge how." However, this definition or explanation is very misleading because the *knowledge how* that is being referred to is knowledge how to perform *cognitive* activities not overt behavioral activities (see Anderson, 1990). Knowledge about how to perform activities in a real or behavioral sense, that can be verbalized, is *declarative* knowledge. Thus, in the lower section of Table 3.1, we see that declarative knowledge includes such things as "how to bake a cake," "how to change a flat tire," how to set an alarm clock," and so on. We can (potentially) tell someone or verbalize how to accomplish these activities. Because this knowledge can be accessed and stated explicitly, it is classified as declarative knowledge, or *declarative knowledge of procedures* for performing activities. To measure a person's declarative knowledge, the typical method is to simply ask in one way or another.

Declarative knowledge is most frequently assumed to be internally rep-

---

TABLE 3.1   Examples of Declarative Knowledge

Apple is a fruit
Apples have a skin
You eat apples
You eat an apple by biting into it
Sacramento is the capitol of California
The functions of a car
The parts of a car
The causal interconnections of parts on a car
How to bake a cake
How to change a flat tire
What happens when we are late for our own wedding

---

resented in the form of networks (Anderson, 1983; Collins and Loftus, 1975; Graesser and Gordon, 1991; Kintsch, 1988; Sharkey, 1986). These networks vary in their representational syntax, but all have the form of nodes and links, where nodes represent pieces of information and links represent the relationship between those pieces of information. Figure 3.2 shows an example of a simple declarative network depicting taxonomic (definitional) knowledge.

Declarative knowledge structures contain both general semantic knowledge and episodic memory (E. D. Gagne, 1985; Squire, 1987; Tulving, 1985). Semantic knowledge is the type we have been discussing; knowledge of general facts, concepts, and relations, without regard to specific landmarks or contextual events. Episodic memory refers to memory for specific past experiences or events in a person's life including the environmental and temporal cues associated with those events.

Procedural knowledge is knowledge about how to perform cognitive activities; the word cognitive is critical here. Putting together a grammatically correct sentence is a cognitive activity. Retrieving your friend's phone number from memory is a cognitive activity. The ability to perform cognitive activities is knowledge in a *dynamic* sense. But it is not symbolic knowledge that we can directly verbalize. No one can tell you HOW they know if a sentence is grammatically correct, or how they construct sentences. The ability to construct language is procedural knowledge, it is implicit and not accessible for verbalization. Table 3.2 gives some examples of procedural knowledge.

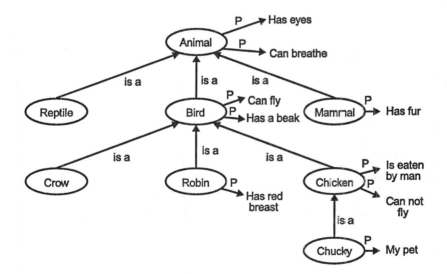

Figure 3.2   Example of a taxonomic knowledge structure. P denotes a Property arc (Figure courtesy of Arthur Graesser).

TABLE 3.2 Examples of Procedural Knowledge

Recognizing that + is a plus sign
Distinguishing a chair from a couch
Detecting tumors in an X-ray
Knowing how to retrieve knowledge from memory
Knowing how to smoothly shift from first to second
Being able to make inferences
Knowing how to proceed in solving physics problems

To evaluate a person's procedural knowledge, the person must perform some task (E.D. Gagne, 1985).

Because of the dynamic nature of procedural knowledge, it is often represented in theory and computer simulation by IF-THEN production rules (e.g., Neves and Anderson, 1981; Singley and Anderson, 1989). It is usually conceived of as the *direct association* between conditions perceived by the person (as a result of environmental or internal stimuli) and the person's response.

Subtypes of Declarative
and Procedural/Skill Knowledge

Anderson (1983, 1990) does not discriminate between types of declarative knowledge or types of procedural knowledge. Declarative knowledge is considered to be represented by propositional networks, and all types of procedural knowledge are represented, at least theoretically, in terms of associative production rules. However, many other researchers discriminate between different types of declarative knowledge, and between different types of procedural knowledge. Therefore, before discussing the remaining taxonomies, I will now describe the knowledge breakdown listed on the far left side of Figure 3.1 that I am proposing as the most basic or molecular level of knowledge types. All other taxonomies can be described as various combinations of these molecular knowledge types.

Declarative knowledge. Declarative knowledge can represent a wide variety of relationships such as taxonomic, causal, goal hierarchy, functional, structural, and spatial. For example, Graesser and colleagues have developed a model of knowledge and comprehension that assumes world knowledge can be broken down into four major classes; taxonomic (defining) structures, spatial region hierarchies, causal networks, and goal hierarchies (for a brief review, see Graesser & Gordon, 1991). Figure 3.3 shows a declar-

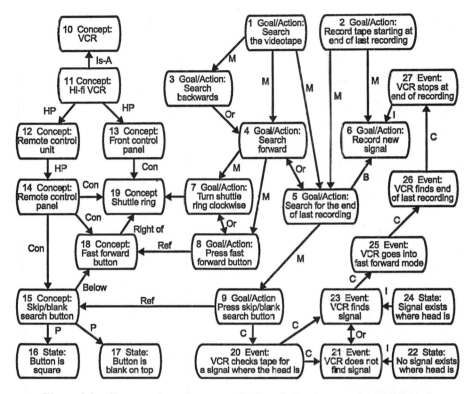

Figure 3.3 Conceptual graph structure for knowledge about a particular type of hi-fi VCR. The abbreviations for arc categories are: B = Before, C = Consequence, Con = Contains, HP = Has Part, I = Initiate, M = Means, P = Property, Ref = Refers-to.

ative knowledge network containing all four types of knowledge. Taxonomic knowledge is represented by the interrelationships between nodes 10 through 17; spatial region knowledge is represented within nodes 13–15, 18, and 19; causal knowledge is shown in nodes 9, and 20–27; and goal hierarchy knowledge is shown in nodes 1–9. It can be seen that these different types of knowledge are all interconnected within a network structure.

While these distinctions between types of knowledge structures can be useful for certain activities, training and instruction often does not require such a fine distinction. A more frequently useful differentiation can be obtained by distinguishing between only two types of declarative knowledge. This distinction is often implicitly made in training and instructional design research. This distinction is between the declarative "knowledge that" and "knowledge how." That is, in Figure 3.3, all of the taxonomic, causal, and spatial knowledge would be considered knowledge about things or systems;

knowledge that. We will refer to this type of knowledge as **conceptual** knowledge.

Conceptual knowledge essentially includes all types of declarative knowledge other than how to do things, including both episodic and semantic knowledge. Thus, declarative conceptual knowledge is knowledge about what things are, how they are related, causal relationships, functional relationships, and so forth. When we learn about computers, for example, we often get this type of information as background material, before we go on to learning specific operating procedures.

The goal hierarchy knowledge in Figure 3.3, or knowledge about how to do things, is "knowledge how." We will refer to this type of declarative knowledge as **rule** knowledge, or alternatively, as **explicit procedural** knowledge. Declarative rule knowledge is often referred to in the literature as goal structures or goal hierarchies. This is because, while we actually perform actions in a serial one-thing-at-a-time fashion, there is evidence that this knowledge is psychologically represented in a hierarchical fashion such as the information shown in Figure 3.3 (Bower, Black, & Turner, 1979; Galambos & Rips, 1982; Graesser, 1981; Graesser & Gordon, 1991; Lichtenstein & Brewer, 1980). We will see later that representing rule knowledge in terms of goal hierarchies is a convenient tool for task analysis.

Procedural/skill knowledge. Nonverbalizable **procedural** knowledge has been described in many ways, and there doesn't seem to be an all-inclusive listing anywhere of the various types of procedural knowledge. As noted earlier, I have included enough types in the far left column of Figure 3.1 to accommodate the more popular models. I have referred to this type of knowledge as procedural/skill knowledge because many people in the fields of training and human factors refer to this as skill rather than procedural knowledge (possibly because of the problem with the use of the word "procedure").

1. Automated rules. When we try something and it works, we store a declarative rule that can be directly retrieved the next time we are in the same situation. These declarative rules are then used when a new situation matches the characteristics of the rule(s). When we use a rule enough times, it becomes "automated." That is, we can use it without any deliberate effort or thinking. Thus, our activity, over time, comes to be dependent on rules that are automatically accessed and applied without any cognitive effort.

An example of an automated rule can be seen in learning to drive a manual stick-shift on a car. When we are first taught, someone might tell us, "to put the car in first gear, move the shift lever over to the left and then up." We must access and apply this rule in a conscious and deliberate fashion for quite a few times when we are first learning. However, with enough practice, the association becomes automatic; the goal of putting the car in first gear is associated with the movement of the shift left and up.

2. Perceptual skills. There are certain types of cognitive processing that seem to take place at a level outside of our symbolic, declarative memory system. Perceptual skills are one of those types. Perceptual learning is the process of learning to respond to a particular pattern of stimuli in a particular manner. The set of stimulus elements is processed at a basic, "nonconscious" level and appropriate responses generated through a process of trial and error. Over time, we learn to recognize certain patterns as meaning certain things. To quote Rasmussen (1986):

> Survival has most probably been granted to those individuals who were best—by direct pattern recognitions—at quickly identifying cues in the environment related to immediate goals, such as approach, escape, eat, and don't eat. . . . The subconscious main processor may therefore be assumed to have evolved into a data-processing system with particularly high capacity in functions needed in the control of the body. (p. 77)

3. Domain-general strategies. Much of our knowledge has to do with basic cognitive processing skills, such as storing knowledge, searching for and retrieving knowledge, transforming knowledge through inference processes, etc. This type of procedural knowledge or skill will be referred to as domain-general strategies. It is procedural knowledge acquired over a lifetime that allows us to process knowledge, that is, attend, learn, remember, and think. There are a variety of domain-general strategies, and many taxonomies have been suggested, some directly related to instructional design. For example, Weinstein and Mayer (1986) distinguish between rehearsal strategies, elaboration strategies, organizing strategies, comprehension monitoring strategies, and affective strategies (strategies used to focus attention, control anxiety, and manage learning time effectively).

4. Domain-specific strategies. Domain-specific strategies are cognitive processing skills that are developed and used within a particular field or domain. Examples would be working backwards to solve geometry problems. If a person learned to solve geometry problems in this manner, they might not realize that this domain-specific strategy could be used in other fields or domains as well. If they did not generalize this knowledge, it would remain a domain-specific strategy. Other domain-specific strategies include formulating plans (e.g., in chess), methods of skimming through an article (for scientific research), etc.

5. Psychomotor skills. Finally, there are certain tasks that rely more on refinement of motor skills than on cognitive processing per se. Examples are playing the piano, learning to type on a typewriter or computer, riding a bicycle, and so forth. Because psychomotor skills are often one part of a training program, they are included in our taxonomy. In terms of learning mechanisms, they are viewed as being similar to other procedural skills. They are acquired through relatively extensive practice where the system makes successively more correct responses and fewer incorrect responses. This process will be described in more detail under the section on controlled vs. automatic processing.

All of these subtypes of procedural knowledge have much in common. They are dynamic, implicit structures that are acquired through experience. Because they are similar in more respects than not, and are acquired in basically the same way, I will rarely refer to specific types. However, when it is necessary to segregate these subtypes out, I will use the terminology given on the left side of Figure 3.1.

Rasmussen's Levels of Cognitive Processing

Jens Rasmussen (1980, 1983, 1986) has studied the processes and strategies used by people to control complex systems. As a result of this work, he has developed a model of human information processing that outlines the mental processes that occur during task performance (see Rasmussen, 1986). As shown in Figure 3.1, he divides processing into three categories: skill-based processing, rule-based processing, and knowledge-based processing. A person is assumed to move fluidly back and forth between these three levels depending on the nature of the task and the person's experience.

Knowledge-based processing. When faced with a relatively unfamiliar task, a person uses knowledge-based processes. Knowledge is defined as a conceptual, structural model of concepts. In knowledge-based behavior, sensory input is first transformed into conceptual symbols which are then used for reasoning about the task in processes such as goal formulation, plan selection, plan evaluation, etc. Knowledge-based behavior is exhibited in novel situations for which actions must be planned at the time, using conscious analytical processes and stored knowledge. It is behavior necessitated by a lack of relevant rules or skills. This type of processing appears to be similar to the use of what we have called declarative conceptual knowledge.

Rule-based processing. Rule-based behavior is goal-oriented also, but is based on a "feedforward" control mechanism, that is, a stored rule. The composition of a sequence of subroutines in a familiar situation is directly controlled by stored rules or procedures that have been created through previous experience. Rule-based processing is similar to the use of what we are calling declarative rules or explicit procedural knowledge.

Skill-based processing. As similar occasions are repeatedly experienced, the use of rule-based behavior is discarded for the more efficient skill-based behavior. In skill-based behavior, features of the situation are directly associated with automated sensorimotor patterns. It can be seen that skill-based processing seems to be similar to procedural knowledge. However, it is difficult to determine where strategy knowledge fits into Rasmussen's scheme. It is probably safest to map skill-based processing onto our automated rules and psychomotor skills categories, and assume that domain-general and domain-specific strategies are involved in the hidden processes governing knowledge-based processing.

Controlled vs. Automatic Processing

Schneider and Shiffrin (1977) and Shiffrin and Schneider (1977) described two types of cognitive processing, controlled and automatic. **Controlled processing** is when a person accesses information, brings it into working memory, performs transformations upon the information, generates responses, etc. Controlled processing is a deliberate, effortful use of knowledge requiring significant amounts of attentional and other cognitive resources.

Alternatively, **automatic processing** is fast and demands no attentional resources. Automatic processing develops slowly and only as a result of extensive practice at a task (meaning hundreds of similar experiences). Automaticity is assumed to develop only under conditions where there is a *consistent* mapping between the stimulus elements and the response. Once a task has become automated, it is extremely rapid, performed automatically, and demands virtually no resources. This means that the person's cognitive resources are freed to perform other controlled tasks.

With regard to our taxonomy shown in Figure 3.1, it is probably not possible to perfectly map the controlled vs. automatic dichotomy onto our basic levels of knowledge listed on the left side. However, we can tentatively assume that controlled processing is essentially the use of declarative knowledge, and also the beginning stages of psychomotor skills. Automatic processes are automated rules and also psychomotor skills that have received a great deal of practice.

Gagne's Five Kinds of Learned Capabilities

Working in the field of instructional theory, Gagne (R. M. Gagne, 1985; Gagne, Briggs, & Wager, 1988) has developed a category scheme in which five qualitatively different learning outcomes may result from a training or instructional program. He also suggests that there are certain environmental conditions that are necessary for each of these outcomes to occur.

The five outcomes are regarded as different learned capabilities, and four of them are what we might call "types of knowledge." The fifth, attitudes, is included in this discussion because training programs sometimes have as a goal to change trainee attitudes.

1. Verbal information. According to Gagne et al. (1988), verbal information is "the kind of knowledge we are able to *state*. It is *knowing that,* or "*declarative knowledge*" [sic]. Gagne further states that the method to use for measuring verbal knowledge is to ask the learner questions. It can be seen from this definition that this category of knowledge corresponds to what has been called declarative knowledge.

2. Intellectual skills. Intellectual skills are processes which "allow the individual to interact with their environment in terms of symbols or conceptualizations" (Gagne et al., 1988). Intellectual skills range from reading, writing, and mathematics to advanced technical skills of science and engineering. Intellectual skills are described as being the same thing as Anderson's (1983) procedural knowledge.

3. Cognitive strategies. Cognitive strategies are described as being "special and very important kinds of skills. They are the capabilities that govern the individual's own learning, remembering, and thinking behavior" (Gagne et al., 1988). Cognitive strategies are assumed to grow slowly over long periods of time spent studying, learning, and thinking. Gagne also differentiates between domain-specific strategies and domain-general strategies, a dichotomy that was retained in the basic taxonomy column on the left of Figure 3.1.

4. Motor skills. Motor skills are physical activities which require practice before smooth performance is achieved. Examples include riding a bicycle, driving a car, drawing a circle or straight line, etc. Gagne et al. (1988) point out that motor skills are often components of a task that enter into further learning. For example, students must learn the skill of printing letters in order to move on to writing language.

5. Attitude. As noted before, attitudes are not a class of knowledge per se. But they are a "cognitive" entity that can be acquired through training, and are therefore potentially important. R. M. Gagne (1985; Gagne et al., 1988) defines attitudes as an affect toward things, persons, and situations. The effect of an attitude is to "amplify an individual's positive or negative reaction. . . . a persisting state that modifies the individual's choices of action."

Our view of this topic comes from the Theory of Reasoned Action developed by Fishbein and Ajzen (1975) and Ajzen and Fishbein (1980). The theory assumes that people make reasonably good use of the information available to them. It further assumes that pieces of declarative knowledge, termed beliefs, have a direct and causal effect on one's attitudes. That is, one's beliefs about computers determine one's attitudes toward computers. And one's beliefs about *using* a computer determine one's attitude toward using a computer. Since most of the attitudes we are concerned with in training programs are really trainee attitudes toward performing some behavior, this behavioral attitude is of central importance.

Ajzen and Fishbein (1980) have developed and tested a model which outlines the factors determining behavior. Figure 3.4 shows the organization of the causal model, and how the components ultimately determine behavior. It can be seen that behavior is predominantly a function of the individual's intentions. Behavioral intentions are themselves a function of two factors, the individual's attitude toward the behavior, and their subjective norm.

Attitude toward the behavior is the person's feeling of favorableness or unfavorableness toward performing the behavior (Ajzen & Fishbein, 1980). **Subjective norm** is the person's overall perception concerning whether other people who are important to them desire them to perform the behavior. This part of the model has important implications for training and its ultimate ef-

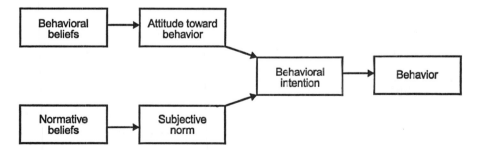

Figure 3.4   Structural model of Ajzen and Fishbein's theory of reasoned action.

fectiveness. An employee may complete a training program and have a very good attitude toward a particular behavior (such as use of a safety device), and yet perceive that others don't wish him/her to perform that behavior (subjective norm). Depending on the work environment, subjective social norms may have a very powerful effect.

The view taken in this book comes from research in social psychology showing that changing a person's attitude can effectively be accomplished by making a change in his/her belief system (Ajzen & Fishbein, 1980). Since beliefs are declarative knowledge, changing a trainee's attitude would be accomplished by imparting declarative knowledge in a manner that is accepted by the trainee.

Knowledge, Skills, and Abilities (KSAs)

In the field of training and instructional design, a knowledge taxonomy that has held a place of prominence for several years is a scheme developed by Prien (1977) and made well known by Goldstein (Goldstein, 1986; Goldstein, Macey, and Prien, 1981). I will only briefly review these terms because they can be defined using concepts that we have already covered extensively.

Knowledge.   Knowledge refers to "an organized body of knowledge usually of a factual or procedural nature, which, if applied makes adequate job performance possible" (Goldstein, 1986). It can be seen that this category maps onto what has been termed declarative knowledge.

Skills.   Skills refer to "the capability to perform job operations with ease and precision." According to Goldstein (1986), use of the term *skill* usually refers to psychomotor types of activity. For this reason, I have cross-referenced the term skill to the generic category of psychomotor skill.

Ability.   Ability is used to refer to cognitive capabilities "necessary to perform a job function." They cannot function alone, but require the application of some (presumably declarative) knowledge base. Examples given in Goldstein (1986) include:

- Ability to shift priorities in response to a change in supply conditions
- Ability to evaluate the capabilities of subordinates for promotion
- Ability to recognize the usefulness of information supplied by others

I have assumed that *abilities* map generally onto what we are calling procedural or skill knowledge, with the exception of motor skills.

Terminology To Be Used in This Book

It can be seen that the various taxonomies all deal with largely the same types of knowledge, but cluster or define them in differing ways. In this book, I will use the terminology described on the left-hand side of Figure 3.1 because it is consistent with the majority of the taxonomies covered, and does not use terms that are ambiguous such as "knowledge," "abilities," or "procedural" knowledge.

I make major distinctions between declarative and procedural/skill knowledge, because the acquisition of these two types of knowledge is fundamentally different (E. D. Gagne, 1985). I also distinguish between use of declarative *conceptual* knowledge and declarative *rule* knowledge (rule knowledge may occasionally be referred to as explicit procedural knowledge because use of the word "procedure" is such standard terminology in the training industry). The distinction between conceptual and rule declarative knowledge is retained because there are major differences in how these two types of information are presented in instructional materials and how they are represented and used by learners.

I will usually refer to the training and acquisition of procedural/skill knowledge without reference to the specific types. Occasionally it may be necessary to refer to specific types of skills, such as psychomotor skills. For example, some types of skills can be taught more efficiently by explicitly stating procedural rules, while others such as domain-general inference strategies cannot be taught directly. The next two sections will review two additional topics that are also cited widely in the research literature relevant to training, mental models, and novice vs. expert performance.

MENTAL MODELS

The concept of **mental model** (Johnson-Laird, 1983; Gentner & Stevens, 1983) is becoming a popular tool in much of the system interface design community and among training analysts as well (e.g., Converse, Cannon-Bowers, & Salas, 1991; Kieras & Bovair, 1984; Rouse & Morris, 1986; Wilson & Rutherford, 1989). For this reason, I will briefly review the concept and its possible relationship to the types of knowledge discussed earlier.

Definition of a Mental Model

In their review of the use of mental models as a construct within psychology and human factors, Wilson and Rutherford (1989) note that "the literature contains a number of allusions to what mental models might be, but very few formal or explicit definitions exist" (p. 618). To illustrate this, here are a few of the diverse descriptions:

1. According to Toffler (1970): "Every person carries within his head a mental model of the world—a subjective representation of external reality."

2. Wilson and Rutherford (1989) suggest that a mental model is: "a representation formed by a user of a system and/or task, based on previous experience as well as current observation, which provides most (if not all) of their subsequent system understanding and consequently dictates the level of task performance."

3. Johnson-Laird (1981, 1983) states that "A model represents a state of affairs and accordingly its structure. . . . Its structure mirrors the relevant aspects of the corresponding state of affairs in the world" (1981, p. 174), and that mental models "enable individuals to make inferences and predictions, to understand phenomena, to decide what action to take and to control its execution, and above all to experience events by proxy" (1983, p. 397).

Most descriptions of mental models tell us what they can do (e.g., allow understanding and prediction of future events) rather than what they are. Perhaps the most widely used conceptualization is that of a mental representation of some system, such as a VCR or "the world" (as in physics). This view is represented by Kieras' (1988) suggestion that a mental model is the basic conceptualization of some system. It is a dynamic model that can be "run" to derive predictions of future behavior. Analogously, Rasmussen (1986) describes a mental model as "the mental representation of the anatomic or functional structure of the system," and Lindgaard (1987) views mental models as representations that can range from mental images of concrete objects or systems to abstract conceptions of the system.

These views all suggest that we have some internal representation of a thing or system or place with definite boundaries. As an example, we have a mental model of our television set, what the buttons are, how it works, what will happen if we do certain things, and so on. This representation provides an internal system that we can use to generate problem solutions, running the model in a hypothetical manner to see what will happen under certain conditions. In addition, most researchers agree that a person may have several mental models regarding a single system (e.g., Wilson & Rutherford, 1989), where each model represents different information about the system, or different "views" of the system. For example, Rouse and Morris (1986) suggest that separate mental models could relate to different aspects of a system such as state, function, purpose, or form. It is not clear whether these different mental models are actually simply subparts of a single complex mental model.

Rasmussen (1986) assumes that the cognitive procedures used to transform, or *run* the model are not actually part of the model itself. This is in contrast with most psychologists who feel that the operational procedures are also an important part of the model (e.g., Rouse & Morris, 1986). In fact, this may be the only distinguishing factor between the currently fashionable term of "mental model," and an older concept, "schema." That is, mental models are the use of (schema) information in a computationally dynamic manner.

In applying the concept to the design of systems for ease of use, Norman (1983, 1986) makes a useful distinction between the designer's *conceptual model* of a system, the *system image* that is projected, and the *user's mental model* of the system. The designers' conceptual model is the idea of the system that they have in their head as they design it. This model is almost never given explicitly to the user. Instead, there is a system image that is projected through its outer features, displays, documentation, controls, and operations. Users develop their own internal mental model of the system based on their interaction with its features. This mental model might be fuzzy, incomplete, and inconsistent in its parts. And the user's mental model might bear very little resemblance to the designer's mental model. The goal of human factors design is to create a simple and elegant design model, and then create a system image that conveys the design model to the user. If users are able to acquire a complete and accurate model of the system, they are much more likely to be able to successfully use the system.

Relationship Between Mental Models and Types of Knowledge

Johnson-Laird (1983) suggested that parsing (creation) of verbal propositions creates a mental model that is structurally analogous to the represented world. This suggests that the mental model is the "big picture" of whatever world is being modeled; the parts, functions, causal relations, spatial relations, etc. Most researchers also agree that the model includes computational means to transform or *run* the model. This may sound to the reader much like all types of knowledge discussed previously, excluding motor skills, that happen to be in regard to some particular system. This is, in fact, the way that I view mental models. There seems to be no clear distinction between the concept of mental model and the types of knowledge we described earlier. If one considers the types of knowledge structures described (taxonomic, spatial, causal, etc.) and *combine* those structures with operating procedures, this looks very much like a mental model.

Application of Mental Models to Training

Why is the concept of mental model relevant to design of a training program? There are two fundamentally different ways that the concept is important to the designer.

Domain mental models. One application is the need to consider trainees' mental models relevant to the task domain itself. For example, if one is teaching about basic principles of physics, then the analyst should be concerned with any mental models of the task domain, this being the "world," brought in by the learner. These models may be correct or incorrect, and may be consistent or inconsistent with models to be trained in the instructional program. In addition, the designer may want to specifically and explicitly represent the models of experts in the field, and try to determine ways to convey those models to learners. There is also some evidence that novices often do not directly acquire the models of experts but move through intermediate models. This means that the trainer may have to identify sequences of models moving from novice to intermediate to expert, and find ways to bring the novice through these different models.

To summarize this point, before developing a training program, we will want to measure the mental models of novices, intermediates, and experts in the task domain. Then, we attempt to develop a training program which optimally brings the learner's mental model to more closely resemble the intermediate and expert model. This idea can be expanded to include the training of a *common* mental model for groups of people who must work together in some task domain (e.g., Cannon-Bowers, Salas, & Converse, in press; Converse et al., 1991).

Training system user models. The other reason the concept of mental models is important for training design is that training is increasingly being accomplished via computers. Learners have limited cognitive resources to devote to a set of tasks (Wickens, 1980), and if their resources are being allocated to understanding how to use the training system, they will not be used for learning the domain knowledge. The training system interface should be designed so as to convey a simple and usable mental model for the user. That is, the user should understand the parts of the training system, their interrelationships, their functions, and most especially, how to accomplish the learner's goals. This interface design issue will be addressed more directly in Chapter 11.

NOVICE AND EXPERT PERFORMANCE

Now that we have examined the different types of knowledge used in task performance, we can look for basic differences between novice and expert performance. By knowing, at some general level, the typical differences between experts and novices, we can more soundly design our methods for task analysis (next chapter). We can also make more informed decisions regarding what we can reasonably expect of a given training program, and how to measure the success of that program. In this section, I will briefly summarize Anderson's (1983) view of how people change as they become experts in a task.[1] Then I will summarize some of the empirical findings among researchers studying expert-novice differences.

Anderson's Three-Stage Model of Expertise

Anderson (1983) updated a theory originally proposed by Fitts and Posner (1967) suggesting that people pass through three stages in the process of learning a task and becoming an expert. In the first *cognitive* stage, people are primarily involved in accumulating relevant declarative knowledge (conceptual and perhaps some rule). When a person must perform a task, relevant pieces of declarative knowledge are retrieved from memory and operated upon by domain-general procedures. In this first stage, decision making and problem solving tend to be slow, tedious, and prone to error.

By the second *associative* stage, repeated use of declarative knowledge has begun to result in domain-specific procedures being compiled. These are direct associations between specific conditions and the resultant action. Thus, the need for slow, analytical processing of declarative conceptual knowledge becomes more and more bypassed.

With enough practice, the individual moves into the third, *autonomous,* stage. In this stage, activities have been repeated so many times that the procedures have become automated. Putting it briefly, the stimulus-response associations grow in number, existing associations become stronger with use, and as the number of associations increases, they become more highly specialized. After enough practice, the associations are strengthened to the point where no effort is required, and the use of declarative knowledge is relegated to the occasional novel situation.

[1]Rasmussen (1986) has presented a similar model.

Empirical Evidence for Models

Several studies have been conducted to test Anderson's theory and similar ideas (see Gordon, 1992, for a review). Without going into details, the conclusions we can draw at this point in time are that (1) there is evidence for separate and distinct processing mechanisms corresponding to the use of declarative conceptual types of knowledge, and implicit, nonverbalizable procedural/skill knowledge; and (2) there is not evidence that the sequence of stages is invariant. While there is some evidence that people *often* pass through the stages in the order described, there is also evidence that it is not always the case (e.g., Lewicki, 1986; Lewicki, Czyzewska, & Hoffman, 1987; Willingham, Nissen, & Bullemer, 1989). That is, there is evidence that learning can take place at the subconscious level without declarative knowledge involvement, right from the beginning (e.g., Sanderson, 1989; Willingham et al., 1989). There has been a relatively large number of studies yielding evidence that we are able to learn complex perceptual tasks without any conscious knowledge of what cues we are using to determine the correct response (see Gordon, 1992, and Squire, 1987 for a review). That is, learning can take place in an unconscious procedural system without analytic use of declarative knowledge.

Interestingly, Willingham et al. (1989) found that although subjects *could* learn to respond correctly to perceptual sequences without any explicit declarative knowledge of the underlying rules, if participants *did* have an awareness of the rules, their performance was enhanced. This finding has implications for training which will be discussed later.

There is also some indirect evidence that individuals rely more on implicit skill knowledge as they become adept. Designers building expert systems in the field of artificial intelligence have frequently noted that it is difficult to get experts to verbalize their knowledge (see Gordon, 1989). Given the view of knowledge that we have built in the previous sections, this is quite understandable. Because of their extensive experience, experts rely predominantly upon implicit skill knowledge, and this is the type that is difficult if not impossible to consciously describe. Only when they are faced with a novel task would they rely on analytic use of conceptual knowledge, making it more likely that they could explain what they are thinking about and how they are performing the task.

Characteristics of Expert Performance

Many researchers have questioned and observed novices and experts in order to categorize the differences in knowledge and strategies between the two groups. Much of this work is reviewed in Chi, Glaser, and

Farr's (1988), *The Nature of Expertise*. A few of the major findings are summarized below:

1. Experts are especially good at thinking within their own particular domain, but are not generally any better than novices at performing tasks within another domain.
2. Experts have a great deal of domain-specific declarative knowledge.
3. The declarative knowledge of experts is more well-organized, and it is organized around different superordinate concepts. Experts organize their knowledge around fundamental and critical concepts and principles within the field, while novices tend to group or associate their knowledge in terms of more superficial or unimportant features.
4. Experts have a larger repertoire of domain-specific strategies that are easily accessible during task performance or problem solving.
5. Related to point 4, experts have a very large number (10,000 to 100,000) of situation-specific production rules (procedures) that are automatically invoked during task performance.
6. The existence of the large number of situation-specific implicit procedures results in experts being able to perform tasks very, very quickly and with low error.
7. Experts tend to have more difficulty expressing their knowledge; not knowledge of facts and rules, but providing explanations for their performance.
8. Experts perceive large, meaningful patterns in the stimuli. That is, novices see all pieces individually and rather equally. Experts have learned to chunk individual stimuli into groups that have some meaning.
9. Experts spend more time developing the initial representation of the problem. It is assumed that they are building more complete and elaborate models, adding constraints that will reduce the problem solution (or subtask) search space.

In summary, we can say that experts have an enormous amount of well-organized declarative and procedural/skill knowledge. The problem for training program design is that we must somehow identify a great deal of that knowledge in order to train others. The next chapter describes a variety of methods for performing this task. We will see that there are no shortcuts—there is no way to quickly extract knowledge from an expert and plant it into a trainee's head.

REFERENCES

ANDERSON, J. R. (1983). *The architecture of cognition.* Cambridge, MA: Harvard University Press.

ANDERSON, J. R. (1990). *Cognitive psychology and its implications* (3rd ed.). New York: W. H. Freeman.

AJZEN, I., & FISHBEIN, M. (1980). *Understanding attitudes and predicting social behavior.* Englewood Cliffs, NJ: Prentice Hall.

BOWER, G. H., BLACK, J. B., & TURNER, T. J. (1979). Scripts in memory for text. *Cognitive Psychology, 11,* 177–220.

BROWN, J. S., COLLINS, A., & DUGUID, P. (1989). Situated cognition and the culture of learning. *Educational Researcher, 18,* 32–42.

CHI, M. T. H., GLASER, R., & FARR, M. J. (eds.). (1988). *The nature of expertise.* Hillsdale, NJ: Lawrence Erlbaum Associates.

CANNON-BOWERS, J. A., SALAS, E., & CONVERSE, S. A. (in press). Shared mental models in expert decision making. In N. J. Castellan (ed.), *Individual and group decision making.* Hillsdale, NJ: Lawrence Erlbaum Associates.

COLLINS, A. M., & LOFTUS, E. F. (1975). A spreading activation theory of semantic processing. *Psychological Review, 82,* 407–428.

CONVERSE, S. A., CANNON-BOWERS, J. A., & SALAS, E. (1991). Team member shared mental models: A theory and some methodological issues. *Proceedings of the Human Factor Society 35th Annual Meeting* (vol. 2, pp. 1417–1421). Santa Monica, CA: Human Factors Society.

FISHBEIN, M., & AJZEN, I. (1975). *Belief, attitude, intention, and behavior: An introduction to theory and research.* Reading, MA: Addison Wesley.

FITTS, P. M., & POSNER, M. I. (1967). *Human performance.* Belmont, CA: Brooks Cole.

GAGNE, E. D. (1985). *The cognitive psychology of school learning.* Boston, MA: Little, Brown.

GAGNE, R. M. (1985). *The conditions of learning* (4th ed.). New York: Holt, Rinehart and Winston.

GAGNE, R. M., BRIGGS, L. J., & WAGER, W. W. (1988). *Principles of instructional design* (3rd ed.). New York: Holt, Rinehart, and Winston.

GALAMBOS, J. A., & RIPS, L. J. (1982). Memory for routines. *Journal of Verbal Learning and Verbal Behavior, 21,* 260–281.

GENTNER, D., & STEVENS, A. L. (Eds.). (1983). *Mental models.* Hillsdale, NJ: Lawrence Erlbaum Associates.

GOLDSTEIN, I. L. (1986). *Training in organizations: Needs assessment, development, and evaluation* (2nd ed.). Monterey, CA: Brooks Cole.

GOLDSTEIN, I. L., MACEY, W. H., & PRIEN, E. P. (1981). Needs assessment approaches for training development. In H. Meltzer and W. R. Nord (eds.), *Making organizations humane and productive*. New York: John Wiley & Sons.

GORDON, S. E. (1989). Theory and methods for knowledge acquisition. *AI Applications in Natural Resource Management, 3*, 19–20.

GORDON, S. E. (1992). Implications of cognitive theory for knowledge acquisition. In R. Hoffman (ed.), *The cognition of experts: Psychological theory and empirical AI* (pp. 99–120). New York: Springer-Verlag.

GRAESSER, A. C. (1981). *Prose comprehension beyond the word.* New York: Springer-Verlag.

GRAESSER, A. C., & GORDON, S. E. (1991). Question answering and the organization of world knowledge. In W. Kessen, A. Ortony, & F. Craik (eds.), *Memories, thoughts, and emotions: Essays in Honor of George Mandler* (pp. 227–243). Hillsdale, NJ: Lawrence Erlbaum Associates.

JOHNSON-LAIRD, P. N. (1981). Mental models in cognitive science. In D. A. Norman (ed.), *Perspectives on cognitive science* (pp. 147–191). Hillsdale, NJ: Lawrence Erlbaum Associates.

JOHNSON-LAIRD, P. N. (1983). *Mental models.* Cambridge, England: Cambridge University Press.

KIERAS, D. E. (1988). What mental models should be taught: Choosing instructional content for complex engineering systems. In J. Psotka, L. Massey, and S. Mutter (eds.), *Intelligent tutoring systems: Lessons learned* (pp. 85–112). Hillsdale, NJ: Lawrence Erlbaum Associates.

KIERAS, D. E., & BOVAIR, S. (1984). The role of a mental model in learning to operate a device. *Cognitive Science, 8*, 255–273.

KINTSCH, W. (1988). The role of knowledge in discourse comprehension: A construction-integration model. *Psychological Review, 95*, 163–182.

LEWICKI, P. (1986). Processing information about covariations that cannot be articulated. *Journal of Experimental Psychology: Learning, Memory, and Cognition, 12*, 135–146.

LEWICKI, P. CZYZEWSKA, M., & HOFFMAN, H. (1987). Unconscious acquisition of complex procedural knowledge. *Journal of Experimental Psychology: Learning, Memory, and Cognition, 13*, 523–530.

LICHTENSTEIN, E. H., & BREWER, W.F. (1980). Memory for goal directed events. *Cognitive Psychology, 12*, 412–445.

LINDGAARD, G. (1987). *Who needs what information about computer systems: Some notes on mental models, metaphors and expertise* (Branch Paper 126). Clayton, Australia: Telecom Australia Research Laboratories.

NEVES, D. M., & ANDERSON, J. R. (1981). Knowledge compilation: Mechanisms for the automatization of cognitive skills. In J. A. Anderson (ed.), *Cognitive skills and their acquisition.* Hillsdale, NJ: Lawrence Erlbaum Associates.

NORMAN, D. (1983). Some observations on mental models. In D. Gentner and A. L.

Stevens (eds.), *Mental models* (pp. 7–14). Hillsdale, NJ: Lawrence Erlbaum Associates.

NORMAN, D. (1986). Cognitive engineering. In D. A. Norman and S. Draper (eds.), *User-centered system design* (pp. 31–61). Hillsdale, NJ: Lawrence Erlbaum Associates.

PRIEN, E. P. (1977). The function of job analysis in content validation. *Personnel Psychology, 30,* 167–174.

RASMUSSEN, J. (1980). The human as a system component. In H. T. Smith and T. R. Green (eds.), *Human interaction with computers.* London: Academic Press.

RASMUSSEN, J. (1983). Skills, rules, knowledge: signals, signs, and symbols and other distinctions in human performance models. *IEEE Transactions on Systems, Man, and Cybernetics, 13(3),* 257–267.

RASMUSSEN, J. (1986). *Information processing and human-machine interaction: An approach to cognitive engineering.* New York: Elsevier.

ROUSE, W. B., & MORRIS, N. M. (1986). On looking into the black box: Prospects and limits in the search for mental models. *Psychological Bulletin, 100,* 349–363.

RYLE, G. (1949). *The concept of mind.* San Francisco, CA: Hutchinson.

SANDERSON, P. M. (1989). Verbalizable knowledge and skilled task performance: Association, dissociation, and mental models. *Journal of Experimental Psychology: Learning, Memory, and Cognition, 15,* 729–747.

SCHNEIDER, W., & SHIFFRIN, R. M. (1977). Controlled and automatic human information processing: I. Detection, search and attention. *Psychological Review, 84,* 1–66.

SHARKEY, N. E. (1986). A model of knowledge-based expectancies in text comprehension. In J. A. Galambos, R. P. Abelson, and J. B. Black (eds.), *Knowledge structures* (pp. 49–70). Hillsdale, NJ: Lawrence Erlbaum Associates.

SHIFFRIN, R. M., & SCHNEIDER, W. (1977). Controlled and automatic human information processing: II. Perceptual learning, automatic attending and a general theory. *Psychological Review, 84,* 127–190.

SINGLEY, M. K., & ANDERSON, J. R. (1989). *The transfer of cognitive skill.* Cambridge, MA: Harvard University Press.

SQUIRE, L. M. (1987). *Memory and brain.* New York: Oxford University Press.

TOFFLER, A. (1970). *Future shock.* London: Bodley Head.

TULVING, E. (1985). How many memory systems are there? *American Psychologist, 40,* 385–398.

WEINSTEIN, C. E., & MAYER, R. E. (1986). The teaching of learning strategies. In M. C. Wittrock (ed.), *Handbook of research on teaching* (3rd ed.). New York: Macmillan.

WICKENS, C. D. (1980). The structure of attentional resources. In R. S. Nickerson (ed.), *Attention and performance VIII.* Hillsdale, NJ: Lawrence Erlbaum Associates.

WILLINGHAM, D. B., NISSEN, M. J., & BULLEMER, P. (1989). On the development of procedural knowledge. *Journal of Experimental Psychology: Learning, Memory, and Cognition, 15,* 1047–1060.

WILSON, J. R., & RUTHERFORD, A. (1989). Mental models: Theory and application in human factors. *Human Factors, 31(6),* 617–634.

ADDITIONAL RESOURCES

BERGER, D. E., PEZDEK, K., & BANKS, W. P., EDS. (1987). *Applications of cognitive psychology: Problem solving, education, and computing.* Hillsdale, NJ: Lawrence Erlbaum Associates.

GENTNER, D. (1983). A theoretical framework for analogy. *Cognitive Science, 7,* 155–170.

GLOVER, J. A., RONNING, R. R., & BRUNING, R. H. (1990). *Cognitive psychology for teachers.* New York: Macmillan.

GOODSTEIN, L. P., ANDERSON, H. B., & OLSEN, S. E. (1988). *Tasks, errors, and mental models.* London: Taylor & Francis.

GRAESSER, A. C., & CLARK, L. F. (1985). *Structures and procedures of implicit knowledge.* Norwood, NJ: Ablex Publishing.

HOLLNAGEL, E., MANCINI, G., & WOODS, D. D., EDS. (1988). *Cognitive engineering in complex dynamic worlds.* San Diego, CA: Academic Press.

KLAHR, D., & KOTOVSKY, K. (Eds.) (1989). *Complex information processing: The impact of Herbert A. Simon.* Hillsdale, NJ: Lawrence Erlbaum Associates.

MANDLER, G. (1985). *Cognitive psychology: An essay in cognitive science.* Hillsdale, NJ: Lawrence Erlbaum Associates.

OSHERSON, D. N., & SMITH, E. E. (eds.) (1990). *Thinking: An invitation to cognitive science volume 3.* Cambridge, MA: The MIT Press.

SCHOENFELD, A. H. (ed.) (1987). *Cognitive science and mathematics education.* Hillsdale, NJ: Lawrence Erlbaum Associates.

TAUBER, M. J., & ACKERMANN, D. (1991). *Mental models and human-computer interaction 2.* New York: Elsevier.

4

Procedures:
Task Analysis
and Trainee Analysis

After performing the organizational analysis, the designer next turns to evaluation of the job and its component tasks, a process termed **task analysis.** The analyst breaks a job down into its subtasks, and then evaluates each of those subtasks at a detailed level. This involves activities such as determining how the subtasks are accomplished, what knowledge and skills are used in performing the subtasks, and various task characteristics such as difficulty, frequency, hazards, etc.

Either during or after the task analysis, the designer also evaluates the characteristics of the people who are most likely to be trainees, termed **trainee analysis.** All of this information is then used to determine the appropriate performance solution(s), design the training program, and determine the variables to measure in the final evaluation phase. One reason for performing the task analysis before the trainee analysis is that task analysis allows the designer to determine prerequisite knowledge and skills that will be needed by trainees coming into the instructional program. Then, during the trainee analysis, the analyst can evaluate whether the prerequisite knowledge and skills are adequately possessed by the trainees.

The nature of task analysis has changed over the last five to ten years, as technology has advanced and tasks targeted for training become more complex. The next three sections will very briefly (a) describe traditional procedures for performing task analysis, (b) describe a newer approach as designers realized the need for more cognitively oriented task analyses, and (c) describe the range of task analyses currently performed in human factors for system design. The remaining sections review the task analysis methods most appropriate for design of training and performance support systems.

Readers may notice that this chapter is lengthy. There are two approaches which may be taken. First, a reader who will be performing task analysis frequently should read the methods described and become familiar with many of them. This will provide a battery of tools for various future projects. A reader who will not be performing an extensive cognitive task analysis, or who is not interested in becoming proficient at the various methods, should read the first introductory sections plus the most common and useful techniques. In my own personal opinion, these include:

- General data collection methods of structured interviews, verbal protocol analysis, and observation.
- General data representation methods of matrices, hierarchical networks, and flow charts.
- Specific methods of GOMS, FAST. and conceptual graph analysis.

TRADITIONAL TASK ANALYSIS

Task Lists and Hierarchies

Task analysis is a complete assessment of the tasks (or potential tasks) to be performed by the employee as part of the job or duties under analysis. Traditionally, task analysis has been accomplished by breaking down a job to be trained into a list or hierarchy of components such as duties, tasks, and subtasks (e.g., Goldstein, 1986; Merrill, 1987; Rothwell & Kazanas, 1992). While some authors have defined tasks as either cognitive or actions (Reddout, 1987), in practice the task descriptions tended to be oriented toward *outwardly measurable* behavioral variables.

For an illustration, we can evaluate a simple "job" with which everyone is at least somewhat familiar. Imagine that you have just acquired a new breed of bicycle known as a mountain bike. Now you must be trained to adjust, maintain, repair, and safely ride your mountain bike. Similar to training for many pieces of equipment that people acquire for work or play, training for bike ownership currently takes place by way of an owner's manual, perhaps a salesperson at the shop where the bike is purchased, and lots of owner guesswork and trial-and-error. Now imagine that you must perform a task analysis.

By looking through the owner's manual, you might develop a task list that includes items such as the following:

1. Adjust the seat height
2. Adjust the handlebar height
3. Adjust the placement of the brakes
4. Check the inflation of the tires
5. Check chain tension
6. Check break cable tension
7. Check for frame fractures
8. Inflate tires to proper degree
9. Change flat tire
10. Get on the bicycle
11. Ride the bicycle on pavement
12. Ride the bicycle on a dirt trail
13. Ride the bicycle over obstacles
14. Shift between 18 gears
15. Use breaks
16. Ride safely

While this is not a complete list, the reader can see that something as simple as "riding and upkeep of a bicycle" can be a lengthy and complex set of tasks.

Task analyses often break jobs down into hierarchical structures with successively more specific levels of description. This hierarchical analysis can take place above the basic task level, below the task level, or both. As an example of a hierarchical analysis that takes place above the task level, Pearlman (1980) describes a hierarchical taxonomy that breaks a job down into categories moving from **job family** down to **occupation/job type** (e.g., bookkeeper), **job, person,** and finally the **tasks** themselves (see Figure 4.1). As an example of a hierarchical breakdown above and below the task level, Phillips et al. (1988) describe a method for performing task analysis that specifies two levels of "activities" above the tasks and one level, termed task elements, below the task level (see Table 4.1).

Sometimes a set of task specifications can be derived most easily by using a matrix approach. This method requires a definable set of tasks to be performed under varying conditions or with varying objects/people in the environment. For example, a maintenance job may involve a limited set of operations (align, adjust, etc.) which are performed on a specifiable set of hard-

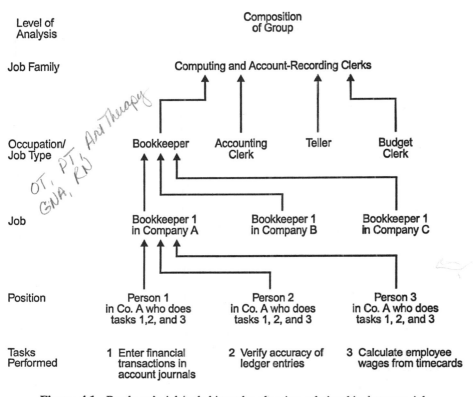

Figure 4.1 Pearlman's job/task hierarchy showing relationship between job family and tasks. (From K. Perlman, "Job families: A review and discussion of their implications for personnel selection," *Psychological Bulletin, 87,* 1–18, 1980. Reprinted with permission of the American Psychological Association.)

ware items. The task specification can be structured by using a matrix of tasks by system components. After the matrix is set up (one dimension on each axis), the analyst can evaluate each cell to determine whether it is a task performed for the job being evaluated.

   In addition to the actions, training analysts also frequently specify the inputs and outputs of task activity, and the conditions for performance. As Goldstein (1986) suggests, each task statement would ideally specify WHAT the worker does, HOW the worker does it, to WHOM/WHAT, and WHY" (p. 40). This list could also include specifying WHEN, that is, under what conditions it is performed. In short, the analyst will often want to specify not only the tasks and subtasks required for a particular job, but also other information or *characteristics* related to each task. Some examples of this type of information are listed in Table 4.2, segregated into general categories.

TABLE 4.1   Example of Activity/Task Breakdown in "Concept of Operations" Methodology

ACTIVITY DECOMPOSITION

ACTIVITIES

1.0   Perform Situation Monitoring

2.0   Resolve Capacity Problems

. . . . .

SUB-ACTIVITIES

1.1   Checking and Evaluating Traffic Flow

2.1   Resolving Transmission Backlog

. . . . .

TASKS

1.1.1   Observe Absence of Overdue Message Alerts

1.1.2   Review Display for Potential Backlogs

. . . . .

2.1.1   Detect Message Backlog

2.1.2   Receive Overdue Message Alert

2.1.3   Determine Validity of Message Backlog/Alert

2.1.4   Determine Appropriate Action to Resolve Backlog/Alert

. . . . .

TASK ELEMENTS

1.1.2.1   SCAN Message__Data__Entry

1.1.2.2   COMPARE significant Message__Data__Entry to global parameters

1.1.2.3   ESTIMATE projected message handling times

. . . . .

(From Phillips et al., "A task analytic approach to dialogue design." In M. Helander, *Handbook of human-computer interaction*, 1988. Reprinted with permission.)

Declarative Knowledge and Skills

An additional class of information that is frequently identified for each task is the knowledge and skills required for a person to successfully perform the task. This analysis is commonly referred to as **Knowledge, Skill, and Ability (KSA) analysis,** following Prien's (1977) categorization. That is, the designer identifies the declarative knowledge that must be possessed by an individual to do the task, and if any of the task is skill-based, they must identify what skills are necessary. For example, Rothwell and Kazanas (1992, p. 104) define task analysis as involving the following activities:

TABLE 4.2   Types of Information Obtained for Task Analysis

CONTEXT

Initiating circumstances or cues (when you do the task)

Equipment, tools, or materials involved in the task

Controls, and critical values

Displays, and critical values

Coordination, communication with others

Effects or results of performing the task

RELATION TO JOB

Importance of the task to attainment of job goals or requirements (e.g., is task essential or optional)

Frequency of task performance

Average time spent performing the task

Concurrent tasks

Severity of consequence if task not performed correctly

PSYCHOLOGICAL FACTORS

Attitude toward the task (positive or negative)

Importance of the task (to the worker)

Difficulty of performing the task

Difficulty of learning the task

1. Determine components of competent performance.
2. Identify activities that may be simplified or otherwise improved.
3. Determine precisely what a worker must know, do, or feel to learn a specific work activity.
4. Clarify conditions needed for competent performance.
5. Establish minimum expectations.

It can be seen that the focus is on description of the tasks, expected performance levels, and the knowledge that must be possessed in order to perform the tasks.

COGNITIVE TASK ANALYSIS

Because an increasing number of jobs have a large number of cognitive components, the traditional behavioral task analysis performed in the workplace is increasingly being augmented by **cognitive task analysis.** Ryder, Redding, and Beckshi (1987) offer the following insights into the distinction:

> Traditional task analysis involves subdividing a job into a hierarchy of components—duties, tasks, subtasks and activities—which are performed as part of the job. The analyst then identifies the inputs, actions and outputs required for each task, as well as conditions for performance and proficiency requirements. . . . The goal of cognitive task analysis is to delineate the mental processes and skills needed to perform a task at high skill levels. By examining the change in knowledge structures and mental processes as one progresses from novice to expert, we can better understand the learning of a new skill and determine how it might be expedited . . . methods for conducting cognitive task analysis are needed which identify the conceptual and procedural knowledge and mental models used to support task performance. (Ryder, Redding & Beckshi, 1987, p. 1263)

While traditional task analysis focused on the knowledge and skills required for each individual subtask, cognitive task analysis goes far beyond that, analyzing the knowledge base for an entire job or set of tasks, looking at concepts and procedures, but also at their interrelationships. Why is cognitive task analysis necessary or important for training program design? One answer is that analysis and training of only the task-specific procedures and task-relevant knowledge is often insufficient for a complex domain. When they encounter novel or difficult problems, workers must have a sound, well-organized, and accessible knowledge base and efficient mental models for task performance (Rouse, 1991). Training which occurs only through sequences of condition-specific procedures is not as effective as training augmented by usable conceptual knowledge (Fisk, 1989; Ohlsson & Rees, 1991). And as Redding (1990) notes, providing the trainee with effective cognitive strategies will help promote transfer of training.

In this book, I will review the most common methods for performing task analysis. Because many of these methods blend behavioral task analysis with cognitive task analysis, I will not divide the methods into these two categories. However, in the remainder of this section, I will provide a brief description of *when* one needs to perform a cognitive task analysis, and an indication of *which methods* are most amenable to *including* a cognitive task analysis.

In the process of performing the initial task analysis, the designer should be sensitive to the following characteristics:

- Complex decision making, problem solving, reasoning, or inferencing from incomplete data
- Large amounts of conceptual knowledge that must be used to perform subtasks
- Large and complex rule structures that are highly dependent on situational characteristics

If any of these characteristics are noted, the designer should strongly consider performing a cognitive task analysis.

There are several recently developed task analysis methods that focus heavily on the cognitive components. Any of these methods would be appropriate for performing a cognitive task analysis:

- Structured interviews
- Verbal protocol analysis
- Critical decision method
- Conceptual graph analysis
- Means and Gott's cognitive task analysis

These methods can also be generalized enough to measure *mental models* as part of the cognitive task analysis. One way to efficiently capture the accuracy and completeness of a person's mental model is to ask them to predict the behavior of a system (Sasse, 1991). This variable can easily be added to any of the methods listed above.

TASK ANALYSIS FOR HUMAN FACTORS DESIGN

Before reviewing the specific methods for task analysis, it is useful to briefly consider task analysis as it is performed in the field of human factors. The main purpose of this section is to introduce readers to the wide range of task analysis methods used in that field. While some of these methods may not be appropriate for training program design per se, knowledge of their existence may prove useful on some future project.

One of the best reviews of task analysis methods used in human factors is a book edited by Kirwan and Ainsworth (1992), *A Guidebook to Task Analysis*. This text categorizes and describes 41 different methods of task analysis. In addition, the book contains ten case studies organized in a very usable fashion. Kirwan and Ainsworth (1992) divide task analysis methods into five categories, with one category devoted entirely to safety. There is also a class of methods used to evaluate the adequacy of task support components. The training analyst may at some point need these types of tools, and should be aware of their existence.

METHODS FOR PERFORMING TASK ANALYSIS

The task analysis methods described in this section are divided into three categories: general methods for data collection, general methods for data representation, and specific task analysis methods. It can be seen that two categories are generic methodologies, the third is specific commonly used methods. Some methods, such as conceptual graph analysis, combine-traditional and cognitive task analysis.

For each category, I will describe the predominant approaches or methods along with the major advantages and disadvantages. I will also provide matrices showing how the methods are interrelated. Other reviews of task analysis and cognitive task analysis methods include Kirwan and Ainsworth (1992), and Jonassen, Beissner, and Yacci (1993). The three categories used in this book and their associated methods are shown in Table 4.3 to provide the reader with an overview of the next three sections.

General Methods for Data Collection

This section describes most of the commonly used general approaches for actually collecting information about jobs, tasks, concepts, equipment, task environments, etc. Most of these are based either on eliciting verbal output from task performers or observation of task performance.

Document and equipment analysis.   One important source of information at the initial stage, for both behavioral and cognitive task analysis, is evaluation of pre-existing documentation. This is especially true if the task pertains to the operation of some piece of equipment. A second type of analysis is an evaluation of the equipment itself, describing general physical characteristics, controls, displays, etc.

There are a variety of types of documents that will be available for analysis, depending on the job and tasks being analyzed. An owner's manual may contain a wealth of information relevant to task analysis. This information will usually be reorganized and structured as various types of knowledge including goal hierarchies for performing tasks (containing the sequences of actions required to perform a task), and a description of the system.

Document analysis cannot effectively be conducted without first identifying a particular *format* for the data as it is pulled from the document. Document analysis is not identified with any one particular method of data representation, and most of the methods identified in the next section are appropriate (e.g., outlines, hierarchies, flow charts, etc.). Table 4.4 shows a matrix of the general data collection methods crossed with the general data representation methods. Stars in a cell indicate normal or reasonable combinations of data collection and representation methods. In Table 4.4, it can be seen that data collected from a document might be represented in outline, matrix, network, hierarchy, or flow chart form. In fact, it is common for data

TABLE 4.3 Taxonomy of General and Specific Task Analysis Methods

General Methods for Data Collection
- Document and equipment analysis
- Unstructured interviews
- Structured interviews
- Group interviews
- Sorting and rating
- Questionnaires
- Verbal protocol analysis
- Observation
- Task simulation with questions

General Methods for Data Representation
- List and outline
- Matrix (cross-tabulation table)
- Structural network
- Hierarchical network
- Flow chart
- Timeline chart

Specific Task Analysis
- Methods Position Analysis Questionnaire (PAQ)
- Delphi and Focus groups
- Controls and Displays Analysis
- Hierarchical Task Analysis
- Extended Task Analysis Procedure (ETAP)
- The GOMS model
- Critical Incident Technique
- Functional Analysis System Technique (FAST)
- Cognitive Task Analysis (Means & Gott)
- Conceptual Graph Analysis
- Associative scaling algorithms
- Activity Sampling
- Functional flow diagrams

from documents to be represented in more than one format, depending on the type of data.

As in the organizational analysis, documents can almost always yield useful information. They are a good place to start, as they often contain basic information. However, it is also almost always the case that the information is

TABLE 4.4   Commonly Used Combinations of Task Analysis Data Collection and Data Representation Methods

Data Collection	Data Representation Data					
	List and Outline	Matrix	Structural Network	Hierarchical Network	Flow chart	Timeline Chart
Document and Equipment Analysis	*	*	*	*	*	
Unstructured Interview	*		*	*	*	
Structured Interview	*	*	*	*	*	
Group Interview	*	*	*	*	*	
Sorting and Rating	*	*	*			
Questionnaire	*	*		*	*	
Verbal Protocol	*			*	*	
Observation	*	*		*	*	*
Task Simulation with Questions	*			*	*	

incomplete. For that reason, one of the other methods described below should be used in conjunction with this method.

Summary of Advantages and Disadvantages.   Document analysis provides a good starting point because it usually includes what one or more experts have considered to be basic material necessary for a job or task. This method provides a good introduction and overview of the job and its subtasks to the analyst. Document analysis is a fairly efficient way to obtain a large amount of information without a lot of extraneous material. The major disadvantage is what you *won't* find in documentation. For some jobs, such as trade jobs like logging or carpentry, there will be almost no information concerning goal structures (what to do and under what circumstances). As might also be expected, the procedural/skill kinds of knowledge developed by experts is almost totally lacking.

Unstructured interviews.   Another good place to start for a task analysis is by talking to experts about their job, and how they perform the tasks. The analyst must somehow get an overview or feel for the task domain, and as a supplement to document analysis, this can be accomplished through unstructured initial interviews.

Interviews conducted to begin the task analysis should be relatively short; sometimes as little as 15–20 minutes is appropriate for the initial interview (see Gordon & Gill, 1992). These interviews are usually conducted with experts; this allows the task analysis to reflect how the job *should* be done. However, there are also times where it is wise to interview a novice. The reason is that the task analysis will have to describe the basic goals and tasks themselves, *as well as* the prescribed method for accomplishing the tasks. As an example, consider the case of developing a training program (or job aid) for a library search system. There are really two parts to this task. The first part has to do with all the goals and tasks with which the students walk in. The second part has to do with using the computerized search system to accomplish those goals and tasks. Interviews with students would yield the first type of information, while interviews with an expert on the search system would yield the second type of information.

As a general rule, the training analyst introduces the interviewee to the work being performed, and asks the expert to provide an overview of the tasks or subtasks. The interview is said to be unstructured because the expert is free to verbalize whatever he/she wishes, and there are no formal questioning guidelines for the analyst. Questions are kept to a minimum, and the analyst determines subjectively at the time what questions are appropriate.

Summary of Advantages and Disadvantages.   The biggest advantage of the unstructured interview is that it takes little to no training. Anyone can ask the interviewee to tell them about their job. But this flexibility is also the biggest disadvantage of unstructured interviews. There is little control over what information is elicited, and it is easy to get off on long tangents. Among other problems, it is difficult for the analyst to know *what* questions to ask, to know *which* trails of discussion to pursue further, and to know *when* various areas of knowledge have been adequately covered (Gordon & Gill, 1989). The expert is usually required to remember what he/she has already verbalized (or risk repetition), and may have difficulty knowing exactly what it is that the analyst is looking for.

In addition, the knowledge elicited will be declarative conceptual and rule knowledge that the expert was able to access in a relatively abstract setting without a specific task context. If we assume that experts rely heavily on tacit procedural skills in performing their job, we must also assume that there is likely to be much knowledge not captured by unstructured interviews (Gordon, 1989, 1992a).

In summary, because of the serious problems with using unstructured interviews, they should be constrained to use for a short introduction to the domain. After that, the analyst should use other more efficient and effective methodologies.

Structured interviews.   Interviews can provide a wealth of information if conducted systematically. A number of methods for structuring interviews have been developed including laddering (Diederich, Ruhmann, & May, 1987; Grover, 1983), goal decomposition (Grover, 1983), repertory grid techniques (Boose, 1986, 1988), question answering (Graesser, Lang, & Elofson, 1987), FAST (Creasy, 1980; Fewins, Mitchell, & Williams, 1992), and question probes (Gordon & Gill, 1989, 1992; Moore & Gordon, 1988). Some of these will be described later in the chapter.

In structured interviews, the analyst usually conducts multiple interviews with one or more experts, and perhaps one or more novices. Before each interview, the analyst prepares some type of guideline for questions to be given during the interview. Usually the interview is kept to an hour or less, and is tape-recorded. The interview should be conducted in a private place, preferably close to where the interviewee works. A table and blackboard are often helpful so that the respondent can illustrate points. If the task involves use of special hardware or software, it is helpful to conduct the interview with those items available.

Interview guidelines might be a list of types of information to be acquired, or might be a set of questions for each of several items on a checklist. As an example of the most common approach, imagine that we begin with the list of tasks described earlier for the procedures associated with mountain bike ownership (a list developed using document analysis). For each item in the list, we could ask "How do you perform this activity?" and "Why do you perform this activity?" *How* and *why* questions are common among specific techniques because they get at the backbone of a goal hierarchy. Other questions can elicit the various types of information shown in Table 4.2.

After the interview, the analyst usually takes some period of time to add the information elicited from the interviewee to the data base. This information then provides the basis for a new interview. For example, each procedure that was elicited becomes the basis for a new set of how and why questions. Successive interviews are conducted until the analyst moves outside of the domain of interest (e.g., you move into discussions of the physical materials of a system, the specific hand motions required to press a key, etc.).

Summary of Advantages and Disadvantages. Structured interviews are both useful and widely used. They have the advantage of being relatively simple and familiar, give one the opportunity to elicit the types of information sought, and there is enough flexibility to make sure that the answers are complete. In addition, they provide guidelines to the analyst and interviewee; it is more clear what the analyst is expecting, and the structure tends to keep discussion focused. After a period of time, the interviewee develops a mental model of the kinds of information being sought, and begins to shorten and focus his/her answers in that direction.

There are, however, certain disadvantages to the use of structured interviews. First, they require time from both the person being interviewed and the analyst, and it takes considerable time to transcribe the audio (or video) tapes and transform the data into whatever format is being used. Second, even with structuring tools, it is often difficult to make sure that certain information hasn't been overlooked. And third, there is really no way to deal with discrepancies between different interviewees.

A final disadvantage that is the most critical is the fact that structured interviews require the interviewee to abstractly think about and discuss the task. For conceptual declarative types of knowledge, this is not a problem. But the method assumes that the person is able to adequately remember and verbalize how and why he/she performs procedures. This is often an unrealistic expectation. As we discussed in the previous chapter, it may be too difficult to verbalize procedural/skill types of knowledge. For this reason, it is a good idea to use both structured interviews and some type of data collection based on task performance, such as performance observation. In addition, for some jobs, the analyst must make sure to focus on unusual or "critical" situations as well as the normative ones.

Group interviews. There are times when collecting data for the task analysis is best performed by getting people to talk together in groups. This will usually be done for one of two reasons. First, asking people to discuss their tasks and knowledge in a group can result in a greater body of information than when the analyst only works with people individually. People can retrieve only so much knowledge when trying to remember information on their own. However, research has shown that people are able to access much more knowledge if there are external cues to trigger that knowledge (Anderson, 1990). The statements verbalized by other experts talking in a group can act as those triggers. Also, having experts gathered together to discuss different approaches or methods may bring out the various procedures for a task, the rationale behind procedures, and the conditions under which one might choose one procedure over another.

The second reason has to do with resolving inconsistencies; there will be times when the information obtained from various sources do not agree. For example, several workers may perform a given job using different procedures, or in a different sequence. In performing the task analysis, the designer may wish to identify a single best set of procedures. Getting the workers to converge on a single set of procedures or other types of knowledge can be helpful to the analyst.

Summary of Advantages and Disadvantages. The advantages of group interviews (when structured) are the same as those for structured individual interviews. In addition, they provide a methodology to obtain consensus from a number of people. Disadvantages are the same as for structured interviews, with the additional disadvantage that they are generally more difficult to conduct.

Sorting and similarity ratings.

This category refers to two methods of identifying how the concepts related to tasks being analyzed are organized in a person's knowledge structures.

Sorting. It is sometimes useful to determine how a person categorizes or applies meaning to various concepts related to their tasks. Many task analysis techniques try to identify the way we structure knowledge by having people sort cards containing the relevant concepts (Hirshman & Wallendorf, 1982; Miller, 1969; Stein, Baxter, & Leinhardt, 1990). These concepts can be single ideas, such as "computer," but they can also be tasks, such as "turn on the computer."

A typical card sorting task will present a person with a series of cards. Each card contains one word or phrase representing a unitary concept. The person is asked to sort the cards into piles where the cards in each pile are similar to one another in some way (Miller, 1969). Sometimes the person is asked to provide a label for each stack of cards, and/or sort the groups into subgroups. Jonassen et al. (1993) provide detailed instructions for conducting a card sort. Results from card sorts can be represented in list, outline, or graphical form by using associative scaling methods (described below).

Card sorts are usually done with two classes of people. First, they are done with several experts in order to determine how they organize the knowledge in their field of expertise. There are techniques, such as associative scaling algorithms, for combining card sort results from several such experts. The second example is where card sorts are done with "novices," that is, those people who will be trainees. The purpose of this task is to determine how the people represent their task in general—not necessarily how they would perform the specific tasks, but how they represent the general ideas and goals in their job. This information can be useful in designing system interfaces, understanding the basic differences between novices and experts, and designing training programs to fit the way trainees think about the task.

Similarity Ratings.   In this method, people are presented with pairs of concepts, and asked to rate their similarity (Brown & Stanners, 1983; Johnson, Cox, & Curran, 1970; Jonassen et al., 1993). Usually, all possible pairs are given to the respondent, and the ratings are summarized in some type of distance matrix showing the degree of similarity between the concept pairs. The other method for summarizing the results of similarity ratings is to depict the concepts as a network, where the graphical distance between nodes reflects the similarity between concepts. The associative scaling algorithms described in the next section will show examples of this display method.

Summary of Advantages and Disadvantages.   The major advantages of card sorting and similarity ratings are that they are simple, straightforward tasks for both analyst and trainee (or expert). In addition, they result in simple, objective data. Their ease of use makes these tools popular when the basic structure of knowledge needs to be identified.

There are a number of relatively serious disadvantages to the use of sorting and similarity rating scale techniques. First, distances derived from direct similarity ratings have been found to be unreliable (Reitman & Rueter, 1980). In addition, the data yielded is quite limited. For example, it is often not particularly helpful in task analysis to know the similarity between a set of job-related concepts. Knowledge structures contain a number of relationships that one might argue to be more important than similarity. The goal, taxonomic, causal, and spatial structures described in Chapter 3 illustrate the importance of capturing the qualitative relationship between concepts. The use of sorting and similarity rating is most useful in those circumstances where the "clustering" of concepts or procedures is central, for example, in designing a software display screen.

Questionnaires.   Questionnaires are hardcopy or computer-based documents containing a limited set of written questions, usually in a fixed format (Cooper, 1986). Questionnaires are given to both prospective trainees as well as expert job performers. As discussed in Chapter 2, questionnaires may contain open-ended and/or closed-ended questions. The most commonly used types of items on questionnaires for job and task analysis are:

- Multiple choice items
- Rating scales

- Rankings
- Matrices to be filled in
- Checklists
- Short answer or open-ended

Questionnaires are a useful follow-up tool after interviews and observation. They provide external validity, allowing the analyst to assume that he/she has adequately captured the task for the general population of people performing the job. The general guidelines regarding questionnaires given in Chapter 2 are applicable for task analysis questionnaires.

Summary of Advantages and Disadvantages. The advantages of questionnaires include familiarity, ease of deployment, low cost, simplicity and straightforward nature of the data collected, the number of potential respondents, and the relatively large amount of data that can be collected. These features make questionnaires an attractive method for task analysis. The major disadvantages are the difficulty of developing a good questionnaire and the limited nature of the data that can be collected. On this second point, imagine trying to get a person who works on an assembly line putting together computer components to fill out a questionnaire explaining how he or she performs their job. The nature of the instrument is very linear and constrained. The nature of most tasks is complex, where subtasks are interrelated in a variety of ways, and each subtask depends on a complex set of preceding conditions. This type of data is often difficult to capture in a questionnaire. In addition, the method suffers from those described under structured interviews; it is often difficult to verbalize or explain certain types of knowledge such as perceptual or skill knowledge.

Verbal protocol analysis.   Many researchers perform task analysis, and particularly cognitive task analysis, by having the worker or expert "think out loud" as they perform a task. The assumption is that the person is expressing the knowledge and reasoning behind *specific task performance.* The verbalizations regarding the performance of a task are termed **verbal protocols** or **think-aloud protocols.** Analysis of such protocols, termed **verbal protocol analysis,** leads to insights into the knowledge and skills being used (Ericsson & Simon, 1993; Means & Gott, 1988; Rasmussen, 1986).

Verbal protocols are usually one of three types: concurrent, retrospective, or prospective (Tolbert & Bittner, 1991) . To obtain a **concurrent** protocol, the expert is asked to perform one or more tasks and verbalize both the knowledge and strategies responsible for his/her task activities. Sometimes the experts are given specific focal points such as identification of stimuli that are important or responsible for their responses, decision strategies, or "why" they made a certain response (Klein, Calderwood, & Clinton-Cirocco, 1986).

Concurrent protocols can sometimes be difficult to collect. If the task takes place quickly, or is too cognitively demanding, verbalization is inappropriate. In these cases, a **retrospective** protocol can be collected (Ericsson & Simon, 1994). For a retrospective protocol, an expert is asked to review his/her own or someone else's task performance, most frequently by watching a videotape. They are asked to remember (or generate) the knowledge and reasoning responsible for the choice of actions. In an empirical evaluation of concurrent vs. retrospective protocol for software usability testing, Ohnemus and Biers (1993) found that retrospective protocols actually yielded more usable information for analysts than did the concurrent protocols.

A third method for collecting cognitive task analysis data is the use of **prospective** verbal protocol. In this method, the expert is given a scenario and asked to imagine that he/she performs the task. They are asked to verbalize their most likely thought processes and action. This method is useful if task performance is simply too difficult, dangerous, costly, or time-consuming. But like structured interviews, knowledge verbalized may be qualitatively different because of the lack of actual task cues.

Summary of Advantages and Disadvantages. Advantages of the method include the following:

- It is more likely to elicit task-relevant information than having the person respond to interview questions without any task cues.
- It is relatively easy and inexpensive to implement, other than having to pay the expert to take the time to perform the task.
- It tends to provide information of two types, declarative conceptual knowledge and explicit rule knowledge for how to perform subtasks, both of which are important for complete task analysis.
- It has face validity.

The method also has certain disadvantages which are frequently overlooked. First, it is very time-consuming and if done properly, requires the expert to perform a large number of tasks. The expert may have difficulty verbalizing what he/she is doing and why. This is sometimes frustrating for both parties involved. The discussion of knowledge types in Chapter 3 sheds light on this difficulty. Given the distinction between *verbalizable* declarative knowledge and *nonverbalizable* procedural/skill knowledge, we can assume that the data collected from verbal protocol analysis will predominantly be declarative rules (e.g., Greeno, Riley, & Gelman's planning nets, 1984; Ohlsson & Rees, 1991). However, research shows that in most circumstances experts rely predominantly on procedural/skill knowledge, which is not easily verbalized. This causes difficulties in providing a verbal protocol explicitly stating *why* the expert acts in a certain way (Gordon, 1992a; Rasmussen, 1986).

One major assumption of most verbal protocol analysis methods is that the expert is capable either of accessing and verbalizing the *skill knowledge* directly, or at least inducing that skill knowledge from "viewing" the situation and their particular response. There is some evidence that, at least sometimes, this assumption is not warranted. Experts have been known to be completely in error concerning the cues responsible for their behavior.

Observation. To circumvent the question of whether verbalized knowledge reflects the knowledge actually used for performance, some researchers suggest use of behavioral observation. This can yield data of two types. First, it is possible to simply catalog a set of situations and the associated responses. For perceptual learning, this is an appropriate approach. For example, a trainee at a fast food restaurant must learn to recognize when the french fries are exactly the right shade of golden brown. While the expert may be able to verbalize some type of information, a simple knowledge-capturing mechanism is to take photographs of different stimuli (in this case various degrees of french fry brownness), with the response of the expert ("too dark, too light, just right"). Thus, observation can result in capturing perceptual skills and other nonverbalizable knowledge by capturing the stimulus set characterizing the subtask and its associated response, without recourse to any symbolic explanation. When this type of analysis is used, it is critical that the analyst capture the entire spectrum of relevant stimuli and their associated response.

There is some evidence that knowing symbolic (declarative) rules can enhance learning of skill knowledge. For this reason, the analyst may want to make observations of a wide range of behavior and then look for patterns of stimulus-response associations. By doing so, it may be possible for the analyst and/or the expert to induce the underlying associative skills responsible for the behavior. There has not been any research to speak of that addresses the validity or adequacy of such inductive methods, probably because one can never directly determine *what* knowledge has actually been used for task performance.

Summary of Advantages and Disadvantages. The major advantage of observation is that the data collected most closely captures subtask performance. The behavior is most valid in the sense that it is the least disrupted by the data collection process itself. There are times when only observation will yield data relevant to what the expert (or novice perhaps) actually does. The disadvantages are that it is usually costly and time-consuming to set up. There are also circumstances where the task cannot be done, for reasons such as interference or safety. In addition, observing a person's task performance does not *directly* give one information concerning why the person is doing what they are. This information must be inferred, which is usually no small task.

Task simulation with questions. This method is one where the expert is asked to perform the task in a simulated environment, and answer questions during the process (Kirwan & Ainsworth, 1992). The task is performed in this manner when it is not possible to actually do the job in its normal setting while answering interview questions (e.g., flying a jet airplane, etc.). The process is similar to verbal protocol, but the person is responding to specific question probes given by the analyst. Examples of such questions are: "Why are you _____?" "Under what conditions would you not _____?" "What would happen if you _____?" etc. These questions are similar to those used in a structured interview, except they are given in the context of performing a specific task.

Summary of Advantages and Disadvantages. Like other task performance measures, this method provides a rich array of cues for the respondent. The method is usually well-structured and is therefore fairly easy for both analyst and respondent. It also captures thoughts and behaviors closer to the real task than structured interviews. Finally, it is safer and easier than using observation of actual task performance. Disadvantages are similar to those for verbal protocol analysis. However, with the right questions and a relatively small number of scenarios (including both normal and "critical" or unusual events) the method can be very efficient, yielding a knowledge base including both conceptual and procedural types of information.

General Methods for Data Representation

Up until this point, we have discussed methods for eliciting knowledge from experts. In addition to choosing one or more data collection methods, the analyst must identify one or more methods for data representation. To make this choice, it is useful for the analyst to determine what types of knowledge are involved in performance of the particular task(s) being analyzed. Identifying the types of knowledge can usually be done after the initial data collection using document analysis and unstructured interviews. For example, if the domain is tennis, piano playing, or some other field dominated by psychomotor skills, then analysis of conceptual knowledge might not be a key factor (although tennis actually has a fairly large component of cognitive "strategy").

The various types of information that seem to predominate in the task can be identified in terms of types of knowledge discussed earlier. The analyst should make an attempt to identify and explicitly represent *all* types of knowledge that might impact the trainee's learning to perform the task. While some researchers have focused predominantly on skill knowledge (e.g., Anderson, 1983), there is enough evidence that conceptual knowledge and mental models are important for transfer and novel problems that declarative knowledge should not be ignored (Cannon-Bowers, Tannenbaum, Salas, & Converse, 1991; Rouse, 1991; Wickens, 1992).

Once the predominant types of knowledge have been identified, the analyst can choose appropriate representation methods from the list below. A comparison and some guidelines will be presented after the methods have been described.

List and outline.   Probably the most traditional format for task analysis data is the use of lists, including outlines or hierarchical lists such as the one shown in Table 4.1. Hierarchical outlines are extremely useful because they can easily convey the structure of tasks and subordinate subtasks, that is, goal hierarchies. Table 4.5 shows an example of a goal hierarchy for powering on a computer in outline form. For a large database, the format is more easily edited and stored than large hierarchical graphs. The outline or table format is so familiar and easy to work within that it is almost always used at one point or another during the task analysis even when other formats such as graphical hierarchies are also being utilized.

Matrix.   A format closely related to the list is that of a cross-tabulation table or matrix (Jonassen et al., 1993; Kirwan & Ainsworth, 1992). A matrix lists the concepts or procedures relevant to the task down the left side of the matrix. Across the top are listed any attributes or dimensions that the analyst feels are appropriate. Instances of such dimensions include actions performed upon system components, the knowledge and skills required to perform the tasks, relevant controls and displays, communications, and consequences of errors (see Table 4.6).

Creation of task matrices is sometimes termed task decomposition. Matrices are useful because, when set up properly, they can provide a sound organization for the data and prompt acquisition of the relevant knowledge for all of the concepts and procedures. That is, the cells act as a cue for collecting data.

**TABLE 4.5   Part of Task Breakdown Into Subtasks
for Powering On a Computer**

Step 1. Insert the operating system diskette in drive A.
    a. Identify the operating system diskette.
        (1) Identify a diskette.
        (2) Identify the diskette label.
        (3) Know discriminating information appearing on the operating system diskette label.
    b. Identify drive A.
        (1) Identify a diskette drive.
        (2) Know the position of drive A.
    c. Open the diskette drive door.
        (1) Identify the latch.
        (2) Know how to open and close the latch.
    d. Orient the diskette correctly before insertion.
        (1) Determine the top of the diskette.
            (a) Locate the label.
        (2) Determine the leading edge of the diskette.
            (a) Identify read/write opening.
    e. Insert the diskette.
    f. Close the diskette drive door.
Step 2. Turn the computer's power switch to "on."
......

(Reprinted with the permission of Macmillan College Publishing Company from THE DESIGN, DEVELOPMENT, AND EVALUATION OF INSTRUCTIONAL SOFT-WARE by Michael J. Hannafin and Kyle L. Peck. Copyright 1988 by Macmillan Publishing Company, Inc.)

Structural network.   It is becoming increasingly common to see knowledge and procedures represented in graphical format (e.g., Gordon & Gill, 1992; Zaff, McNeese, & Snyder, 1993). In this section, the term structural network will refer to a graphical format of nodes and links that describe factual knowledge as concepts and their interrelationships. In the terminology of conceptual graph structures described in Chapter 3, it refers to taxonomic knowledge, spatial knowledge, and causal knowledge. This knowledge is represented in a network form which might be somewhat hierarchical in nature but probably not strictly so. Goal structures are central to task analysis and are considered separately under the next section, *hierarchical networks*.

Structural networks can be developed using any syntax that the analyst wishes. Many researchers use a method developed in educational psychology,

TABLE 4.6   Example of Tabular Task Analysis Format

Step no.	Description	Displays	Required Action	Feedback	Comms.
A	**RACK FROM STORAGE TO PLANT**				
A1	**RACK HANDOVER**				
1	Accept control of transfer machine from storage at Control Station 1 (CS1)	*CONTROL OFFERED* Light illuminated	ACCEPT push- button	*CONTROL OFFERED* light goes out, and *CONTROL ACCEPTED* is illuminated	Confirmation of control with Storage Control Room
2	Read rack number	CCTV monitors	Camera controls		
3	Input rack number to computer for checking	Previous reading Computer prompt	Input number using keyboard	Computer response— *continue or re-enter*	
A2	**DELIVER RACK TO C1**				
4	Open Gate A and the building door		2 push-buttons	None	
5	Release travel locks		*RELEASE LOCKING BOLT* push-button	*LOCK OPERATING* light will go out	

(From Kirwan & Ainsworth, *A guide to task analysis,* 1992, Taylor and Francis Inc. Reprinted with permission.)

concept mapping, where one uses whatever concepts and labels that seem appropriate for the domain under study (e.g., Klinger & Gomes, 1993; Zaff et al., 1993). Figure 3.2 is an example of a concept map. There are also several graph syntaxes that have been developed as a way to standardize development of structural networks, one of which is **conceptual graph structures** (see Figure 4.2).

The advantages of using a network format over lists or tables is that it

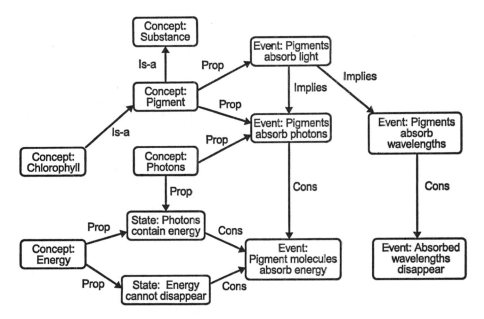

Figure 4.2   Structural network for concepts related to photosynthesis.
Cons = Consequences, Prop = Property.

not only makes interrelationships between ideas explicit, but it visually conveys those interrelationships in a very useful manner (Jonassen et al., 1993;
Gordon, Schmierer, & Gill, 1993). This ability to see the structural relationships of a knowledge base make the data collection task easier.

The major disadvantage of using networks for representation is the un-
wieldiness of the databases. Graphs must be drawn on paper or developed
using special computer software (e.g., *SemNet*™ produced by the SemNet
Research Group in San Diego; see Fisher, 1990). Both methods have their
advantages. Graphs on paper are visually very useful and can act as a guide
during interview processes. It is easier to see the big picture than with the
computer format because computers cannot display very many nodes at one
time. However, computer networks have advantages such as efficient storage
of large graphs, ease of navigation, and ease of adding to or changing the
graphs. SemNet™ also allows the analyst to view and print the graphs in list
format.

Hierarchical network.   This format is very frequently used to represent task or goal hierarchies. Hierarchical networks consist of upside-down
tree structures, with nodes and links between the nodes. The links may or may
not be labeled. An example of a hierarchical network is presented in Figure 4.3.

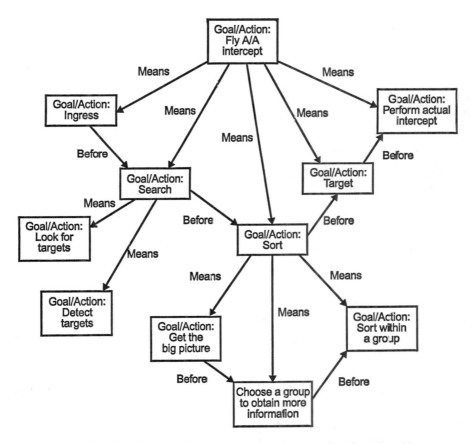

Figure 4.3 Part of hierarchical network for tasks performed during an Air-to-Air intercept in an F-16.

Any graphical syntax can be used in creating hierarchical networks, with the one requirement that some set of "lower" or subordinate nodes must support or allow the procedure above them. Like structural networks, there are certain graph syntaxes that have been developed to standardize the development of hierarchical networks, two of which are the GOMS model and Conceptual Graph Structures. The benefits of such standardization are listed below. Hierarchical networks are probably one of the most useful methods in task analysis. Like structural networks, they convey both content and structure of the task components.

The disadvantages of using hierarchical networks to represent goal structures include the fact that they generally do not capture the time and sequential elements as do flow charts and timeline charts. And like structural networks, they can be cumbersome for large databases. However, as noted previously, there are now software programs such as SemNet™ that allow the analyst to toggle between a list and graphical format of the same hierarchy.

Flow chart. Another graphical notation frequently used for task analysis is a flow chart format. Flow charts capture the chronological, linear aspects of a task. A flow chart usually shows the sequence of subtasks as they are normally performed, along with certain juncture or decision points. They often indicate alternative paths to perform the same task, and categorize different types of activities such as decisions, information transmission, and operator actions. The graph may also indicate how the worker relates to other people or equipment. Sometimes a flow chart specifies the optimal path, other correct paths, and even common incorrect paths. Figure 4.4 shows an example of one type of task flow chart termed an Operational Sequence Diagram. This particular diagram was developed by Phillips et al. (1988) for the task of resolving transmission backlogs. Figure 4.5 shows a flow chart indicating various options open for performing a set of subtasks for scanning an image into

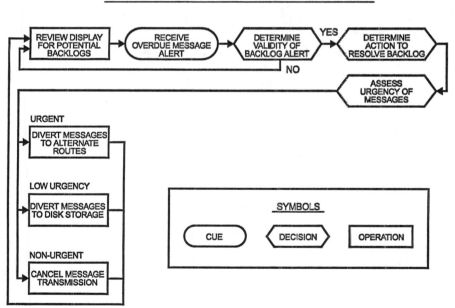

Detection and Resolution Function: resolving transmission backlog

Figure 4.4   Task analysis in flowchart format with decision points. (From Phillips et. al., "A task analytic approach to dialogue design," In M. Helander, *Handbook of human-computer interaction,* 1988. Reprinted with permission.)

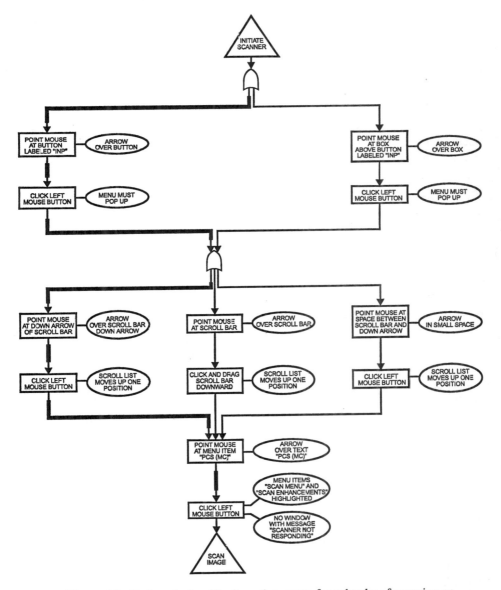

Figure 4.5 Task analysis with alternative routes for subtasks of scanning an image into a PC. (From a course project completed by Bill Brown at University of Idaho. Used with permission.)

a PC. There are a great many types of flow charts; the section below will describe only a few of the most applicable.

Summary of Advantages and Disadvantages.   In general, formal flow chart methods and syntaxes have been developed to provide a manner for representing task information that is easy to understand and is more rigorous and precise than text (Kirwan & Ainsworth, 1992). Those factors are the major advantages of using flow charts, they convey information in a precise format that visually shows the interrelationship among tasks.

Disadvantages have to do with constraints imposed by the graphical format. First, there is considerable evidence that people represent task information cognitively as goal hierarchies rather than strictly linear sequences of information (Jonassen et al., 1993). Trying to impose a flow chart sequence on a set of tasks may artificially impose a sequence where there doesn't have to be one. In addition, certain types of information tend to be left out. This information includes conditions that initiate tasks, results of task performance, and conceptual knowledge associated with tasks. Nevertheless, flow charts provide valuable information, and are especially useful when combined with other methods.

Timeline chart.   Timeline charts are used for timeline analysis. According to the American National Standards Institute, a timeline analysis is "an analytical technique for the derivation of human performance requirements which attends to both the functional and temporal loading for any given combination of tasks." In other words, each task must be described along with the amount of time necessary to perform it. Timeline charts are used for this purpose. Obviously, if an analyst is evaluating tasks where time is relatively unimportant, this analysis would not be worth pursuing.

A timeline chart usually depicts the task on a vertical axis and time units across the horizontal axis. Lines, bars, or boxes then show the time requirements for each task or subtask. An example from Kirwan and Ainsworth (1992) is shown in Figure 4.6.[1] In this particular example, additional information is conveyed by using different pattern coding in the boxes.

Timelines require the use of certain types of behaviorally based data collection methods, such as verbal protocols, or more ideally, a less intrusive observation method. Usually the analyst makes videotapes of the task performance for later detailed analysis. From this description, one can see certain disadvantages to the method. It is tedious and time-consuming to create timeline charts. Other disadvantages include the very restricted nature of the data

---

[1]The Kirwan and Ainsworth (1992) text is an excellent reference for this type of format.

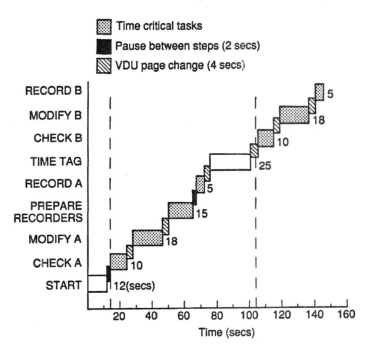

Figure 4.6   Example of timeline chart. (From Kirwan & Ainsworth, *A guide to task analysis,* 1992, Taylor and Francis Inc. Reprinted with permission.)

(temporal relationships only). Obviously, timelines do not provide much qualitative information, such as the relationship between subtasks. The advantages of timeline charts are that they are conceptually very simple to develop without special training. They have high face validity in that they represent what they are supposed to represent. When time factors are important features of the task analysis, they are an effective tool.

Specific Task Analysis Methods

In this section, I will very briefly review some of the more specific methodologies that have been developed over the years to perform task analysis. Some of these are somewhat general instances, such as hierarchical task analysis, while others are very specific techniques and/or data representation procedures. For each method, I will first list the generic methods used in the technique (if any are specified) and then give a brief text description. For additional information, readers are referred to Jonassen et al. (1993) and Kirwan and Ainsworth (1992).

Position Analysis Questionnaire

- *Data collection method(s)*: Questionnaire
- *Data representation method(s)*: Matrix

The Position Analysis Questionnaire was developed to identify general features pertaining to a job (McCormick, Jenneret, & Machan, 1969). In this technique, employees are given a questionnaire containing a list of job "elements" organized around categories such as mental processes, work output, job context, etc. They are asked to identify those job elements that apply to their particular position.

Delphi and focus groups

- *Data collection method(s)*: Group interview
- *Data representation method(s)*: None specified

The Delphi method is considered a structured group interview method used to bring individual opinions to some type of convergence (Dalkey, 1969; Jonassen, Hannum, & Tessmer, 1989). The focus group is an unstructured group interview method designed to foster open communication, discussion, generation of ideas, and problem solving.

Delphi Method.   The Delphi method is a systematic process of obtaining opinion from several people iteratively until some kind of consensus is reached. It does not necessarily have to be conducted within the context of an actual meeting between the people. As Jonassen et al. (1989) state:

> Delphi involves an anonymous, independent, noncompetitive survey of experts to obtain consensus without necessarily involving group meetings. The technique essentially entails a series of surveys using the same experts, each survey dependent upon the responses of the previous one. (p. 397)

Jonassen et al. (1989) suggest the following steps to perform a Delphi analysis:

1. Select the individuals who are to participate in the analysis. The group should consist of at least 7 and no more than 100 individuals.
2. Write an initial question and distribute as a questionnaire to all of the group members. This question is usually a relatively simple and open-ended one, such as "How do you accomplish the task of X?"
3. Clarify responses from the first questionnaire. This can be done through personal interviews of the group members, or through distribution of a second questionnaire.

4. Write and distribute third questionnaire. In the third questionnaire, all of the information or items from the respondents is listed in some type of structured format. All members are asked to rank or somehow rate the items.

5. Summarize responses to the third questionnaire and distribute subsequent questionnaires. A summary of responses is tabulated and presented to respondents. New ratings or rankings are also collected. Each time a questionnaire is sent out, the members will be able to see where divergence and convergence is occurring. More convergence will be obtained through each round. However, at some point the analyst will be able to see that further convergence is not likely to occur.

6. Report the results to group members and management.

The Delphi method has certain advantages and disadvantages, and like any other method must be used for appropriate purposes. The major advantages are that it avoids many of the problems encountered in groups which are actually gathered together. Members can act anonymously; it weights everyone equally so that all types of people can voice an opinion, and it keeps the assertive individuals from dominating the discussion. However, the method is certainly more work, and is more time-consuming than a group discussion. It also may not work well for people who have poor reading skills, or do not like having to take the time to read over material presented in this fashion.

Focus Group.   A focus group is conducted by bringing the individuals together to talk. Workers will, in general, find this a more natural and easier method of sharing data. In a focus group, the analyst usually begins by asking the members to introduce themselves and state their backgrounds. It is standard procedure to tape-record the session, but you must be sure to obtain the members' permission before doing so. The analyst presents concepts, rules, principles, or procedures as topics for discussion. After that point, the analyst should provide only enough input to the group to keep discussion on target, and clarify any ambiguities. The analyst should not verbalize his or her own opinions on the topics. If the group seems to need more time after two hours, another meeting should be scheduled.

After the focus group has met, the analyst will transcribe the tape and compile the data. This can be done in several ways. One is to simply make lists of relevant information, such as task elements, subtask contingencies, knowledge, etc. Another is to perform the analysis and combine it with other methods such as a GOMS analysis.

Controls and displays analysis

- *Data collection method(s)*: Document and equipment analysis
- *Data representation method(s)*: List, Matrix

This is a method developed to evaluate the interface between human and system. A task analysis often requires the full evaluation of a set of tasks, including tasks to operate a system. Such an analysis cannot be completed without some level of analysis of those parts of the system with which the person must interact. The controls and displays analysis provides one method of doing this.

In this method, the analyst identifies all controls and displays relevant to task performance. In addition, all relevant characteristics are determined, including the range of values for both controls and displays, critical values for given tasks, their physical characteristics, absolute and relative location, exactly what is being controlled or displayed, etc.

The information collected is usually listed in tabular format, and used for two purposes. First, it can provide data necessary to determine that all controls and displays are necessary to perform each task. Obviously, this requires analyzing the tasks themselves to determine what type of information is necessary for each task, as well as what type of control input. Second, as a training program is developed, the information relevant to controls and displays can be integrated into the instructional content.

Hierarchical task analysis

- *Data collection method(s)*: None specified
- *Data representation method(s)*: Hierarchical network or outline

Hierarchical task analysis (HTA) is a general method for identifying tasks and subtasks to attain job-related goals. The analysis results in a hierarchy of **operations** required to perform a task and **plans,** statements of the conditions necessary to undertake the operations. Figure 4.7 shows an example of a hierarchical task analysis. The advantage of this method of graphing goal structures is that the analyst is cued to look for initiating circumstances or conditions that must be met for the operations to be performed. One disadvantage of the method is that goal structures are usually many layers deep. Using the HTA syntax requires the analyst to arbitrarily determine which level is to be labeled as "operations" and which level is to become "plans."

Extended task analysis procedure (ETAP)

- *Data collection method(s)*: None specified
- *Data representation method(s)*: Flow chart

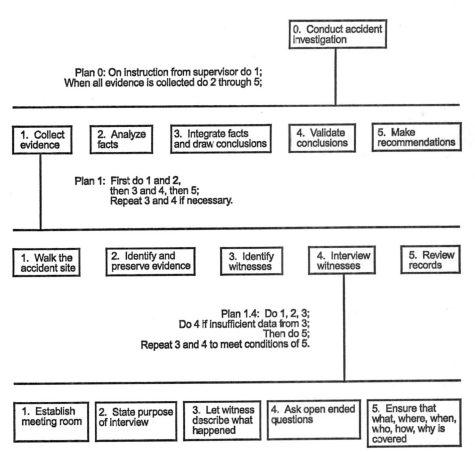

Figure 4.7 Hierarchical Task Analysis for conducting an industrial accident investigation. (Tasks from analysis conducted by David McCallister for a course project at the University of Idaho. Used with permission.)

This method is a three-phase approach to task analysis developed by Reigeluth and Merrill (1984). In this method, the task is first analyzed via an information-processing approach using the following steps (Merrill, 1987):

1. Identify the operations and decision steps of the procedure.
2. Sequence the steps in the order in which they would be performed.
3. Prepare a flow-chart representation of the sequenced steps.
4. Validate the flow-chart, using several different initial inputs.

In the flow chart, the sequence of steps is indicated by arrows, the operations represented by rectangular boxes, and the decision points represented by diamonds.

The second phase essentially consists of performing the same analysis in more detail. That is, the steps identified in the information-processing analysis are broken down into more refined steps. The third phase, *knowledge analysis*, consists of identifying any prerequisite concepts or facts which the trainee must know in order to learn a given step in the flow chart.

The GOMS model

- *Data collection method(s)*: None specified
- *Data representation method(s)*: Hierarchical network

The GOMS model is a representational format for specifying the steps and procedures for performing tasks (Card, Moran, & Newell, 1983). It was developed to analyze the operation of some system, and the type of knowledge that is evaluated reflects this emphasis. Kieras (1988) provides a very readable explanation of how the GOMS model can be used for task analysis.

Essentially, the GOMS model decomposes tasks into four types of node: goals, operators, methods, and selection rules. **Goals** are defined similarly to previous definitions in this chapter; things the operator of a system wants to accomplish. A goal is described as an action-object, usually in the form of <verb noun>. Examples include "adjust bicycle seat," or "move a file." Tasks are decomposed into goal hierarchies consisting of goals and their subgoals. Goals are *not* executable actions.

Operators are the actions that the user executes to accomplish a goal. While both take the form of action-object, they are conceptually distinct; a goal is something to be accomplished, while an operator is something that is actually executed. Kieras (1988) notes that defining a particular action-object such as "move file" as goal or operator is an *arbitrary* and subjective judgment on the part of the analyst. Note that this is similar to the arbitrary discriminations required for the Hierarchical Task Analysis. According to Kieras (1988), operators can be external operators such as read text from screen, turn a page, pressing a key, or mental operators such as storing or retrieving items from long-term memory.

A **method** is a sequence of operators that accomplishes a goal. As such, it is not a conceptually and qualitatively distinct component of the model so much as a collection of the operator component of the model. Figure 4.8 illustrates this idea. Methods will "call" a sequence of actions termed operators, but can call other methods as well.

The fourth component, selection rules, specify which method will be chosen, given that there is more than one method available for accomplishing a goal. For example, in the Windows™ software environment, there are two ways of selecting "OK" when asked if the user is ready to print. The first

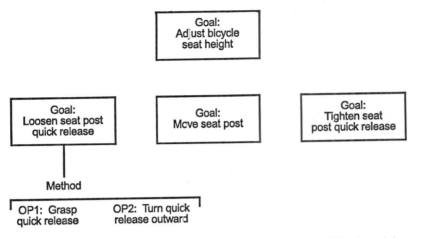

Figure 4.8　Example of goal, method, operator components of GOMS model.

includes the operator(s) of using the mouse to click on OK, the second is pressing the Enter key. Selection rules specify the conditions under which each alternative method will be chosen.

Readers can see that the knowledge represented is procedural in nature, either explicit rule knowledge or implicit skill knowledge. As such, it is a convenient representational syntax for capturing the knowledge directly associated with tasks and subtasks. However, when more general, conceptual knowledge is evaluated, other formats must be used to augment the method. The GOMS method is sometimes considered cognitive task analysis because it analyzes tasks in terms of cognitive goal hierarchies. However, the model does not address conceptual knowledge or mental models, making it a restricted sort of cognitive task analysis.

The technique described in this section consists of a specific graph syntax for performing task analysis. A more complete technique combining a set of data collection methods with the conceptual graph syntax is described under the section below, *Conceptual Graph Analysis*.

Critical incident technique

- *Data collection method(s)*: Structured interview and retrospective verbal protocol
- *Data representation method(s)*: None specified, but usually list or tabular

There are several people who have suggested a variation of the critical incident technique (Flanagan, 1954; Hoffman, 1987; Klein et al., 1986; Means & Gott, 1988). In particular, Klein and collaborators have used a variation of this method, termed **Critical Decision method,** to identify cognitive processes involved in decision making (Klein et al., 1986; Klein & Thordsen, 1988; Klinger & Gomes, 1993).

In its most general form, the critical incident technique identifies various features of important incidents—critical events that represent important variations from normal task performance. Sometimes critical incidents are defined as errors or near-miss situations (e.g., Fitts & Jones, 1947). Other times critical incidents simply mean those situations that are different from normal and also difficult for one reason or another (e.g., Hoffman's, 1987, "tough cases").

Several researchers have tried to provide some guidelines for use of this technique. For example, Flanagan (1954) suggests that somewhere between 50 and 100 incidents should be analyzed. However, Kirwan and Ainsworth (1992) state that for a complex task, it is unknown how many incidents are really required. Meister (1985) has provided guidelines for application of the method. He suggests that analysts do the following:

1. Determine the general aims and goals of the activity.
2. Specify the criteria for effective or ineffective behaviors.
3. Collect the data, asking for detailed information about critical or memorable events. A variety of methods can be used, including individual interviews, group interviews, questionnaires, and checklists. In general, the interview seems to be the most popular.

Advantages to this method include the fact that it may be the only way to identify factors leading to errors and accidents. In addition, it is often a good idea to identify how experts would approach different and difficult scenarios. However, there are certain disadvantages as well. First, analysts will need to identify a particular methodology and practice their skills at applying the method. In addition, Kirwan and Ainsworth (1992) point out that the method is very narrow in terms of data collection techniques and types of data resulting from the analysis. For this reason, it should not be used except as a supplement to other methods.

Functional Analysis System Technique (FAST)

- *Data collection method(s)*: None specified
- *Data representation method(s)*: Hierarchical network

This method was developed to be a relatively quick and easy way to perform task analysis (Creasy, 1980; Fewins, Mitchell, & Williams, 1992). It is similar to HTA and GOMS; however, it is somewhat more restricted in scope. In this technique, the analyst develops a task hierarchy by successively working upward and downward. The analyst works upward by asking WHY a task is performed (thus identifying the superordinate goal), and works downward by asking HOW a function is achieved in operational terms. When all of the "how" and "why" questions can be answered by the hierarchy, the analysis is complete.

Obviously this type of analysis focuses on development of only the tasks and their subtasks. For certain types of situations this type of approach will be adequate (e.g., developing a job aid for a relatively simple and straightforward task). For more complex analyses, the FAST approach can be a good starting point.

Means and Gott: cognitive task analysis

- *Data collection method(s)*: Verbal protocol, structured interview
- *Data representation method(s)*: Flow chart

There are several different methods for performing cognitive task analysis that depend on using questioning methods with a specific network or flow chart syntax; this section and the next present two of them.

Means and Gott (1988) developed this particular method to analyze how experts perform difficult troubleshooting tasks. Once this implicit knowledge is made explicit, intelligent tutoring systems can be developed for training programs. This technique consists of several phases which are briefly summarized below.

1. Specify and categorize major types of problems in the domain.
2. Have experts generate representative problems for each category.
3. Analyst works with experts to identify the knowledge and skills required for that problem.
4. The expert is asked to describe the particular fault within the system.
5. The expert is then asked to describe how he or she would troubleshoot the fault. This phase could really be considered the beginning of the task analysis per se. The troubleshooting steps are outlined in enough detail that a technician could follow them. In this step, the analyst identifies the overall hypothesis or goal(s), the plan guiding various actions, the actions themselves, the consequences of each action, and what the consequences mean to the troubleshooter.
6. During step 5, a flow chart is developed to represent the information. This flow chart shows the preferred solution path, along with alternative solution paths. This represents a sort of *effective problem space*. Any move outside of that space is considered to constitute error.

In some instances, the approach can be modified to have one expert make up the problem and the analyst would then give the problem to another expert to solve. In addition, the problems can be given to novices to solve. This serves several purposes. First, it supports trainee analysis (described below) in that it helps the analyst determine what the trainees do and do not know regarding the task and its associated knowledge base. Second, it helps the analyst identify the novice user's mental model, including any misconceptions. This is also helpful in developing a training program.

The advantages of the method are that it provides a rich body of data for developing instructional systems. Unlike many other task analysis methods, it captures the underlying knowledge and skills used by experts in problem solving. It also provides a data representation format that is easy to translate into computer-based training programs and simulations. The major disadvantage is the fact that like most cognitive task analysis procedures, it is time-consuming and requires a certain amount of expertise to implement.

Conceptual graph analysis

- *Data collection method(s)*: Document analysis, structured interviews, observation
- *Data representation method(s)*: A specific structural and hierarchical network format termed Conceptual Graph Structures

Gordon et al. (1993) suggest use of the conceptual graph structure syntax to represent knowledge obtained through a complementary set of data collection methods including document analysis, structured interviews, and inductive analysis based on behavioral observation. The method has been used in a variety of domains (e.g., Gordon, 1992b, in press; Gordon & Gill, 1992; Gordon, Kinghorn, & Schmierer, 1991; Moore & Gordon, 1988) and has been empirically tested as a method of cognitive task analysis for instructional design (Gordon et al., 1993).

In brief, conceptual graph analysis consists of several methods for iteratively acquiring data and representing the information in the form of conceptual graph structures. If the analyst determines that a training program is required, the graphs are further "engineered" by working with the expert to make the information complete, consistent, and organized for instructional design (Gordon et al., 1993). The syntax will be described in the next section, followed by a brief description of the knowledge acquisition methods.

Conceptual Graph Structures.   Conceptual graph structures consist of nodes that are linked by labeled directional arcs, as shown in Figure 3.3 and reprinted in Figure 4.9 (Graesser & Clark, 1985; Graesser & Gordon, 1991).

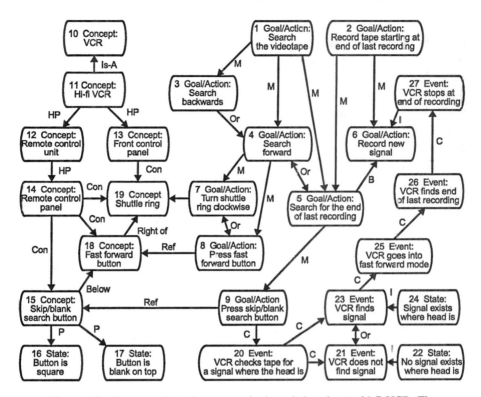

Figure 4.9 Conceptual graph structure for knowledge about a hi-fi VCR. The abbreviations for arc categories are: B = Before, C = Consequence, Con = Contains, HP = Has Part, I = Initiate, M = Means, P = Property, Ref = Refers-to.

In these networks, each node contains specific information in the form of a concept or statement (e.g., "VCR," or "VCR reads signals on tape"), and the category of that concept or statement. The categories include concept, state, event, style, goal and goal/action. **Concepts** are individual entities or ideas usually expressed as nouns or noun-phrases, such as VCR, John Smith's VCR, democracy, or time. **States** are statements of some static condition or ongoing characteristic of a person, object, system or aspect of the environment. Examples would be "Computer is off, John is angry, Animals have eyes, etc." **Events** are statements describing some state change over a period of time, such as Book falls from table, Error message flashes on screen, Earthquake occurs, etc. A **Style** node specifies the speed, intensity, force, or qualitative manner in which an event unfolds (e.g., the pencil falls off the desk quickly, loudly, etc.). **Goal** refers to some state or event that is desired by an agent/person. Graesser's original taxonomy separates goals from events performed in order to accomplish those goals (Graesser & Franklin, 1990). For the syntax described here, **goal/action** nodes are used to represent a node that could be *either* a goal *or* an action that is performed to attain a goal. An example of a goal hierarchy containing goal/action nodes is the subset of nodes in Figure 4.9 labeled 1 through 9. It can be seen that "Search the videotape" might be conceptualized as either a goal or an action performed by the operator. Because of the dual nature of goal nodes, acting as either goal or activity, we have simplified the graphs by using one node to represent both goal and action. However, the analyst can separate these two types of node if desired, as was done in the GOMS model described earlier.

The directional arcs between nodes represent relationships between the nodes, such as "*X is-a Y.*" While there are conceivably hundreds of such relationships in any given knowledge base, CGSs use a restricted set of arc categories to represent these relationships. The legal arc categories are listed in Table 4.7. The table shows that each arc category can only be used to link certain types of nodes. For example, a concept node cannot be linked to another concept node by a **before** arc.

The set of nodes and arcs listed in Table 4.7 acts as a *base* syntax for a given task domain, in that occasionally it is convenient to use additional arc types that are more specific to the particular domain (analogous to a domain having its own technical terms).

Data Collection.   In developing the graphs, data is acquired by using the following combination of activities:

1. Initialize the graphs by using document analysis or unstructured interviews.
2. Expand the graphs by using question probes to structure interviews (described below).
3. Expand the graph goal hierarchies by using observation and rule induction.

A conceptual graph structure is initialized by translating information from documents or unstructured interviews into graph form (Gordon & Gill, 1992). The initial graph is usually a simple overview of the domain; ideally, it is just a simple graph or subgraphs of the tasks being studied. A task list such as the one described for mountain bike ownership will usually suffice.

The initial graph is then used to develop **question probes** for structuring interviews with task experts (or alternatively with novices and intermediate individuals). In essence, for each node on the graph, the analyst asks a particular set of questions. The information obtained is added to the graphs in the appropriate places. This process is iterated until the expert deems the graph complete (see Gordon & Gill, 1992, for a discussion of this issue).

The set of question probes for any given node is generated by combining

TABLE 4.7   Major Arc Categories for Conceptual Graph Structures

Source Node	Arc	Terminal Node	Arc Definition
Concept	**IS-A**	Concept	Source node is a kind or type of terminal node
Concept, Event, State	**EQUIVALENT- TO**	[similar node type]	Source is similar to terminal node, differences are inconsequential
Concept, Event, State	**INSTANCE-OF**	[similar node type]	Source node is a specific instance of the terminal node
Concept	**PROPERTY**	Event, State, Goal, Goal/ Action	Source node concept has the property of the terminal node
Concept	**HAS PART**	Concept	Source node has as a part, the terminal node
Concept	**SPATIAL RELATION, e.g. Contains Left-of**	Concept	Specifies spatial relationship between source and terminal nodes
Event, State, Style	**IMPLIES**	Event, State, Style	Source nodes implies existence of terminal node, overlap in time
Event, State, Style	**CONSE- QUENCE**	Event, State, Style	Source node causes or enables terminal node
Event	**MANNER**	Event, Style	Terminal node specifies the manner in which the source event occurs
Style	**MANNER**	Style	Terminal node specifies the manner in which the style node occurs
Event, State	**BEFORE/ AFTER/ DURING**	Event, State	Specifies the temporal relation between source and terminal node
Goal, Goal/ Action	**MEANS**	Goal/Action	Terminal node is means of achieving source node
Event, State, Style	**INITIATE**	Goal, Goal/ Action	Source node initiates or triggers the terminal node
Goal, Goal/ Action	**BEFORE/ AFTER/ DURING**	Goal, Goal/ Action	Specifies the temporal relation between source and terminal nodes

(*continued*)

TABLE 4.7 Major Arc Categories for Conceptual Graph Structures (cont.)

Source Node	Arc	Terminal Node	Arc Definition
Goal/ Action	**MANNER**	Goal/Action, Style	Terminal node specifies the manner in which the source node occurs
Goal/ Action	**CONSE- QUENCE**	Event, State, Style	Source node causes or enables terminal node, or precedes terminal node
Goal	**OUTCOME**	Event, State	Terminal node specifies whether source node (goal) is achieved
Goal, Goal/ Action Event, State, Style	**REFERS-TO**	Concept	Source node has a compo- nent that refers to the ter- minal concept
Goal, Goal/ Action Event, State, Style	**AND/OR**	[similar node type]	Both source and terminal nodes exist Either source or terminal nodes exist

the contents of the node with a generic set of questions developed specifically for this purpose. The generic set was designed to elicit all possible types of information that can "legally" be related to a given type of node, based on the conceptual graph structure theory. The generic set of question probes for the various node types is given in Table 4.8.

Thus, for each concept in the graph, the analyst would ask all of the questions related to concepts, such as "what is X, what are the properties of X," etc. In using question probes to structure interviews, the expert and ana- lyst can look at the graph sketch during the questioning process. As the expert talks, the analyst adds the answers to the graph. In this way, the graph acts as a cue for the question probes, and also provides a short-hand notation system for recording the verbalized information.

The question probe method elicits verbalizable conceptual and rule knowledge. However, most experts use a large amount of procedural/skill knowledge that is difficult to verbalize in structured interviews. One method to alleviate this problem is *observation of performance*. The expert performs a representative set of problems (as in the Means & Gott approach above), and

TABLE 4.8   Generic Question Probes for Node Categories

Node: Question probes

For a Concept:
What is _____?
What are the types of _____?
What are specific instances or examples of _____?
What are the properties of _____? OR What are the properties of _____
    that distinguish it from _____?
What are the parts of _____?
(any domain-specific questions)

For an Event:
What happens before _____?
What happens after _____?
What are the consequences of _____ occurring?
Why does _____ occur? (Optional because it covers the same information
    as previous two)

For a State:
What happens before _____?
What happens after _____?
What is the consequence of _____ occurring?
What causes or enables _____?
Why is _____? (Optional because it covers the same information as
    previous two)

For a Style:
What are the consequences of _____?
What causes or enables _____?
How _____?
What happens if not _____?

For a Goal:
What happens before having the goal of _____?
What happens after having the goal of _____?
How is the goal of _____ attained?
What state or event initiates the goal of _____?
What is the outcome of _____?
Why is the goal _____? (Optional because it covers the same information
    as previous two)

For a Goal/Action:
How do you [how does a person] _____?
What do you [what does a person] do before _____?
What prevents you from being able to _____?

(continued)

TABLE 4.8   Generic Question Probes for Node Categories (cont.)

Node: Question probes
What do you [what does a person] do after _____?
What states or events cause or enable you [a person] to _____?
Why do you [why does a person] _____?
What are the consequences of _____?
What happens if you do not _____?

task performance is observed and/or videotaped. After this process, the analyst and expert review the tapes, identifying the initiating cues for each task. If it is not possible to determine a set of rules that can account for the behavior, the knowledge must still be captured somehow. One way of doing this is to index the examples into the graph directly. The set of examples can be used directly for training at the proper time.

The advantages of this method are that it captures all types of knowledge with one graphical syntax. That is, a goal hierarchy can be constructed, and all related taxonomic, spatial, and causal knowledge tied in at appropriate points. Advantages of using question probes are that it takes the burden off both analyst and expert to determine the direction of questioning, and the method acts as a "press" into all areas of the knowledge structures.

The disadvantages are that the use of conceptual graph structures requires learning of a graphical syntax; like learning a language, this takes some time. In addition, the process of developing the graphs is itself time-consuming. Like other cognitive task analysis methods, it should be used when the task is complex (e.g., large and intricate goal structures) and/or task performance depends on a large and complex conceptual knowledge base (e.g., Gordon, 1992b).

Associative scaling algorithms

- *Data collection method(s)*: Sorting and Similarity Rating
- *Data representation method(s)*: Structural network

Scaling algorithms are a general set of procedures performed on chunks of information, such as concepts or statements, that have been ranked, sorted, or judged in some fashion by experts (e.g., Cooke & McDonald, 1987; McDonald & Schvaneveldt, 1988; Schvaneveldt, 1990; Zubritsky & Coury, 1987). The methods use mathematical procedures such as multidimensional scaling to infer the underlying cognitive structure responsible for a given pattern of data (e.g., see Goldsmith, Johnson, & Acton, 1991). As an example, for cognitive task analysis, the designer may ask several experts to identify the major concepts, activities, and so forth in the domain. Experts are then asked to sort these into piles, or judge all possible pairs in terms of their similarities. The similarity judgments are subjected to a mathematical analysis that infers an underlying multidimensional space or network that can account for the judgments (or sortings).

The analysis results in a structure (typically one per expert) that is a "map" of the knowledge, where topics or concepts are laid out in topological relationship indicating the "closeness" or degree of psychological association between the concepts. These are often drawn as graphs where nodes differ with respect to how closely they are linked to one another. Figure 4.10 shows an example of such a graph for the domain of statistics. The network was based on the ratings of five students in a course on research techniques.

The structures derived through multidimensional scaling are appealing because they are "empirically" derived. That is, once the items are sorted or judged, there is only one mathematically best solution or fit. Most other cognitive task analysis methods have at least some elements of subjectivity, such as how to represent the knowledge, or how fine a level of detail to go into. However, the method does have certain drawbacks. If our goal is to identify the conceptual, procedural, and skill knowledge of experts to the point where it can be taught to trainees, we must have very explicit and fine-grained representations of that knowledge.

Multidimensional scaling is not only a very indirect method of evaluating general cognitive structure (Redding, 1989), but it also results in a significant loss of information. That is, only clustering of key concepts is obtained, with no information on the relationships between the nodes. Nor has any type of cognitive strategy been adequately represented using this method.

Stated another way, while this type of analysis may be useful for interface design or evaluating the general organizational structure of novices and experts, the knowledge provided by multidimensional scaling methods is often inadequate for training program design.

Activity sampling

- *Data collection method(s)*: Observation
- *Data representation method(s)*: Table

Activity sampling is an instance of methods that attempt to identify the proportions of time spent on various job tasks. The technique is performed by sampling an employee's behavior at certain intervals, with data including the type, frequency, and duration of activities (Christensen, 1950).

Obviously, this type of analysis is only possible for tasks that are behav-

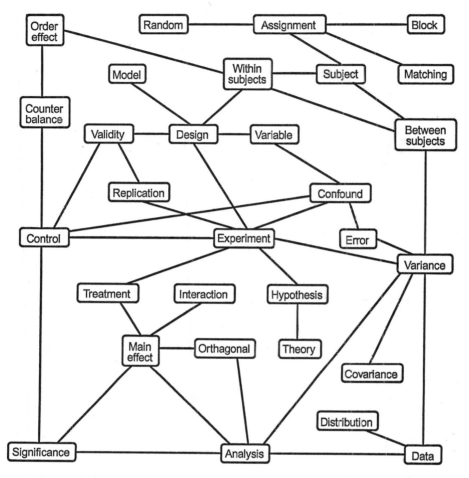

Figure 4.10 Example of network produced using a network scaling algorithm. (From Goldsmith, Johnson, & Acton, "Assessing structural knowledge," *Journal of Educational Psychology, 83*, 88–96, 1991. Copyright 1991 by the American Psychological Association. Reprinted with permission.)

ioral and therefore measurable. Sometimes the tasks occur so rapidly that videotaping is necessary. Kirwan and Ainsworth (1992) suggest that sampling should "be continued for sufficiently long to sample the full range of activities, and should yield in the region of 1000 sampling points."

Data collected from activity sampling is usually simply listed in a table with frequency tallies. Advantages of the method are that it collects data under realistic circumstances and tends to be nonintrusive. However, it requires a rather large amount of work to determine the tasks and transform the data from the tapes. The method should only be undertaken when relative percentages of task time are important.

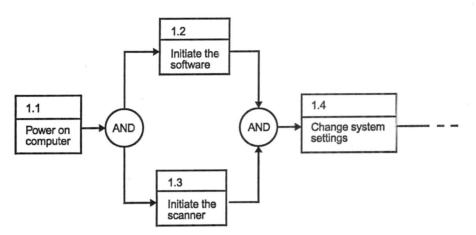

Figure 4.11   Example of part of a functional flow diagram.

Functional flow diagrams

- *Data collection method(s)*: None specified
- *Data representation method*: Flow chart

Functional flow diagrams are flow charts that use blocks to depict sequential interrelationship between various system functions (Geer, 1981). While they are often used to show the flow of functional activities within a physical system, such as a power plant, they can also be used to show the functional flow of human activity. An example of such a flow chart is shown in Figure 4.11. The activity is usually placed in a block, with numbers indicating the typical sequence. All blocks are linked by AND or OR gates.

CHOOSING TASK ANALYSIS METHODS

As we have seen, there are many methods for identifying knowledge and tasks, and many methods for representing that information. No one method has been shown to be the ideal technique, and based on the evidence to date, a combination of methods will probably be best. The representation of knowledge is typically done via lists, knowledge graphs, or task flow charts representing the flow of behavioral and cognitive processes. A training analyst should be familiar with all three of these formats for two reasons. First, each of these representational formats is more useful in some circumstances than others. For example, goal lists or hierarchies will be important for training procedures, whereas semantic networks can be used to identify and organize conceptual knowledge. Unfortunately, the newness of the field has resulted in a definite lack of guidelines as to the efficiency and appropriateness of any method for given types of applications. Second, even within one training application, the designer will frequently shuttle back and forth between different types of representation.

We can say that each of the representation methods is more appropriate for certain types of data than others. In general, when the information involves declarative rule knowledge, that is, procedures performed to attain goals, appropriate and popular formats include hierarchical networks and flow charts. When the time to perform a number of subtasks is also critical, a timeline chart is appropriate. If the hierarchy of procedures is quite large, it is often useful to represent the information in list or outline form (or both).

Procedural/skill knowledge can be handled in one of two ways. If the underlying rules cannot be induced, the specific instances of stimuli and associated responses can be represented in list (or graphic set) form. If the rules can be induced, the information can then be represented similar to declarative rule knowledge.

Explicit rule and implicit skill knowledge are most likely to be represented as IF-THEN production rules, or some type of goal hierarchy or planning net. Kieras (1988) argues that IF-THEN production rules are good computational representations if the analyst is planning to run simulations, but not good for general design purposes. Similarly, a tutorial might ultimately be based on production rules, but for the cognitive task analysis stage, other more readily understandable formats are desirable.

Declarative knowledge can be represented as lists (such as system components) or structural networks. The advantage of structural networks is the ability to directly capture the interrelationships between concepts, entities, and characteristics (Jonassen et al., 1993). Lists or outlines tend to leave out the relationships between items. However, there are times when declarative knowledge can be most efficiently represented as lists, tables, or matrices.

At some point, the designer must ask how deeply the analysis should delve into the details of the knowledge domain. That is, tasks and related information can be broken down to a more molecular level infinitely. Where to cut off the process is essentially an arbitrary decision. Most analysts agree that there is a level at which the designer will realize that enough is enough. For example, "open the drive" on a computer could be broken down into small physical actions such as lift hand, move hand toward computer, etc. However, there is usually not any point in carrying out this type of analysis for training design purposes.

SUMMARY OF TASK ANALYSIS

The goal of task analysis is to break the task down into component subtasks and identify all relevant information pertaining to those tasks. The information that is relevant will change from project to project. *In general,* we can say that the analyst should:

1. In a preliminary fashion, identify the *types* of knowledge relevant to the job or task being analyzed. For example, if the task relies heavily on decision making using knowledge of some physical system (e.g., troubleshooting), the database should include conceptual and causal knowledge. Tasks that are time-critical with multiple task-sharing will need a representation that captures the time characteristics.

2. Develop a priority list of the types of knowledge representation most useful for the project. The first item in the priority is almost always a task hierarchy in either list/outline or network format. Such a task hierarchy should include subtasks, their initiating conditions, and the consequences.

3. Determine the appropriate representational formats, such as structural networks for causal knowledge, flow charts for sequential information, etc.

4. Collect data beginning with document and equipment analysis.

5. Continue collecting data using appropriate methods such as a combination of interview and observation or protocol analysis.

6. Represent data as it is collected if possible. This allows you to identify gaps and also to determine when the analysis is relatively complete.

TRAINEE ANALYSIS

By this point, the training analyst should have performed a significant portion of the task analysis. For most jobs or tasks, this will have resulted in breaking down the behavioral and cognitive components until the level of prerequisite knowledge has been reached. For example, a terminal node on a goal structure might be "use wrench to loosen bolt." There is no merit in breaking down this activity any further if it is assumed that the trainee will know how to perform the activity, and will know what is meant by wrench, loosen, and bolt. For each task or activity, analysis stops at the level where it is assumed that the trainee has the necessary knowledge to perform the activity.

Each of these *terminal sites* should be evaluated and the assumed prerequisite knowledge and skills identified. These assumptions of pre-existing knowledge and skills should then be verified *before* development of the training program. Once a list is made of the assumed trainee knowledge and skills, a group of representative trainees are evaluated to determine the extent of the prerequisite knowledge and skills.

In addition to determining levels of prerequisite knowledge and skills, the analyst should determine any basic conceptual models (mental models) that the trainees bring with them. For example, researchers in the area of physics have found that novices bring in simplistic and incorrect models of the world that actually interfere with the acquisition of more complex and correct models. Any pre-existing mental models on the part of learners are important for development of a training program, to either build upon or correct if necessary. Many of the task analysis methods described earlier can be used for this process.

Training programs should be tailored as much to individual learners as possible. For this reason, we also determine relevant *demographic* characteristics such as range and mean age, predominant language, interests, background or job history, etc. It also means determining the extent of any necessary physiological conditions (e.g., heart capacity), or physical capabilities (e.g., strength, dexterity, etc.). In some cases, it may be desirable to classify trainees according to their predominant *learning style* (Kolb, 1984). Several questionnaires have been published for this purpose (Hagberg & Leider, 1982; Stephen, 1987).

And finally, trainees will bring with them certain attitudes toward their organization, toward their job, toward specific tasks, and toward a potential training program. They often have strong feelings about what would and would not be desirable means for performance support or training. The analyst should be aware of these beliefs and attitudes ahead of time in order to effectively overcome any negative attitudes and make use of any positive attitudes. If this information has not been obtained through the organizational analysis, it should be done at this point. Fishbein and Ajzen (1975) and Ajzen and Fishbein (1980) provide lengthy descriptions of how to measure beliefs and attitudes. For example, Likert-type rating scales may be used in questionnaires to assess attitudes toward various factors. If necessary, the beliefs and attitudes can be assessed in one interview or questionnaire along with demographics and knowledge.

REFERENCES

ANDERSON, J. R. (1983). *The architecture of cognition.* Cambridge, MA: Harvard University Press.

ANDERSON, J. R. (1990). *Cognitive psychology and its implications* (3rd ed.). New York: W. H. Freeman.

AJZEN, I., & FISHBEIN, M. (1980). *Understanding attitudes and predicting social behavior.* Englewood Cliffs, NJ: Prentice Hall.

BOOSE, J. H. (1986). *Expertise transfer for expert system design.* New York: Elsevier.

BOOSE, J. H. (1988). Uses of repertory grid-centred knowledge acquisition tools for knowledge-based systems. *International Journal of Man-Machine Studies, 29,* 287–310.

BROWN, L. T., & STANNERS, R. F. (1983). The assessment and modification of concept interrelationships. *Journal of Educational Psychology, 52,* 11–21.

CANNON-BOWERS, J. A., TANNENBAUM, S. I., SALAS, E., & CONVERSE, S. A. (1991). Toward an integration of training theory and technique. *Human Factors, 33(3),* 281–292.

CARD, S., MORAN, T., & NEWELL, A. (1983). *The psychology of human-computer interaction.* Hillsdale, NJ: Lawrence Erlbaum Associates.

CHRISTENSEN, J. M. (1950). The Sampling Technique for use in activity analysis. *Personnel Psychology, 3,* 361–368.

COOKE, N. M., & MCDONALD, J. E. (1987). The application of psychological scaling techniques to knowledge elicitation for knowledge-based systems. *International Journal of Man-Machine Studies, 26,* 533–550.

COOPER, M. A. R. (1986). *Fundamentals of survey measurement and analysis.* London: Collins.

CREASY, R. (1980). Problem solving, The FAST way. *Proceedings of Society of Added-Value Engineers Conference,* 173–175. Irving, TX: Society of Added-Value Engineers.

DALKEY, N. C. (1969). *The Delphi method: An experimental study of group opinion.* Santa Monica, CA: Rand Corporation (RM-5888-PR).

DIEDERICH, J., RUHMANN, I., & MAY, M. (1987). KRITON: A knowledge-acquisition tool for expert systems. *International Journal of Man-Machine Studies, 26,* 29–40.

ERICSSON, K. A., & SIMON, H. A. (1993). *Protocol analysis: Verbal reports as data* (rev. ed.). Cambridge, MA: MIT Press.

FEWINS, A., MITCHELL, K., & WILLIAMS, J. C. (1992). Balancing automation and human action through task analysis. In B. Kirwan & L. K. Ainsworth (eds.), *A guide to task analysis* (pp. 241–251). London: Taylor & Francis.

FISHBEIN, M., & AJZEN, I. (1975). *Belief, attitude, intention, and behavior: An introduction to theory and research.* Reading, MA: Addison-Wesley.

FISHER, K. N. (1990). Semantic networking: The new kid on the block. *Journal of Research and Science Teaching, 27,* 1001–1018.

FISK, A. D. (1989). Training consistent components of tasks: Developing an instructional system based on automatic/controlled processing principles. *Human Factors, 31(4),* 453–463.

FITTS, P. M., & JONES, R. E. (1947). Analysis of factors contributing to 460 "Pilot Error" experiences in operating aircraft controls. Reprinted in H. W. Sinaiko (ed.), *Selected papers on human factors in the design and use of control systems* (pp. 332–358). New York: Dover Books.

FLANAGAN, J. C. (1954). The critical incident technique. *Psychological Bulletin, 51,* 327–358.

GEER, C. W. (1981). *Human Engineering Procedure Guide.* Wright-Patterson Air Force Base, Ohio, US: Rept No. ARAMRL-TR-81-35.

GOLDSMITH, T. E., JOHNSON, P. J., & ACTON, W. H. (1991). Assessing structural knowledge. *Journal of Educational Psychology, 83,* 88–96.

GOLDSTEIN, I. L. (1986). *Training in organizations: Needs assessment, development, and evaluation* (2nd ed.). Monterey, CA: Brooks/Cole.

GORDON, S. E. (1989). Theory and methods for knowledge acquisition. *AI Applications in Natural Resource Management, 3(3),* 19–20.

GORDON, S. E. (1992a). Implications of cognitive theory for knowledge acquisition. In R. Hoffman (ed.), *The psychology of expertise: Cognitive research and empirical AI* (pp. 99–120). New York Springer-Verlag.

GORDON, S. E. (1992b). *Cognitive task analysis for segment defensive counter air mission.* Technical report submitted to Northrop Corporation. University of Idaho, Moscow, ID.

GORDON, S. E. (in press). Eliciting and representing biology knowledge with conceptual graph structures. In K. M. Fisher and M. R. Kibby (eds.), *Biology knowledge: Its acquisition, organization, and use.* New York: Springer-Verlag.

GORDON, S. E., & GILL, R. T. (1989). Question probes: A structured method for eliciting declarative knowledge. *AI Applications in Natural Resource Management, 3,* 13–20.

GORDON, S. E., & GILL, R. T. (1992). Knowledge acquisition with question probes and conceptual graph structures. In T. Lauer, E. Peacock, and A. Graesser (eds.), *Questions and information systems* (pp. 29–46). Hillsdale, NJ: Lawrence Erlbaum Associates.

GORDON, S. E., KINGHORN, R. A., & SCHMIERER, K. A. (1991). Representing expert knowledge for instructional system design: A case study. *Proceedings of the Human Factors Society 35th Annual Meeting* (pp. 1412–1416). Santa Monica, CA: Human Factors Society.

GORDON, S. E., SCHMIERER, K., & GILL, R. T. (1993). Conceptual graph analysis: Knowledge acquisition for instructional system design, *Human Factors, 35,* 459–481.

GRAESSER, A. C., & CLARK, L. C (1985). *Structures and procedures of implicit knowledge.* Norwood, NJ: Ablex.

GRAESSER, A. C., & FRANKLIN, S. P. (1990). QUEST: A cognitive model of question answering. *Discourse Processes, 13,* 279–304.

GRAESSER, A. C., & GORDON, S. E. (1991). Question answering and the organization of world knowledge. In G. Craik, A. Ortony, and W. Kessen (eds.), *Essays in honor of George Mandler* (pp. 227–243). Hillsdale, NJ: Lawrence Erlbaum Associates.

GRAESSER, A. C., LANG, K. L., & ELOFSON, C. S. (1987). Some tools for redesigning system-operator interfaces. In D. E. Berger, K. Pezdek, and W. P. Banks (eds.), *Applications of cognitive psychology: Problem solving, education, and computing* (pp. 163–181). Hillsdale, NJ: Lawrence Erlbaum Associates.

GREENO, J. G., RILEY, M. S., & GELMAN, R. (1984). Conceptual competence and children's counting. *Cognitive Psychology, 16,* 94–143.

GROVER, M. D. (1983). A pragmatic knowledge acquisition methodology. *International Joint Conference on Artificial Intelligence, 83,* 436–438.

HAGBERG, J., & LEIDER, R. (1982). *The inventurers.* Reading, MA: Addison-Wesley.

HANNAFIN, M. J., & PECK, K. L. (1988). *The design, development, and evaluation of instructional software.* New York: Macmillan.

HIRSHMAN, E. C., & WALLENDORF, M. R. (1982). Free-response and card-sort techniques for assessing cognitive content: Two studies concerning their stability, validity and utility. *Perceptual and Motor Skills, 54,* 1095–1110.

HOFFMAN, R. (1987). The problem of extracting the knowledge of experts from the perspective of experimental psychology, *AI Magazine, 8,* 53–64.

JOHNSON, P. E., COX, D. L., & CURRAN, T. E. (1970). Psychological reality of physical concepts. *Psychonomic Science,* 19(4), 245–246.

JONASSEN, D. H., BEISSNER, K., & YACCI, M. (1993). *Structural knowledge: Techniques for representing, conveying, and acquiring structural knowledge.* Hillsdale, NJ: Lawrence Erlbaum Associates.

JONASSEN, D. H., HANNUM, W. H., & TESSMER, M. (1989). *Handbook of task analysis procedures.* New York: Praeger.

KIERAS, D. E. (1988). Towards a practical GOMS model methodology for use interface design. In M. Helander (Ed.), *Handbook of human-computer interaction.* The Netherlands: Elsevier.

KIRWAN, B., & AINSWORTH, L. K. (Eds.) (1992). *A guide to task analysis.* London: Taylor & Francis.

KLEIN, G. A., CALDERWOOD, R., & CLINTON-SCIROCCO, A. (1986). Rapid decision making on the fire ground. *Proceedings of the Human Factors Society 30th Annual Meeting* (pp. 576–580). Santa Monica, CA: Human Factors Society.

KLEIN, G. A., & THORDSEN, M. (1988). Use of progressive deepening in battle management. *Proceedings of the 11th Biennial DoD Psychology Conference,* Colorado Springs, CO.

KLINGER, D. W., & GOMES, M. E. (1993). A cognitive systems engineering application for interface design. *Proceedings of the Human Factors Society 37th Annual Meeting* (pp. 16–20). Santa Monica, CA: Human Factors Society.

KOLB, D. (1984). *Experiential learning: Experience as the source of learning and development.* Englewood Cliffs, NJ: Prentice Hall.

MCCORMICK, E. J., JENNERET, P. R., & MACHAN, R. C. (1969). *A study of job characteristics and job dimensions as based on the Position Analysis Questionnaire.* Rept. No. 6, Occupational Research Center, Purdue University, West Lafayette, Indiana.

MCDONALD, J. E., & SCHVANEVELDT, R. W. (1988). The application of user knowledge to interface design. In R. Guindon (ed.), *Cognitive science and its applications for human-computer interaction* (pp. 289–338). Hillsdale, NJ: Lawrence Erlbaum Associates.

MEANS, B., & GOTT, S. P. (1988). Cognitive task analysis as a basis for tutor development: Articulating abstract knowledge representations. In J. Psotka, L. D. Massey, and S. A. Mutter (eds.), *Intelligent tutoring systems: Lessons learned.* Hillsdale, NJ: Lawrence Erlbaum Associates.

MEISTER, D. (1985). *Behavioral analysis and measurement methods.* New York: John Wiley & Sons.

MERRILL, P. F. (1987). Job and task analysis. In R. M. Gagne (ed.), *Instructional technology: Foundations* (pp. 141–173). Hillsdale, NJ: Lawrence Erlbaum Associates.

MILLER, G. A. (1969). A psychological method to investigate verbal concepts. *Journal of Mathematical Psychology, 6,* 169–191.

MOORE, J., & GORDON, S. E. (1988). Conceptual graphs as instructional tools. *Proceedings of the Human Factors Society 32nd Annual Meeting* (pp. 1289–1293). Santa Monica, CA: Human Factors Society.

OHLSSON, S., & REES, E. (1991). The function of conceptual understanding in the learning of arithmetic procedures. *Cognition and Instruction, 8(2),* 103–179.

OHNEMUS, K. R., & BIERS, D. W. (1993). Retrospective versus concurrent thinking-out-loud in usability testing. *Proceedings of the Human Factors Society 37th Annual Meeting* (pp. 1127–1131). Santa Monica, CA: Human Factors Society.

PEARLMAN, K. (1980). Job families: A review and discussion of their implications for personnel selection. *Psychological Bulletin, 87,* 1–28.

PHILLIPS, M. D., BASHINSKI, H. S., AMMERMAN, H. L., & FLIGG, C. M., JR. (1988). A task analytic approach to dialogue design. In M. Helander (Ed.), *Handbook of human-computer interaction* (pp. 835–857). The Netherlands: North-Holland.

PRIEN, E. P. (1977). The function of job analysis in content validation. *Personnel Psychology, 30,* 167–174.

RASMUSSEN, J. (1986). *Information processing and human-machine interaction: An approach to cognitive engineering.* New York: Elsevier.

REDDING, R. E. (1989). Perspectives on cognitive task analysis: The state of the state of the art. *Proceedings of the Human Factors Society 33rd Annual Meeting* (pp. 1348–1352). Santa Monica, CA: Human Factors Society.

REDDING, R. E. (1990). Taking cognitive task analysis into the field: Bridging the gap from research to application. *Proceedings of the Human Factors Society 34th Annual Meeting* (pp. 1304–1308). Santa Monica, CA: Human Factors Society.

REDDOUT, D. (1987). What is a task? *Performance and Instruction, 26(1),* 5–6.

REIGELUTH, C. M., & MERRILL, M. D. (1984). *Extended task analysis procedure (ETAP): User's manual.* Lanham, MD: University Press of America.

REITMAN, J. S., & RUETER, H. H. (1980). Organization revealed by recall orders and confirmed by pauses. *Cognitive Psychology, 12,* 554–581.

ROTHWELL, W. J., & KAZANAS, H. C. (1992). *Mastering the instructional design process: A systematic approach.* San Francisco, CA: Jossey-Bass.

ROUSE, W. (1991). *Design for success: A human-centered approach to designing successful products and systems.* New York: John Wiley & Sons.

RYDER, J. M., REDDING, R. E., & BECKSHI, P. F. (1987). Training development for complex cognitive tasks. *Proceedings of the Human Factors Society 31st Annual Meeting* (pp. 1261–1265). Santa Monica, CA: Human Factors Society.

SASSE, M. HOW TO T(r)ap users' mental models. In M. J. Tauber and D. Ackermann (eds.), *Mental models and human-computer interaction 2* (pp. 59–79). Amsterdam: North-Holland.

SCHVANAVELDT, R. W. (1990). *Pathfinder associative networks: Studies in knowledge organization.* Norwood, NJ: Ablex.

STEIN, M. K., BAXTER, J. A., & LEINHARDT, G. (1990). Subject matter knowledge and elementary instruction: A case from functions and graphing. *American Educational Research Journal, 27,* 639–663.

STEPHEN, L. (1987). Assessing your learning style. In R. Bard, C. Bell, L. Stephen, and L. Webster (ed.), *The trainer's professional development handbook.* San Francisco, CA: Jossey-Bass.

TOLBERT, C. A., & BITTNER, A. C. (1991). Applications of verbal protocol analysis during the system development cycle. In W. Karwowski and J. Yates (eds.), *Advances in Industrial Ergonomics and Safety III.* London: Taylor & Francis.

WICKENS, C. D. (1992). *Engineering psychology and human performance* (2nd ed.). New York: Harper Collins.

ZAFF, B. S., MCNEESE, M. D., & SNYDER, D. E. (1993). Capturing multiple perspectives: A user-centered approach to knowledge acquisition. *Knowledge Acquisition, 5,* 79–116.

ZUBRITSKY, M. C., & COURY, B. G. (1987). Multidimensional scaling as a method for probing the conceptual structure of state categories: An individual differences analysis. *Proceedings of the Human Factors 31st Annual Meeting* (pp. 107–111). Santa Monica, CA: Human Factors Society.

ADDITIONAL RESOURCES

BLANK. W. (1982). *Handbook for developing competency-based training programs.* Englewood Cliffs, NJ: Prentice Hall.

CRAIG, R. (Ed.) (1987). *Training and development handbook: A guide to human resource development* (3rd ed.). New York: McGraw-Hill.

DIAPER, D. (Ed.) (1989). *Knowledge elicitation: Principles, techniques, and applications.* Chichester, UK: Ellis Harwood.

FOSHAY, W. R. (1983). Alternative methods of task analysis: A comparison of three methods. *Journal of Instructional Development, 6(4),* 2–9.

FOSHAY, W. (1986). Choosing the best alternative technique for task analysis. In M. Smith (ed.), *Introduction to performance technology.* Washington, D.C.: National Society for Performance and Instruction.

FREDERIKSEN, N., GLASER, R., LESGOLD, A., & SHAFTO, M. G. (1990). *Diagnostic monitoring of skill and knowledge acquisition.* Hillsdale, NJ: Lawrence Erlbaum Associates.

HERSCHBACK, D. R. (1976). Deriving instructional content through task analysis. *Journal of Industrial Teacher Education, 13(3),* 63–73.

KENNEDY, P., ESQUIRE, T., & NOVAK, J. (1983). A functional analysis of task analysis procedures for instructional design. *Journal of Instructional Development, 6(4),* 10–16.

OPPENHEIM, A. N. (1992). *Questionnaire design, interviewing, and attitude measurement.* Pinter Pubs, UK: Saint Martin's Press.

SANDERSON, P. M., VERHAGE, A. G., and FULD, R. B. (1989). State-space and verbal protocol methods for studying the human operator in process control. *Ergonomics, 32/11,* 1343–1372.

ZEMKE, R., & KRAMLINGER, T. (1982). *Figuring things out: A trainer's guide to needs and task analysis.* Reading, MA: Addison-Wesley.

5

Procedures: Training Need and Resource Analyses

DETERMINING THE NEED, FINDING THE BEST ANSWER

Once the analyst has conducted the organizational analysis, task analysis, and the trainee analysis, this information is evaluated to determine the best method of enhancing performance. This is termed the *training need analysis*. Once it has been established that the best solution is a training or performance support system, the designer conducts an analysis to identify the resources available for program development, the *resource analysis*.

If the presenting problem is one of low performance levels, the first step is to rule out any external or context variables that could be the source of the problem. This means evaluating the results of the analyses in search of physical or social factors that could be causing the lower performance.

EXAMPLES: Workers on a small parts assembly line are not adequately performing the task. Evaluation of the assembly site shows that wrist supports are incorrectly positioned, and the support frame for the assembly task is unstable.

Employees at a paper mill walk into hazardous areas and don't wear steel-toed boots. Analysis reveals that the general attitude among employees and management alike is that this is acceptable behavior as long as one is "careful."

Once the analyst has determined that there are not physical or social environment factors leading to the low (or unsafe) performance, the next task is to determine what should be done to enhance task performance. The general priority of design solutions considered for implementation should be the following:

1. **Primary Design.** Design or redesign the job/task, and its associated products and systems, so that training and/or performance support is not needed.
2. **Support.** Design an auxiliary performance support system or job aid so that training is not needed.
3. **Train.** Design an effective training program (possibly with associated job aids or performance support systems).

This priority system is applied for all of the duties and tasks under consideration. It is possible that different tasks will result in different solutions. Examples of these possibilities are given below.

Design Approach #1: System Design or Redesign

The priority system listed in the last section means that the analyst, and other design team members, should first look at the job tasks and their current performance levels to determine whether redesigning either the task itself or a system that supports the task would increase job performance to acceptable levels. If the job is a new one, the associated system components should be designed to support performance with as little training as possible. Examples of systems that should be usable without explicit training are household fixtures (such as water faucets), household appliances (such as microwaves, stoves, televisions, VCRs and video recorders, gas grills, etc.), public systems such as telephones and automatic teller machines, basic operation of automobile dashboard controls and displays (such as radio, air, windshield wipers, etc.), safety devices such as seat belts, and so forth. Obviously, a parent would have to teach operation of these systems to a child, but an adult should require no product-specific training.

Example of a System Redesign Solution

You have accepted responsibility for training the forklift operators working for your company. You have decided to first focus on general safety issues. You develop a long list of declarative and procedural knowledge needed by operators for safe forklift operation. You have also found a videotape that covers many of the topics. Your next job should be to determine whether training on each of these topics, and use of the videotape, is the most appropriate solution.

One of the items you evaluate is the process of lifting material on the forklift in front of the vehicle. Because of the leverage effect of a forklift, raising heavy loads too high on the front will cause the rear wheels to come off the ground. You know that you want to train your employees to know when the forklift load is being raised too high. Analyzing the causal factors, you realize that there is a complex relationship between the load weight, the center of gravity of the vehicle, the height of the load on the forklift, and the force of the wheels on the ground. You realize that you must train your employees to understand this relationship well enough that they can evaluate the load weight and only lift it high enough to still keep pressure on the rear wheels.

The problem with training employees to perform this perceptual task is that it will often be difficult to determine the weight of the loads. Materials handled by forklifts may not be marked with respect to weight, and they are often much too heavy to judge "by hand." Therefore, the problem with trying to train this perceptual task is that it *really cannot be done.* One alternative would be to try to train operators to only lift loads to an acceptable level for even the heaviest loads. Unfortunately, this would probably not be a creditable solution for operators.

In this particular case, a better solution would be to change the design of the equipment itself. In other words, the danger is in the rear wheels lifting off the ground. This danger could be addressed by warning the operator whenever it began. For example, one might equip the vehicle with a display and/or warning system to inform the operator when the load was at an appropriate height, and especially when the load was at a dangerous height. Such a system could be designed by combining a device to measure force on the rear wheels with a visual display and/or an auditory warning signal.

Design Approach #2:
Performance Support Systems and Job Aids

There are many circumstances where the task and its associated equipment or environment cannot be changed. Examples include such things as a software package that is used by your company, a telephone with conference call and other functions, and formal procedures that have been mandated by government or your own particular firm. In these cases, the analyst must choose some combination of training, performance support, and job aids. In general, it is desirable to provide performance support without specific prior training when possible. That is, the advantage of well-designed performance support systems and job aids is that they provide training at the time it is needed.

What types of job are appropriate for performance support systems or job aids, without prior training?

- Jobs or tasks that allow sufficient time for a person to look up the necessary information.

> EXAMPLE: Obtaining information from a potential medical patient. An example of a task that would *not* be appropriate for *only* a performance support system would be performing CPR.

- Jobs requiring use of large amounts of information, and/or complex judgments and decisions.

> EXAMPLE: A sales representative from a computer chip manufacturer must provide product information to a client.

- Jobs or tasks that won't suffer from the person reading instructions or looking at diagrams.

> EXAMPLE: A person reads a procedure list for using a credit card to purchase gasoline.

- Job or task requires a great number of steps that are difficult to learn and remember.

> EXAMPLE: Procedures for flight check.

- Jobs or tasks where safety is a critical issue, and there are no repercussions or ill effects of relying on a job aid.

> EXAMPLE: Routine startup and control of a nuclear power plant. An example of a task that would not be appropriate would be the skill of interacting appropriately with a business associate or client (the client would probably wonder why the sales representative needed coaching on how to ask about the client's needs).

- Jobs where a novice performs the task, or a person performs the task infrequently.

> EXAMPLE: A freshman in college goes to the library to find references.

It is *not* appropriate to substitute this solution for training of tasks where experts perform the task frequently and/or need to be highly skilled or automatic, such as operating a vehicle.

- Jobs where the employees have difficulty obtaining training (due to distance, time, etc.).

> EXAMPLE: A researcher wants to submit an article to a particular journal, and needs to know the requirements and procedure.

What type of jobs and tasks are not good candidates for performance support systems? *In general* we can say that training is better for tasks with the following characteristics:

The task consists of steps performed quickly and/or in rapid succession.
The task is performed frequently.
The task must be learned in order to perform necessary higher level tasks.
The person wishes to perform the task unaided.
The person is expected for some reason to perform the task unaided.
Performance of the task would be hindered by attending to some type of aid.
The task is perceptual and there is no good way to develop a job aid.

When performance *can* be supported with just-in-time materials such as performance support systems and job aids, employees are usually happy about not having to spend (waste) time attending special training programs.

Example of an Electronic Performance Support System Solution

Intel Corporation has difficulties providing up-to-date information and training to sales and support personnel scattered over the globe. A solution to their problem was to disseminate files over computer networks every time product updates were issued (Raybould & Towle, 1993). In this case, an in-depth task analysis was conducted. The duties and tasks of sales representatives were identified, as well as the normal data gathering methodologies used to support the tasks. A performance support system was then developed that had several functions, including the following:

- Provide product and market information
- Provide simple to use interactive job aids that facilitate feature comparisons between any chosen Intel product and one or more competitor product.
- Provide product training modules (to build and update background knowledge)
- Provide training on appropriate means of interacting with and serving clients

The interface of the system was developed to map onto the tasks and "mental model" of the employee (see Figure 5.1).

Example of a Job Aid Solution

The University of Idaho has a computer-based literature search system for references in the field of psychology. The system is used primarily by psychology majors and other students taking undergraduate psychology courses. The interface for the search system is not particularly easy to use, and most students have difficulty using the system, even after the first time. The current method for dealing with this problem is for librarians to take time to help each student, essentially an inefficient tutoring approach. The librarians were interested in solutions to this problem.

Analysis of the organization, task, and student users revealed several important facts:

- The interface software could not be modified.
- The librarians were partial to computer-based training, especially multimedia.
- The students were not interested in computer-based training, or any other type of training that took more than about 2 minutes.
- There was currently no way to provide computer-based training or performance support systems *at the site* of the search computers.
- The tasks performed by most students were relatively few, and consistent across people, although each task could be accomplished by several different means.

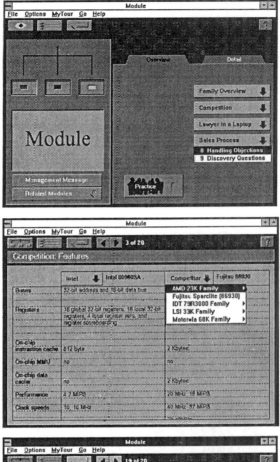

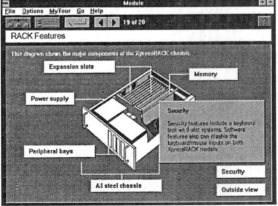

Figure 5.1 Performance Support System interface for Intel sales representatives. (System developed by Ariel PSS Corporation for Intel Corporation. Graphics used with permission.)

The analysts for this project decided that a simple hardcopy job aid "attached" at the site of the literature search computers would be the best design solution for this particular circumstance.[1] Using sound design procedures, they designed a simple fold-open laminated job aid that focused on the most simple and straightforward methods for accomplishing the search tasks. While the librarians were originally biased toward a more technologically sophisticated training program, they are now extremely happy with the final product, and one job aid sits at each computer search site. Final testing has shown that 100% of the students were able to use the job aid without any prior training or on-site help from library personnel. This example shows that the glitzy training options are not always the best design solution.

Design Approach #3: Training Programs

At this point you should have evaluated the potential for system redesign and performance support systems without relying on training prior to job performance. In some cases, you will have concluded that a training component is still needed, perhaps to augment one of these other design solutions. There are also many instances where a basic training program is the most appropriate solution, such as performing CPR, or flying a fighter jet. In the remainder of the book, we will examine methods for developing training programs.

Example of Combined Redesign, Performance Support, and Training Program

Some jobs are complicated, and require both an extensive evaluation and a complex combination of performance support solutions. Such a job is described in Wickens (1992). This job is that of a nuclear power plant process control operator. The job is basically that of controlling and regulating normal processes, and occasionally dealing with malfunctions. This job is sometimes described as "hours of intolerable boredom punctuated by a few minutes of pure hell" (Wickens, 1992). The difficulties of the job have several sources. First, the process itself is highly complex, and difficult to completely comprehend. One could say that it is difficult if not impossible to develop a fully complete and correct mental model of the process. Second, because of this system complexity, the number of controls and displays is astoundingly large; more than 3,000 in 1975 (Wickens, 1992).

---

[1] Analysts in this case were Janice Alves-Foss, Rich Hanowski, and Brett Hunt, students in the University of Idaho Human Factors graduate program.

How is one to prepare an operator to interact with such a large and complex system interface? And equally important, how can we support and train the operators to deal with the vast array of controls and displays when a malfunction suddenly occurs? The solution is a multifaceted one. First, the system and its controls can be redesigned to be more "error tolerant" (Rouse & Morris, 1986). Error-tolerant systems are ones where (a) human error is less likely to occur, (b) actions are monitored for error and appropriate interventions implemented, and/or (c) it is possible for the operator to detect and recover from human error without system disruption or failure.[2]

Second, the interface is redesigned to be more usable and useful to the operator (Wickens, 1992; Woods, O'Brien, & Hanes, 1987; Woods & Roth, 1988). Design solutions include the addition of predictive displays which provide output for one or more variables based on computer models of the power plant. Third, some tasks may be automated, although the decision to automate a task is actually a very complex one (Rouse, 1991). Fourth, when malfunctions occur, special real-time performance support systems can be available to support problem solving and diagnosis. And fifth, effective training provides an accurate and useful mental model of the system as well as appropriate procedural steps (Roth & Woods, 1988; Woods, 1988).

The remainder of this book will proceed through the procedures of developing training programs, job aids, and performance support systems. While the emphasis is on training program development, many of the principles and guidelines are equally applicable to performance support. An example of this general applicability of principles is found in the next section on training resource analysis.

TRAINING RESOURCE ANALYSIS

Once a training program or performance support system has been tentatively chosen as the appropriate design solution, a different sort of organizational analysis is undertaken. This **training resource analysis** is conducted to identify all factors that are relevant to development and delivery of a training program and/or performance support system. This task consists of a complete evaluation of the organization(s) responsible for the development of the pro-

---

[2]An example of this third approach is a software program that asks if you are *sure* you want to delete that file!

gram and also its distribution. If the specific organization is unknown, then all potential development and delivery organizations should be analyzed.

The purpose of this analysis is to gather information that can support instructional system design decisions. Most critical in the analysis is evaluation of resources for program development and program delivery. Factors that should be evaluated include:

- Facilities (space, power, lighting, etc.)
- Financial resources available for the project
- Time
- Available personnel and their skills
- Equipment
- Cost of resources at each prospective development site
- Similar or usable existing programs
- Storage and delivery services
- Dissemination resources
- Costs of delivery or dissemination resources

This information should be gathered by interviewing knowledgeable people and performing site visits. A general description of available resources is the result of the process. The analyst should be sure to obtain estimates of future resources as well as currently existing resources.

Other information will prove useful as well, and can be sought through interviews or examination of organizational documents. For example: Are there certain types of training programs that are "favorites" at the facility? Are there certain types of programs that people would find unacceptable? What are the program priorities, and who has the control over programs and their status?

The designer should perform broad, loosely structured interviewing with the goal of identifying any physical or psychological factors that would constrain or impact the design and success of a training or educational program. Interviews should include a representative sample of potential trainees. This serves two purposes: first, it provides information that can't be found anywhere else. Second, it helps users feel involved, which leads to a greater acceptance of the program after it is fielded. This data collection process can be carried out in an equivalent fashion whether the training will occur in the organization containing the job, a dedicated training organization, or an educational institution.

SUMMARY

The goals of the training need and resource analyses are to:

- Rule out any physical or social factors that might be responsible for low performance
- Move through the priority of design solutions to identify the most effective and appropriate approach (or combination of approaches); in particular, rule out system redesign and job aids before settling on training alone as a design solution
- Collect data regarding program development and delivery resources and constraints

If there is more than one potential training site, the analysis should be carried out for all candidates, to support a final decision as to the site with the lowest cost/benefit ratio (or alternatively, solicit bids).

REFERENCES

RAYBOULD, B., & TOWLE, B. (1993). *Good help is easy to find: Electronic job aids.* Paper presented at Interactive '93, May, Anaheim, CA.

ROTH, E. M., & WOODS, D. D. (1988). Aiding human performance: I. Cognitive analysis. *LeTravail Humain, 51,* 39–64.

ROUSE, W. B. (1991). *Design for success: A human-centered approach to designing successful products and systems.* New York: John Wiley & Sons.

ROUSE, W. B., & MORRIS, N. M. (1986). Understanding and enhancing user acceptance of computer technology. *IEEE Transactions on Systems, Man, and Cybernetics, SMC-16,* 539–549.

WICKENS, C. D. (1992). *Engineering psychology and human performance* (2nd ed.). New York: Harper Collins.

WOODS, D. D. (1988). Commentary: Cognitive engineering in complex and dynamic worlds. In E. Hollnagel, G. Mancini, & D. D. Woods (Eds.), *Cognitive engineering in complex dynamic worlds* (pp. 115–129). London: Academic Press.

WOODS, D. D., O'BRIEN, J. F., & HANES, L. F. (1987). Human factors challenges in process control: The case of nuclear power plants. In G. Salvendy (ed.), *Handbook of human factors* (pp. 1724–1770). New York: John Wiley & Sons.

WOODS, D. D., & ROTH, E. (1988). Aiding human performance: II. From cognitive analysis to support systems. *Le Travail Humain, 51,* 139–172.

6

Background:
Learning
and Motivation

Once we have completed analysis of the context, task, trainees, training need, and resources, it is time to write functional specifications. This document specifies the requirements and constraints within which the system design process will operate. It includes general program goals, specific instructional objectives, and any constraints that will provide a boundary for developing the design concept. It is generally best to specify not only what topics the system will cover, but some general *instructional tactics* as well. For example, we might determine through the organizational and task analyses that certain tasks need to be trained to automation. The fact that these tasks need to be trained to automation should be included in the functional specification. This, in turn, will impact the types of design concepts that are considered in the design phase.

In order to specify instructional tactics, we need to know how people learn different types of information, and what instructional methods support the different types of learning. This chapter provides an overview of how we acquire both declarative and procedural/skill knowledge, and the basic instructional methods that support acquisition of those types of knowledge. We will also review trainee motivation and how instructional methods increase or decrease motivation. Most of the methods discussed are not restricted to any particular delivery system such as computers or live instruction. Rather, they are methods that should be incorporated into the system depending on the types of knowledge being trained, regardless of the media utilized. This chapter will therefore provide basic instructional guidelines; they can be used directly in the system specification document, and also as a reference later during actual program development.

A BASIC INFORMATION-PROCESSING MODEL OF MEMORY

Cognitive psychologists have identified certain characteristics of memory that can most easily be described in terms of a basic information-processing model. Such a model assumes that there are certain components of our cognitive system that treat incoming information in qualitatively different ways (see Lachman, Lachman, & Butterfield, 1979). More specifically, there are at least three different types of processing modes and their associated storage systems for information as it comes in from the environment. The major components of the information-processing model are shown in Figure 6.1. Many variations of this model exist which are more complex and have a greater number of components, but this model is suitable for our purposes.

It can be seen that the model assumes that information comes from the task environment into a **sensory memory** component. From there, certain information is selected for processing in **working memory**, and may also be stored in **long-term memory.** As we discuss the major characteristics of each of these three components, the reader should keep in mind that the model is *not* assuming that the components represent different physiological places to which information is transported and stored. This is especially true for the distinction between working memory and long-term memory. Rather, the components represent states of knowledge and certain associated processes. This point should become more clear through the descriptions below.

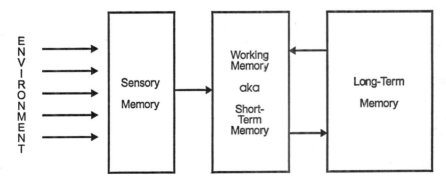

Figure 6.1   Information-processing model of memory.

Sensory Memory

A vast majority of the physical stimuli that impinge upon a person are transformed into internal perceptual representations for further processing. This fact is intuitively reasonable because we don't know what sights and sounds and other sensory input will be important and should receive our attention until we can process them in some preliminary fashion and assign to them some **meaning.** This wide-scale preliminary processing and assignment of meaning occurs in sensory memory. This component is capable of holding a great deal of information, but only for very short periods of time (about one second for visual sensory information and several seconds for auditory information). If the organism needs the information longer than that, it must be "selected" for further processing. This further processing occurs in working memory. The remaining unneeded information is dropped from the sensory memory system (or replaced by new incoming information).

Recall our statement earlier that humans have a very limited amount of cognitive processing capacity. Although a great deal of information gets as far as sensory memory, it takes cognitive resources to attend to subsets of that information and transform it for use in working memory. The limits in our cognitive resources dictate the amount of information we can transform. Sperling (1960) found that we can only transform 4–5 items within the 1-second time span before the information in sensory memory decays or is replaced. The implication is that of all the information a trainee may "see" or "hear" in a training program, he/she can only bring in a small subset of items for actual cognitive processing at any given time.

There are certain factors that tend to determine the information in sensory memory that is attended to and selected for further processing. First, out of all the stimuli represented in sensory memory, the most *salient* will receive attention. For example, certain colors are more salient than others. If items are presented on a computer screen in blue, green, and red, the ones in red will be most visually salient and therefore receive more attention. Also, when a visual stimulus is different than the others, or more centrally located, it tends to be more salient. We know that between visual and auditory information, auditory information tends to be more salient (Sanders & McCormick, 1987; Wickens, 1992).

The second factor that determines what information is attended to is the meaning assigned to the various stimuli. As the incoming pieces of information are given meaning, there may be certain elements that either have *special meaning* to us or that are *incongruent* with our expectations. These things will catch our attention. As an example of information with special (individual) meaning, when we are at a cocktail party, we notice immediately when our name is spoken in a conversation going on behind us, even if we have not been listening to that particular conversation. Certain items catch our attention because they are intrinsically important to us (warning sounds, etc.). The second type of information, incongruous information, consists of elements that would be normal in one information environment but not another. For example, seeing an elephant in a zoo would not catch our attention, but seeing it on a bus probably would.

A third factor that affects attentional resources is the degree to which the information is intrinsically interesting to the person. This will, of course, vary from one person to the next. But there are certain types of stimuli that tend to be interesting to broad ranges of people. For example, children like to watch animations with a lot of color and activity.

Finally, we deliberately choose some items to transform from sensory memory because our current goals and cognitive processes supporting those goals direct us to attend to those items. If we are at a restaurant to meet a friend, we will be visually scanning the environment for the image of that person. There is a large amount of attention that is volitional in this sense.

In summary, people constantly receive a great deal of information from the environment that gets to the preliminary processing stage of sensory memory. However, after that point most of it is lost. In a training environment, the information that gets the most extensive processing will depend on the *amount* of information being presented, the *salience* of various stimuli, the degree to which the information is "*interesting*," and the degree to which the information is called for by short-term or long-term *goals* of the trainee.

Working Memory

Any information that is selected from the sensory memory component for further processing is brought into working memory. Working memory can be thought of as the desktop or work surface of the mind. It is the active knowledge and skills currently being used for cognitive processing. Information that is active in working memory comes from two sources, stimuli from the external environment and knowledge retrieved from long-term memory. An example is the information you are reading that comes from the environment. You combine it with previous knowledge that you activate from long-term memory. Of course, you also have a lot of knowledge that is not being activated, and that information is said to be in long-term memory.

Working memory used to be called short-term memory because when researchers first identified the components of the information-processing model, they based their descriptions and names on the memory characteristics of the components. Without using special tactics such as repetition, short-term memory holds information for a much shorter period of time than long-term memory. Working memory is now a preferred name because it seems more accurate in describing the active knowledge that is being used for processing, and because information *can technically* be held in working memory for an indefinite period of time by repeating or "rehearsing" it.

As with sensory memory, we can talk about working memory in terms of its *holding capacity, time* that information lasts in the system, and *selection processes* for transfer to another component (long-term memory). The capacity of working memory was first defined by Miller (1956) as being approximately 7 ± 2 chunks of information. That is, working memory can only keep around seven items active at a given time. More recent research has shown that if the pieces of information are complex, such as a sentence as opposed to a single digit, the capacity becomes considerably less than seven (Anderson, 1990). Even without putting a number on the capacity of working memory, we can see that the ability of working memory to sustain operations on numerous pieces of information is very limited. This phenomenon again illustrates the existence of limited cognitive resources. Therefore, when we teach knowledge or skills, there will be a critical limit reached by the trainee in terms of how many pieces of information he/she can "hold" in working memory, and the number of transformations that can be performed on those pieces of information.

Once information is activated within working memory, it will quickly decay if it is not somehow kept active. Visual information decays within about seven seconds whereas auditory information lasts a few seconds longer (Anderson, 1990). When information decays in this way, it cannot be reactivated unless it has been transformed into long-term memory format. That is, something must happen to the pattern of activation to make it susceptible to reactivation later. If that process does not occur, the person will not be able to "remember" the information. This process of transforming the information for later retrieval is the process of transferring the knowledge to long-term memory.

Each chunk or piece of information in working memory must be reactivated within 7 to 12 seconds to prevent its dropping out of the working memory system. For example, if the content is a sequence of numbers (such as a phone number), we will sequentially reactivate each number; in some sense you can "hear" yourself mentally repeating the number. This process of voluntarily reactivating information in working memory is known as **rote rehearsal.**

Long-Term Memory

When information is processed in working memory, the *patterns of activation* representing each piece of information are strengthened. This strengthening of patterns makes it more likely that we can later reactivate them. In addition, associations are formed between different items in working memory. These associations are stored along with the individual items in long-term memory.

As we process new information in working memory, the information is comprehended and processed by reactivating previous knowledge from long-term memory. When we add this old information to working memory, new associations are formed among both new and old pieces of information. The tying of old knowledge in with the new is termed **elaborative rehearsal** (Anderson, 1990). This process of forming associations in working memory will turn out to be critical for both storing and retrieving information from long-term memory.

Once information drops from short-term memory, how is it retrieved or reactivated from long-term memory? When we perceive "new" information, it is likely to be something that we have experienced previously. This causes the old pattern for the information to be reactivated. More interestingly, it also causes patterns to be reactivated that were previously **associated** with the information due to coexistence in working memory. So as you receive external information into working memory, those memory patterns are reactivated *along with* a number of related ideas, sounds, and images. This associative retrieval process often occurs automatically and *without effort* (Lachman et al., 1979; Schneider & Shiffrin, 1977; Shiffrin & Schneider, 1977). For example, if we are blindfolded and smell a rose, it causes us to think of the concept "rose" and perhaps imagine an image of a rose (see Figure 6.2).

We also use these associations in memory to *deliberately* retrieve items from long-term memory using controlled search processes (Schneider & Shiffrin, 1977; Shiffrin & Schneider, 1977). If we are trying to recall what we had for dinner last night, we will activate the concepts of "yesterday evening," the "image" of where we were, who we were with, and so forth. This helps us recall what we ate because they are all associated. Such controlled search processes use working memory and require cognitive resources.

What types of association form in working memory that will lead to later retrieval of information? Underwood (1969, 1983) has postulated that there are potentially a dozen *types* of associations that might be formed. These potential associations are called attributes. Such attributes include the acoustical properties of stimuli, the time that it occurred, the emotional state of the indi-

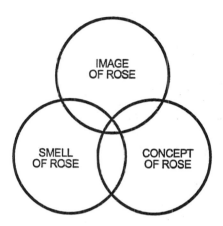

Figure 6.2 Associations among three separate patterns of activity; smell of rose, concept of rose, and image of rose.

vidual, and the context (e.g., you remember seeing the idea on the top right corner of the textbook page).

These associations may also include verbal associative attributes, such as table-chair, key-lock, or rose-flower. Out of the dozen or so potential attributes that might be associated with a given concept or idea, only the one or two most salient of them will actually become associated in long-term memory on a given occasion. These attributes can then act as *retrieval cues* to reactivate associated information (Underwood, 1983).

It can be seen that the associations in long-term memory are very critical because they are the things upon which we rely to cue the original information when we wish to remember it. Thus, a major goal of a training program is to promote the existence of associations among various relevant pieces of knowledge. Gagne (1985) asserts that unless two pieces of knowledge exist together in working memory at one time, they will not be associated in long-term memory. One finding from the field of education is that novices tend to have isolated and fragmented pieces of knowledge, while experts have highly interconnected and cohesive knowledge structures (i.e., lots of associations).

ACQUIRING KNOWLEDGE

Against the background of the information-processing model, we can discuss the general mechanisms by which we learn declarative and procedural knowledge. Knowing these mechanisms allows us to design training systems that maximally promote knowledge acquisition.

Acquiring Declarative Knowledge

The associations formed between declarative pieces of information take place in working memory and can be based on different factors. We may associate two words because they were next to each other on a grocery list. We may associate two ideas because they are different ways of saying the same thing. Or we may associate two words because they share the same phonetic sounds. As noted earlier, some researchers have developed taxonomies of different associations.

What factors increase the likelihood that a particular memory will be reactivated? There are two critical components that have been identified. First, the **strength** of the pattern that represents the item itself will determine the ease with which it is reactivated. All other things being equal, if a particular pattern has been experienced frequently, it will have a higher strength value and will be more easily retrieved than an item that has not been experienced frequently. This leads to the instructional principle that a person should repeat something many times in order to remember it later. An item's strength is also determined by how recently it has been active; the more recently we experience ideas, the more easily they are reactivated.

The second factor that will increase the likelihood that a memory can be reactivated is the strength and number of **associations** with other pieces of knowledge (whether verbal, image, or whatever). Information is more likely to be remembered if it has associations with many other items, and also if the associations are strong. This second retrieval factor is generally much more powerful than the "strength of pattern" effect caused by simple repetition. Much research has shown that the best way to enhance memory for information is to support association of the to-be-learned information with the learner's *previously existing* knowledge that is meaningful and memorable (Anderson, 1990; Gagne, 1985).

The importance of this second factor is pointed out by Jonassen (1988), who states that to maximally enhance learning, instructional programs need to support **generative learning activities.** Generative learning strategies are: "those that require learners consciously and intentionally to relate new information to their existing knowledge rather than responding to material without using personal, contextual knowledge (the classic drill-and-practice paradigm)" (Jonassen 1988, 154). Examples of tutorial programs that *do not* employ generative learning strategies are computer-based training programs where some declarative segments of knowledge are presented, and then subsequent displays test the learner's ability to recall or recognize some subset of that same knowledge.

Notice that in the framework outlined above, access to information in long-term memory largely depends on accessing related information. This related information can vary in how memorable it is, in and of itself. There seem to be certain types of information that are generally more memorable. This includes concrete objects and perceptions (as opposed to abstract ideas and principles), visual images, and anything unusual. Therefore, to maximize learning, the declarative knowledge to be acquired should be associated with concrete instances and vivid visual images, especially if they are interesting or unusual (notice the tie back to *salience* attentional factors).

Finally, we can ask if there are any differences between the acquisition of declarative conceptual knowledge and declarative rule (explicit procedure) knowledge. A review of the literature does not reveal any major differences in the way that we store and retrieve these two types of information. Whether we are being told the name of the first president, or how to change a spark plug, we store and retrieve both types of knowledge through associative networks. And for both types of *declarative* knowledge, the more well-organized the information in memory, the better we will be able to retrieve that information on the basis of information in a new task.

Roles of Declarative
and Procedural/Skill Knowledge

According to Anderson's ACT* model (Anderson, 1983), declarative knowledge is transformed into production rules (procedural skill knowledge) as soon as a person is required to actually perform some task, such as solving an algebra problem. To accomplish this process of **proceduralization,** there are domain-general skills that act on the declarative knowledge to form actions, which are then captured in domain-specific production rules (Singley & Anderson, 1989). Table 6.1 gives several examples of production rules. Once these initial production rules have been formed, subsequent learning occurs through the transformation of production rules (described in the next section).

Given this view that learners move from using declarative knowledge to domain-specific production rules quite early in the learning process, we might ask why one should worry about declarative knowledge at all. Declarative knowledge is very critical because once a person gets into the use of domain-specific skill knowledge, it tends to be *very* situation specific. That is, it does not generalize well to dissimilar problems. Declarative knowledge is the major component that allows the person flexibility to deal with new problems (Singley & Anderson, 1989). Even experts revert to the use of declarative knowledge when faced with a novel problem.

In addition, it is the nature of the declarative knowledge that determines the goodness of the production rules when they are formed. As Singley and

TABLE 6.1 Examples of Production Rules

IF	Encounter WORD that I don't recognize
THEN	Break WORD into SYLLABLEs.
IF	WORD is broken into SYLLABLEs
THEN	Pronounce each SYLLABLE
IF	GOAL is to find LEAST COMMON DENOMINATOR
THEN	Multiply all DENOMINATORS to get PRODUCT
	Identify COMMON FACTORS among DENOMINATORS and PRODUCT
	Divide PRODUCT by COMMON FACTORS

(From E. Gagne, *The cognitive psychology of school learning*, 1985, Little, Brown and Company. Reprinted with permission.)

Anderson (1989) state, "if the underlying declarative knowledge reflects a shallow or even rote understanding of some procedure, then the resulting production rules will have that same character" (p. 228). Thus, if a person's declarative knowledge is complete and well-structured, that is, they have a sound understanding of the deep structure of a domain, they will be more able to create more general and therefore more powerful production rules for task performance.

One phenomenon that researchers have noted is that people are often not very good at using declarative knowledge to solve problems and develop task-specific production rules (Bereiter, 1984; Bransford et al., 1990; Gordon & Gill, 1989; Perfetto, Bransford, & Franks, 1983; Silver, 1987; Simon, 1980; Singley & Anderson 1989). In problem-solving tasks, even when students most definitely have declarative knowledge that is critically relevant to the task, they will fail to retrieve the knowledge and identify its usefulness. This type of knowledge is sometimes referred to as **inert knowledge** (Bransford et al., 1990), a problem first identified many years ago by Whitehead (1929).

In addition to failing to *retrieve* knowledge and identify it as relevant, the actual use of declarative knowledge for the development of task-specific production rules may be difficult. As an example, consider a computer-based tutorial on basic physics that asks the learner to adjust the height of several roller-coaster hills on the computer screen such that the car will stop on top of the last hill. In solving this problem, the learner can go "into the computer" and ask a variety of experts about topics related to the problem (gravity, g forces, etc.). The problem now confronting the learner is knowing (a) which pieces of declarative knowledge are relevant to the task, and (b) how the

knowledge is relevant to the task. Learners have problems making the necessary set of transformations from declarative knowledge to domain-specific
productions because there is too much potentially relevant knowledge to sift
through, they do not know how to make the necessary transformations, and
sometimes because there is just too much of a processing load on working
memory (Singley & Anderson, 1990). Implications are clear for training programs; simply providing the basic conceptual information is not enough,
trainees must be explicitly taught to retrieve and use the information (Thomas
& Rohwer, 1993).

Acquiring Procedural/Skill Knowledge

As suggested by research reviewed in an earlier chapter, skills are production
rules acquired in one of two ways. First, they may be directly acquired
through trial and error. An example is a child's learning of correct grammatical statements. The other method is through first acquiring declarative knowledge and then transforming that knowledge into productions. According to
Anderson, the acquisition and transformation of domain-specific production
rules (as opposed to learning declarative knowledge) is the predominant activity underlying the acquisition of domain expertise (Singley & Anderson,
1989).

Figure 3.1 showed a subclassification that decomposed skills down into
five basic categories. The view to be presented here is that all skills are acquired in essentially the same manner; by repeatedly associating a stimulus
set with some response. Anderson's original work (1983) suggested that
learning within the procedural system takes place through the mechanisms of
generalization and discrimination. While his more recent model does not include these processes (Singley & Anderson, 1989), they are discussed here
because they offer a convenient way of viewing changes in production rules.

Generalization.   When a particular instance of task or subtask performance is encoded into a set of production rules, each production will have
certain conditions and an associated action. As a simple example, if we are
shown a *black square,* we might form a production rule such as:

    IF figure has four sides
        and sides are equal
        and sides are touching on both ends
        and four inner angles are 90°
        and figure is black
    THEN classify as square

This production will correctly recognize other black squares of any size as "squares." What happens when the person is shown a red square as an example of a square? Through the process of analogy (mapping characteristics of the old production onto the new example), the person must drop the condition of "black," in order to accommodate the new instance. When old production rules are too narrow, they are transformed to be more general by dropping conditions that do not apply across the entire range of examples.

There is a very direct implication of this principle for training. Examples of concepts, procedural demonstrations, and practice problems must contain the correct characteristics or conditions and also *vary sufficiently* on unnecessary or irrelevant attributes to prohibit the irrelevant attributes from being included in the condition side of the production. If this is not done, the person will fail to apply the production in instances where it should be applied (underutilization).

Discrimination.   It is also possible to develop production rules that are not specific enough. For example, if we are told that a square has four sides that are equal and touching, we might develop a production rule such as:

> IF figure has four sides
>   and the sides are equal in length
>   and the sides are touching
> THEN classify as square

If the learner is shown the picture in Figure 6.3, he/she will classify it as a square. This error occurs because the production is not adequately specific, that is, it does not have all necessary and sufficient conditions. If a learner is told that Figure 6.3 is not a square, he/she must try to determine the difference between the characteristics of the figure and the characteristics in the production. This will lead to adding of conditions that more strongly restrict application of the production. This process of adding conditions is termed **discrimination**. Let us imagine that as a result of viewing Figure 6.3, the person added a condition to the "square" production rule, such as "and figure is enclosed." The production rule is now closer to being correct for identifying squares, but would still falsely classify Figure 6.4 as a square. It can be seen that much experience with examples or problems is needed to adequately refine production rules.

The process of refining production rules through discrimination is supported instructionally by providing **nonexamples**. A nonexample is an instance that is described as "not fitting a particular production rule." For example, the drawing in Figure 6.3 would be shown and described as something

Figure 6.3 Instance to be classified by learner as square or not square.

that is not a square. A variety of nonexamples should be given so that the CONDITION side of the production rule is specific enough.

To support development of accurate production rules, several instructional approaches might be taken. First, for pattern recognition, many examples and nonexamples can be presented simultaneously or close together in time, where the examples vary on irrelevant attributes. This helps the learner focus on the differences between examples and nonexamples and weed out irrelevant attributes. In more complex tasks and problem solving, a *subtask analysis* should first be conducted to identify perceptual pattern-recognition aspects of the task. These subtask components can then be taught with examples and nonexamples. Note that feedback must be given immediately after each subtask for this process to work. Finally, researchers have shown that explicitly describing the appropriate sequences of actions and the conditions under which they should be performed does help the learner to focus on important task attributes and speeds the learning process (e.g., Fisk, 1989). Even so, it can be seen that this refinement of production rules requires *extensive practice.*

TRANSFER OF TRAINING

Up to this point, we have been evaluating the learning process in general, where learning can be defined as any change in behavior due to repeated experience in a situation (Bower & Hilgard, 1981). However, we also need to look at a more specific type of learning. That is, if we train a learner using

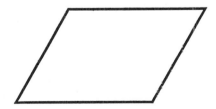

Figure 6.4 Second instance to be classified by learner as square or not square.

certain instances or example tasks, to what extent will he/she perform successfully on other similar but different tasks? A typical goal of training programs, and even formal educational programs, is to teach people in such a way that they will perform successfully on tasks in the post-training environment. What determines how successfully people will **transfer** their learning from the training to the post-training environment?

Identical Elements Theory of Transfer

One theory that has been successfully applied to the transfer of training question is Thorndike's **identical elements theory of transfer** (Thorndike, 1906). Thorndike was one of the first psychologists to oppose the view that the mind can be disciplined through the study of subjects such as Latin or mathematics and subsequently perform better in other unrelated domains. Opposing this "doctrine of formal discipline," Thorndike suggested instead that the mind is simply a collection of countless special operations or functions that are very task-specific.

The identical elements theory assumes that all learning takes place via acquisition of stimulus-response associations. Each learning situation has associated with it certain elements, and learning in one situation will be transferred to a new situation to the extent that the two situations have elements in common. The more elements in common, the more the transfer, meaning that the old response or action will be applied to the new situation. Elements can be either actual stimulus conditions or specific procedures (Hergenhahn, 1988). An example given by Hergenhahn is that looking up words in a dictionary in school may transfer to a variety of situations outside of school, not because the word is the same but because the procedures transfer.

Singley and Anderson (1989) recently argued that the identical elements theory is alive and well. By defining elements as productions, they cogently showed that the number of common elements can strongly predict transfer of training. This **identical productions theory of transfer** is more powerful than Thorndike's original identical elements theory, because the newer version allows very abstract production rules to serve as the common elements.

There is not currently a sufficiently developed theory that specifies what elements are more and less important in transfer. At the most, we know that novices tend to focus attention on superficial concrete elements of problems and retrieve previous problems from memory on the basis of those elements. Experts tend to focus on underlying principles and abstract concepts, retrieving previous examples on the basis of these abstract or relational elements. This means that elements forming the basis of transfer can be either very concrete perceptual elements, or very abstract principles, and one task of the trainer is to move the novice from transferring on the basis of concrete surface elements to transfer on the basis of underlying principles and complex relationships. One way to do this is to vary the concrete specific elements adequately in problems, and to help focus attention on the underlying principles and more abstract elements. When left on their own with only a few problems to solve, novices will not go through this process.

Positive and Negative Transfer

According to the analysis given above, a learner can vary from zero transfer, where there are no elements in common, to strong positive transfer, where there are virtually all elements in common. However, it is more accurate to say that positive transfer will occur when both conditions *and* actions in the production rules are similar. What happens when one side of the production rule is similar and one side is dissimilar? For example, a person learns during training that Stimuli one through five are always associated with response X, but after training the same set of stimuli require a different response. This produces what has been termed **negative transfer,** where training interferes with post-training performance; performance is actually worse than if there had been no training at all.

The situation described above is actually one where previous experience is simply interfering with new learning. That is, you must form a new production rule that has all of the old CONDITIONS plus any specifications denoting the new situation. This production rule may be formed relatively early in experiencing the new situation. However, retrieval of the new production rule is competitive with the old rule. And the more practice a production has received, the stronger it will be. Thus, in the new post-training environment, the old production may be much stronger than the new production, and will accidently be applied for some time until the new production gains adequate strength to compete. It can be seen that the degree of negative transfer will depend on *two factors*: the number of conditions the two situations have in common plus the strength of the original production. This is why people who have many years of experience with a task show strong negative transfer when task stimuli are kept the same but the required action is changed. This situation is much more problematic than one where the response is kept the same but the conditions are changed. Notice that this phenomenon has strong implications for system redesign as well as training. One should not design a system or try to train a person when the task has "old" stimulus elements but a new required response.

PART-TASK TRAINING AND AUTOMATION

Sequencing of Instructional Materials

As we saw in Chapter 4, most job tasks can be broken down into component subtasks. In a training or instructional program, we must ultimately decide whether to train the complex task as a whole, or train the component subtasks individually. This is known as a decision of whole-task versus part-task training. The relevant question here is whether people learn more (or faster) with one method or the other. A related question is, if conceptual knowledge and/ or a task can be decomposed into a hierarchy of components, is it better to teach all of the lower level components before moving on to the higher-level?

Unfortunately, there seems to be no simple answer to these questions. Several researchers have suggested that instructional design should progress from simple to complex components (e.g., Ausubel, 1968; Bruner, 1960; Gagne, 1985; Merrill, 1980). While this approach may be intuitively appealing, research efforts have generally failed to find any evidence supporting this method (VanPatten, Chao, & Reigeluth, 1986). For example, some studies that compared carefully designed curriculum with "scrambled" sequencing of skills found no significant differences in student learning (Payne, Krathwohl, & Gordon, 1967; Niedermeyer, 1968; Pyatte, 1969).

Other researchers studying complex problem-solving tasks have also failed to find significant differences based on curriculum sequencing (e.g., Llaneras, Swezey, & Perez, 1989). To moderate these negative findings, we should point out that one study by Buckland (1968) found that scrambled curriculum significantly impaired learning in less capable students. This suggests that more capable students may be able to make all of the many inferences necessary to learn and bind together the various skills, while less capable students are not.

Reigeluth (1987) also proposed a methodology for sequencing instructional materials, termed **Elaboration Theory.** This method is best understood in terms of a "zoom lens" metaphor, where the learner receives an overview, zooms in to view the parts, then back out to view the whole. Elaboration Theory suggests that the instructional unit start with an overview containing the "epitome." Subsequent material elaborates upon that material in layers (Reigeluth, 1987). This additional material will elaborate on knowledge in one of three ways: *conceptual elaboration* specifying "what," *procedural elaboration* specifying "how," and *theoretical elaboration* specifying "why." Which type of elaboration is performed will depend on the type of knowledge that has been identified for instruction.

There is fairly strong evidence that providing an overview of the material to be learned does promote learning in a variety of domains (Anderson, 1990). Other than this positive effect of what are termed **advance organizers,** there is not substantial research that specifically supports the prescriptions given by Elaboration Theory.

Part-Task Versus Whole-Task Training

A related issue is whether there is any merit in breaking a task down into subtasks and training at least some subtasks separately. For example, should we practice the left hand and the right hand separately when learning a new piece on the piano? There is some evidence that a complex task can be learned more easily if it is broken up into component subtasks and at least some of these are trained individually (e.g., Fisk, 1987; Lesgold & Curtis, 1981; Lesgold & Resnick, 1982; Mane, 1984). On the other hand, reviews of this literature tell us that part-task training is not always superior (Cream, Eggemeier, & Klein, 1978; Wightman & Lintern, 1985), and a few studies have shown superiority for whole-task training (Connelly et al., 1987).

There are many reasons for the lack of superior training with part-task methodologies. The most commonly cited one is that the training analysts did not follow sound psychological theories and principles (e.g. Fisk & Eggemeier, 1988). While this may be true, it doesn't help us understand when to use part-task training and when to use whole-task training. There seem to be two fundamental reasons for the superiority of one versus the other.

First, we have limited *cognitive resources,* and learning of a complex task may sometimes place too much of a burden on those resources and working memory. If some of the subtasks can be learned in isolation, enough working memory is available to form the necessary production rules. Note that part-task training would only be helpful when the entire task required too many cognitive resources. In order to determine whether the entire task to be learned is too workload intensive, the trainer should become familiar with some of the various methods for measuring and predicting cognitive resource workloads (e.g., Andre & Lintern, 1990; Wickens, 1980; Wickens & Kessel, 1980; Wickens, Mountford, & Schreener, 1981).

The second factor affecting the usefulness of part-task training revolves around the nature of the component subtasks. If there is an interaction between the subtasks, where the whole task consists of integrating the many subtasks, there will be an advantage for whole-task training. This is because the learner is able to practice the *skills* of integrating the various subtasks and of allocating attention properly.

Anderson (1990) suggests that if subtasks are independent of one another in total task performance, they are amenable to part-task training. For example, learning to type can be acquired independently of learning to write a research paper on the computer. However, if a complex task requires the integration of many related subtasks, those subtasks should be trained together in the appropriate job sequence. The trainer should use other means (such as memory aiding, or simplifying input procedures) to temporarily reduce cognitive workload imposed by the learning tasks (Andre & Lintern, 1990; Connelly et al., 1987; Lintern & Wickens, 1987).

Automation of Subtasks

In the last section, we saw that the demands of learning a complex task can overload cognitive resources. Sometimes it is the task itself that is resource intensive, not just the learning process. Under these circumstances, researchers have turned to a strategy of automating certain subtasks so that the person can devote more of their cognitive resources to variable tasks requiring controlled processing (Fisk & Eboch, 1987; Gallini & Fisk, 1986).

Fisk suggests that a task be broken down into components using an **elementary component model** matrix (see Gallini & Fisk, 1986). Each cell in the subtask matrix is analyzed to determine the type of task and its consistency of stimulus-response associations over time. The cognitive resources required for the components are estimated as well. Once the components with consistent mapping characteristics are identified, training guidelines based on automatic/controlled processing theory can be applied (Schneider, 1985).

Some preliminary studies have shown that training of subtasks to the point of automation can enhance later task performance (e.g., Fisk and Eboch, 1987). However, not enough research has been conducted to date for us to identify the circumstances under which part-task training for automation of subtasks is advisable. Based on the potentials for negative transfer of training, we can at least say that automation should only be pursued when the analyst can be certain that the stimulus-response associations or mappings will remain constant for future real-world tasks.

MOTIVATION

All of the cognitive processes we have described, from transferring information into working memory to associating new information with previous knowledge, take some amount of cognitive effort. And this effort will not be made by learners unless they are motivated; they must want something and they must get something from their efforts. The goal of this section is to briefly discuss various motivational factors that affect learning, and describe some instructional design strategies that can promote learner motivation.

A great number of learning theories are either directly or indirectly relevant to motivation in an educational or training situation. In fact, there are entire textbooks written on the psychology of motivation. Rather than review all of this literature, I will present two models that bear directly on trainee motivation: Keller's ARCS model (Keller, 1984; Keller & Kopp, 1987; Keller & Suzuki, 1988), and Weiner's (1972, 1980) attribution model. The sections below will present (a) the four basic components of motivation assumed by the ARCS model, (b) Weiner's attribution model which elaborates on the confidence component of ARCS, and finally (c) the specific strategies suggested in the ARCS model.

The ARCS Model of Motivation

Keller's ARCS model integrates ideas and instructional prescriptions from a broad variety of areas including learning theories, environmental theories, attitude theories, attribution theory, equity theory, and cognitive dissonance theory (Keller, 1984; Keller & Kopp, 1987; Keller & Suzuki, 1988). Its primary utility for instructional designers is that it explicitly provides 12 strategies designed to increase learner motivation.

The model first assumes that a person's level of motivation has four components or factors—**attention, relevance, confidence,** and **satisfaction** (hence the acronym, ARCS). Each of these factors plays a part in a learner's overall level of motivation, and each can be enhanced by certain instructional strategies.

Attention.   We have discussed attention as it relates to information processing; attention to certain environmental elements is important for knowledge acquisition. The concern with attention from the motivational perspective is *getting* a learner's attention and *sustaining* it (Keller & Suzuki, 1988). According to Keller, attention can be gained at two levels, the perceptual level of information, and a deeper level. We discussed the perceptual level earlier, pointing out that certain types of information in the environment tend to capture our attention. These are perceptual stimuli that are novel, surprising, incongruous, or uncertain (Kopp, 1982). However, a learner's attention can be also be gained at a deeper "epistemic level" of interest, capturing their curiosity about the nature of some particular phenomenon. Thus, according to Keller and Suzuki (1988), the educator can "introduce startling or unexpected events which arouse a perceptual level of curiosity, or engage the

learners in inquiry oriented behavior that stimulates a deeper level of interest"
(p. 404).

Relevance.   Trainees will be motivated to expend the necessary cognitive resources for learning only if the instructional content has some relevance to their lives. We discussed earlier the fact that information is attended to and brought into working memory if that information is relevant to trainees' goals. This idea can be expanded, any processing will be more likely if it is relevant to learners, either their current goals, future goals, or some aspect of their past or present lives.

Confidence.   Learners will exert the effort necessary to learn and perform some task to the extent that they believe they are capable of succeeding at the task. Stated another way, people are more motivated to perform a task when they believe that they have the necessary knowledge and skills to succeed. Not that success must be absolutely guaranteed, but the likelihood must be within some acceptable limits. This is another way of saying that tasks must be designed so that they are challenging, but the learner still believes there is a relatively high likelihood of success.

Satisfaction.   Once an instructional program has begun and learners have performed some set of tasks, they will remain motivated to a greater degree if they have been satisfied by the process. Satisfaction can come from many sources including external rewards, intrinsic rewards, reduction of curiosity, and meeting of various needs (achievement, self-esteem, etc.). A long line of behavioral research has shown that the use of external reinforcers can have a powerful impact on performance. However, it has also been shown that the use of extrinsic rewards can reduce motivation, especially in those cases where the rewards are under the obvious control of another person (Deci, 1975).

In addition, extrinsic rewards can detract from the effects of intrinsic rewards (Deci, 1975). Intrinsic rewards reflect a general satisfaction derived from doing a task, and doing it successfully. If a person is intrinsically interested in performing the activity, the introduction of extrinsic rewards can alter the person's perception to one where they are now performing the task for external reward rather than inner satisfaction. This is especially problematic when the extrinsic reward system is removed, because now there is no remaining incentive. The reader is referred to Deci and Ryan (1985) for a discussion of the complex interaction between extrinsic and intrinsic reward. In general, it is best to make relatively light use of rewards in an instructional system, and attempt to promote intrinsic satisfaction with task performance.

Finally, there has been research suggesting that people have a need to obtain information relevant to their particular goals and interests, and that gaining this information is reinforcing in and of itself. Incorporating this type of "uncertainty reduction" reinforcement into a training program can be relatively simple and transfer well to the post-training environment.

Weiner's Attribution Theory

In the ARCS model, we saw that confidence in one's ability to perform a task has a substantial impact on a person's motivation to learn and perform within the training program. At any given point in time, the learner's confidence will partly be a function of his/her previous experience, either with the same task or with similar tasks. For example, if the training course content is mathematics and the learner has always done poorly in that domain, he/she may tend to have low confidence and predict that they will do poorly on the training tasks. The resultant lack of motivation will create a self-fulfilling prophecy; the learner makes an inadequate effort and fails at the task.

It would be more precise to say that learners' confidence is some function of how they *perceive* their previous experiences, what they assume to be the cause of their success or failures. If a person perceives his/her previous failures to be caused by a lack of an underlying ability, then he/she will assume that they will not be able to perform the task successfully in the future. When we succeed or fail at a task, we attribute the outcome to some factor, either inside of ourselves or in the external environment.

After evaluating the various research efforts studying causal inference, Bernard Weiner (1972, 1980, 1982) determined that there is a very limited set of causes to which an individual attributes success or failure: ability, effort, task difficulty, luck, mood, fatigue, illness, and help from other people. It can be seen that these factors vary on whether they are internal or external to the person. For example, ability, effort, mood, fatigue, and illness are internal personal causes.

For instructional design, the important issue is the various consequences resulting from causal attributions. Table 6.2 shows the likely consequences of each attribution (summarized from Weiner, 1980, and Bell-Gredler, 1986). Without getting into details, it can be seen that failures that are perceived to be due to external and/or uncontrollable causes do not significantly affect the learner's self-image, although they do often lead to expectations of future failures under similar circumstances (e.g., where causes are task difficulty, and help from others).

On the other hand, when learners attribute their failures to ability, they develop negative affect and self-image, and expect similar outcomes in the

TABLE 6.2 **Consequences of Causal Attributions**

Attribution	Emotion	Self-Image	Expectation for Future
SUCCESS ATTRIBUTED TO:			
Ability	Pride	Confidence, Competence	Same outcome expected
Effort	Pride, activation	Confidence	Same outcome likely but not definite
Task difficulty	Gratitude, disregard	No effect	Same outcome expected
Luck	Surprise	No effect	Same outcome not necessarily expected
Help from others	Gratitude	No effect	Same outcome expected
FAILURE ATTRIBUTED TO:			
Ability	Resignation, apathy	Shame	Same outcome expected
Effort	Guilt	Possibly negative self-image	May still expect future success
Task Difficulty	Depression, frustration	No effect	Same outcome expected
Luck	Surprise, disappointment	No effect	May still expect future success
Mood, fatigue, or illness	Disappointment	No effect	No effect
Help from others	Anger	No effect	Same outcome expected

future. While failure due to effort might generate negative feelings, there is still a possibility of success in the future and thus an "effort" attribution will not decrease motivation to the same degree as a perceived lack of ability. This leads to the educational prescription that students should be encouraged to perceive failure as being due to a lack of effort rather than a lack of ability. Unfortunately, prolonged experiences of failure will still most likely lead the student to move from the "effort" attribution to a more stable cause, such as ability or task difficulty. Both of these will decrease motivation and effort.

Success results in various emotions and expectations depending on the perceived causes of the success. The most desirable causal attributions, from an instructional point of view, are ability and effort; both of these enhance the learners self-esteem and expectancies for future task performance. The implications for training programs should be obvious at this point; care should be taken that all trainees experience at least some moderate degree of success at various subtasks, and that they attribute their success to ability and effort.

ARCS Strategies

Returning back to the ARCS model of motivation, we can now see how variables such as perceived competence, perceived control, and expectancies affect motivation. Learner's will have relatively high motivation if they perceive that with sufficient effort, they have the ability to do the task.

In synthesizing previous research into the ARCS model, Keller developed a set of prescriptions for instructional design (Keller, 1984). These prescriptions are given as a set of 12 strategies designed to enhance learner motivation. The strategies are listed below along with a few of the suggested methods of implementation described by Keller and Suzuki (1988). The reader is referred to Keller and Suzuki for a number of helpful suggestions specific to computer-based applications, and to Keller and Kopp (1987) for one example of an application of the ARCS model.

STRATEGIES TO ENHANCE ATTENTION

Strategy A.1. Perceptual Arousal. Gain and maintain student attention by the use of novel, surprising, incongruous, or uncertain events in instruction. Examples in videotape or computer-based instructional systems would include animation, inverse, flash, special effects, and progressive disclosure of information.

Strategy A.2. Inquiry Arousal. Stimulate information-seeking behavior by posing, or having the learner generate, questions or a problem to solve.

Strategy A.3. Variability. Maintain student interest by varying the elements of instruction. While perceptual arousal usually implies something "new or catchy," the variability element refers to a change among instructional events. For example, instructional segments should be relatively short, and information presentation screens interspersed with interactive segments.

Later in the text we will discuss the need for good human factors in a training program, particularly the interface of a computer-based instructional system. At this point I will just note that variability should never be implemented in courseware such that it results in inconsistency among the interface elements of the system. This creates a system that is more difficult for the trainee to learn and subsequently use, and takes cognitive resources away from the primary learning tasks.

STRATEGIES TO ENHANCE RELEVANCE

Strategy R.1. Familiarity. Use concrete language, and use examples and concepts that are related to the learner's experience and values.

Strategy R.2. Goal Orientation. Provide statements or examples that present the objectives and utility of the instruction, and either present goals for accomplishment or have the learner define them.

Strategy R.3. Motive Matching. Use teaching strategies that match the motive profiles of the students. According to Keller and Suzuki (1988):

> It is much easier to motivate the high achiever, independently of the utility of the instruction, if the teaching strategies provide the opportunity for the learner to set standards, be personally responsible for success, and to receive frequent feedback concerning progress toward goal accomplishment.... However, a person with a high need for affiliation enjoys noncompetitive, cooperative situations.... An option could be provided that would allow two or more students to work together on a program. (Keller & Suzuki, 414)

STRATEGIES TO ENHANCE CONFIDENCE

Strategy C.1. Learning Requirements. Help students estimate the probability of success by presenting performance requirements and evaluative criteria. The program should present objectives and the overall structure of the lesson, prerequisite knowledge and skills, the criteria that will eventually be used to evaluate the learner, and provide opportunities to practice with feedback. When learners know exactly what to expect, they are more likely to be confident in their estimates of success.

Strategy C.2. Success Opportunities. Provide challenge levels that allow meaningful success experience under both learning and performance conditions. The reasoning behind the specification for both learning and performance conditions is that people want some degree of certainty that they will be successful. This means that as people achieve mastery of the skill or content, the difficulty or competitiveness can be increased.

Strategy C.3. Person Control. Provide feedback and opportunities for control that support internal attributions for success. Allow students to have some (but not complete) control over program elements, and provide feedback that leads the learner to attribute success to ability and effort. Most importantly, avoid sequences of instruction where success is impossible or that are likely to result in repeated failures. An example of how to avoid this problem is: as soon as the learner fails at three problems, interrupt activity and provide a different approach (e.g., remedial information or diagnostic questioning).

STRATEGIES TO ENHANCE SATISFACTION

Strategy S.1. Natural Consequences. Provide opportunities to use newly acquired knowledge or skill in a real or simulated setting. The goal should be to allow the learner to apply his/her knowledge in a meaningful environment, be it job, hobby, simulation, case study, or other real or imagined games.

Strategy S.2. Positive Consequences. Provide feedback and reinforcements that will sustain the desired behavior. Extrinsic rewards should be used judiciously. When overdone, they can be distracting, take away from intrinsic reward, annoy the trainee, and take time away from the task. Some simulation/tutorials do nothing when a subtask is accomplished successfully (implying success) but intervene to help when the learner makes an error. The trainer should also take care that failure does not result in a greater "reward" (such as more interesting animations) than does success.

Strategy S.3. Equity. Maintain consistent standards and consequences for task accomplishment. Standards of performance should be applied consistently and fairly over time for the trainee. They should also be applied consistently across different learners to increase each person's sense of equity.

PRINCIPLES AND GUIDELINES
FOR INSTRUCTIONAL DESIGN

In Chapter 3 we learned about different types of knowledge that may be the focus of a training program. In this chapter, we have seen that those types of knowledge are acquired in qualitatively different fashions. Therefore, in designing a training or instructional program, we must first determine what types of knowledge are being acquired (e.g., declarative vs. procedural/skill), and to what degree (e.g., do some tasks need to be automated?). After determining the general content and type of knowledge to be acquired through task analysis, the designer must identify appropriate strategies for instruction.

Based on what we have learned from cognitive, educational, and learning theories, we can list some basic instructional strategies or principles for design. Which strategies should be used will depend on the type of knowledge to be acquired by the learners. In many programs, several types will be acquired, and therefore different strategies should be used in different segments of the program. A variety of principles and strategies are listed below, categorized according to the basic learning mechanisms that they support.

Attention

1. Obtain the learner's attention by using perceptual saliency, novelty, and other qualities known to capture attention.

2. When there are a variety of stimulus elements where some are more critical than others, help focus the learner's attention on important or critical elements. Use color, spacing of elements, motion, flashing, and other attributes that will heighten the saliency of those elements.

3. Use simulations that simplify a complex system to help learners focus on relevant variables.

Acquiring Declarative Knowledge

4. Make declarative information to be acquired meaningful and relevant to the learner's existing knowledge. Either suggest examples of learner's knowledge or experience that might be relevant (if known), or ask the learner to generate aspects of his/her knowledge or experience that is relevant to the material.

5. Tie new information to previously learned information. Use remindings or other prompts to promote remembering of previous instructional information, then show how the new information is related to the old.

6. Provide specific examples or concrete instantiations of abstract concepts and principles. As Ross (1989) has demonstrated, concrete examples and specific instances are often easier to retrieve from long-term memory than more abstract generalizations, although the examples frequently serve as a route to access of the general abstract knowledge.

7. Use memorable types of information to act as retrieval cues. These include items or objects that are concrete, unusual, interesting, etc.

8. Use video and/or animation to present declarative knowledge that involves dynamic systems. This serves two purposes: first, it helps the learner visualize variable relations. Second, it provides more basic concrete "perceptual" information rather than just an abstract description of the processes.

9. Assess and remediate declarative knowledge. Use diagnostic tests to identify holes, misconceptions, and inconsistencies in trainee knowledge structures, and remediate with either (a) correct information, or (b) questioning that guides the learner into identifying and repairing the problem themselves.

10. Before presenting a learning segment, provide a general structure. Such advance organizers should not be a summary of the information, but rather a summary of the simplest, most general ideas and their relationships.

11. Provide instructional objectives to the learner so that they know what they will be expected to do as a function of the program. Such expectancies will help shape goals in working memory; these goals will increase the likelihood that trainees focus on important aspects of the information.

12. Promote active, generative learning. For example, rather than telling a learner some scientific principle, provide experiences that will guide the learner to generate and test the principle. Likewise, ask questions that will cause the trainee to work with declarative knowledge to generate relationships and conceptual structures.

13. When presenting new information that must be stored in long-term memory, do not exceed the learner's capacity to transform the information. Gagne (1985) estimates that a lecture given at a typical speech rate presents 30 propositions per minute. Assuming only half of those are new, the learner must encode 15 propositions per minute. However, Gagne estimates that students are only capable of encoding approximately six new propositions per minute! And if students are actively elaborating upon the information, they will be slowed down further.

14. In teaching definitions of concepts, use example and non-example strategies similar to those for procedural-skill knowledge (#17 below).

15. Strengthen the knowledge and make it more accessible by repeating the information (or asking the learner to generate it) after periods of time where it is not used. This use of distributed practice (as opposed to massed practice) will increase the likelihood that the trainee will recall the information at a later time after training has finished.

16. Related to #15, test for retrieval of declarative knowledge

under conditions similar to the conditions of retrieval in the post-training environment. For example, students often study information immediately before taking a test. Being able to successfully access the information under those circumstances is not at all indicative that the same learner will be able to recall the information after six months without studying. If recall of information after several days without review is important for the post-training environment, this condition should be replicated in the testing environment as well.

As another example, if the trainee will have to later access and use declarative knowledge for problem solving, their ability to do this should be directly trained and tested in the instructional program. One cannot test recall of declarative knowledge and assume that it will be accessed and used under different circumstances.

Acquiring Procedural/Skill Knowledge

17. For basic pattern-recognition or perceptual learning, carefully design examples and nonexamples. These should cover the full range of critical and irrelevant attributes or conditions.

18. Provide examples and nonexamples close together in time, and make the similarities and differences noticeable.

19. For more complex skills, carefully define and provide a range of demonstration examples and problems to solve so that the trainee develops correct production rules.

20. Enhance development of production rules for difficult, abstract, or complex problems by providing explicit descriptions of the conditions and their associated action. For example, point out the characteristics of a mathematical problem that indicate the use of a particular problem-solving strategy.

21. Make important situational attributes and production rule conditions more perceptually salient to focus the learner's attention on critical features.

22. When learners are first acquiring productions, provide relevant declarative knowledge or prompt the learner to retrieve it. The declarative knowledge should be somewhat restricted so that the learner does not have a great deal of difficulty determining what declarative knowledge is relevant to the task and what knowledge is irrelevant.

The instructional program may need to also show how the concepts and relationships within the declarative knowledge lead to the production rule. For example, the declarative knowledge of the gearing subsystem of a bicycle may lead to the production rule that:

IF goal is easier pedaling
   and person is riding bicycle
THEN move right shift lever counterclockwise

However, the "bridging" knowledge between the declarative knowledge and production rule might not be obvious. The translation can be made explicit for the learner if necessary.

23. Train certain subtasks to automation if the post-training environment contains potential for overload of cognitive resources. As a test for automation, the learner should be able to additionally perform a second controlled task (such as carrying on a meaningful conversation) without detrimental effects on the performance of the task being automated. Another way to test for automation is to graph the time required to perform the subtask over trials and look for a point where the slope has significantly leveled out.

24. Use repetition of tasks or subtasks spaced over time to strengthen associations (as in #15 above).

25. Teach and/or test for any required subskills before higher-level skills.

26. Emphasize practice under varied circumstances, requiring more than you probably think is necessary. Skills degrade quickly over time, but research shows that this effect is lessened by overlearning of the task and using varied contexts.

27. Provide feedback after performance, preferably for each subtask being learned. That is, for a mathematics problem, do not wait until the entire problem has been solved to give "correct" vs. "incorrect" feedback. Provide this feedback for each discriminable step. This is more informative for the learner and is more efficient in revising the individual subtask production rules.

28. Where a learner is incorrect on a subtask action, allow them the option of trying a different action, asking for the correct action, and/or receiving additional declarative information bearing on the subtask. This strategy will depend on the type of skill being acquired. It would not be as appropriate for rapid perceptual learning as for complex decision making.

29. Train and test the actual skills required for the post-training

environment, and test them under circumstances similar to those of the post-training environment. As Merrill (1988) suggests, have trainees practice in "experiential situations in which the learner can perform a task similar to that which will be required when the skills are applied in the real world" (p. 95). If being able to shift gears in an actual car is the training objective, it is not appropriate to give a paper-and-pencil test of this knowledge (Bjork & Druckman, 1991).

Cognitive Workload

30. Don't overload cognitive resources. For example, don't require too many items to be held, and/or too many processes to be performed, in working memory all at once. This decreases the resources that can be allocated to learning the task.

31. Related to #30, for complex tasks, use memory aids or other methods to reduce the trainee's cognitive workload as he/she first learns the task.

32. Use part-task training where subtasks are independent and whole-task training would result in too much demand on cognitive resources. Also use part-task training under conditions where training of consistent subtasks is degraded by whole-task performance (see Schneider, 1985).

33. Make use of both visual and auditory channels of communication. Researchers have shown that there are separate cognitive resources for visual and auditory channels, and that we can process more information if the information is split between the two media (Wickens, 1992).

Motivation

34. Use motivational strategies prescribed by Keller (described in the previous section). This will require learning about the knowledge and interests of the students or trainees.

35. When in doubt, ask students and trainees whether *they* find the tasks and information interesting, relevant to themselves, and motivating.

REFERENCES

ANDERSON, J. R. (1983). *The architecture of cognition.* Cambridge, MA: Harvard University Press.

ANDERSON, J. R. (1990). *Cognitive psychology and its implications* (3rd ed.). New York: W. H. Freeman and Company.

ANDRE, A. D., & LINTERN, G. (1990). Attention theory as a guide to part-training for instruction of naval air-intercept control. *Proceedings of the Human Factors Society 34th Annual Meeting* (pp. 1347–1351). Santa Monica, CA: Human Factors Society.

AUSUBEL, D. P. (1968). *Educational psychology: A cognitive view.* New York: Holt, Rinehart & Winston.

BELL-GREDLER, M. E. (1986). *Learning and instruction: Theory into practice.* New York: Macmillan.

BEREITER, C. (1984). How to keep thinking skills from going the way of all frills. *Educational Leadership, 42*, 75–77.

BJORK, R., & DRUCKMAN, D. (1991). How do you improve human performance? *APS Observer, 4(6)*, 13–25.

BOWER, G. H., & HILGARD, E. R. (1981). *Theories of learning* (5th ed.). Englewood Cliffs, NJ: Prentice Hall.

BRANSFORD, J. D., SHERWOOD, R. D., HASSELBRING, T. S., KINZER, C. K., & WILLIAMS, S. M. (1990). Anchored instruction: Why we need it and how technology can help. In D. Nix and R. Spiro (eds.), *Cognition, education, and multimedia: Exploring ideas in high technology* (pp. 115–141). Hillsdale, NJ: Lawrence Erlbaum Associates.

BRUNER, J. S. (1960). *The process of education.* New York: Random House.

BUCKLAND, P. R. (1968). The ordering of frames in a linear program. *Programmed Learning and Educational Technology, 5*, 197–205.

CONNELLY, J. G., WICKENS, C. D., LINTERN, G., & HARWOOD, K. (1987). Attention theory and training research. *Proceedings of the Human Factors Society 31st Annual Meeting* (pp. 648–651). Santa Monica, CA: Human Factors Society.

CREAM, B. W., EGGEMEIER, F. T., & KLEIN, G. A. (1978). A strategy for the development of training devices. *Human Factors, 20*, 145–158.

DECI, E. L. (1975). *Intrinsic motivation.* New York: Plenum Press.

DECI, E. L., & RYAN, R. M. (1985). *Intrinsic motivation and self-determination in human behavior.* New York: Plenum Press.

FISK, A. D. (1987). High performance cognitive skill acquisition: Perceptual/rule learning. *Proceedings of the Human Factors Society 31st Annual Meeting* (pp. 652–656). Santa Monica: CA: Human Factors Society.

FISK, A. D. (1989). Training consistent components of tasks: Developing an instructional system based on automatic/controlled processing principles. *Human Factors, 31(4)*, 453–463.

FISK, A. D., & EGGEMEIER, F. T. (1988). Application of automatic/controlled processing theory to training tactical command and control skills: 1. Background and task analytic methodology. *Proceedings of the Human Factors Society 32nd Annual Meeting* (pp. 1227–1231). Santa Monica, CA: Human Factors Society.

FISK, A. D., & EBOCH, M. M. (1987). Applications of automatic/control processing theory to complex tasks: An encouraging look. *Proceedings of the Human Factors Society 31st Annual Meeting* (pp. 674–678). Santa Monica, CA: Human Factors Society.

GAGNE. E. D. (1985). *The cognitive psychology of school learning.* Boston, MA: Little, Brown and Company.

GALLINI, J. K., & FISK, A. D. (1986). An information-processing approach to instructional system design. *Educational Technology, 26,* 24–26.

GORDON, S. E., & GILL, R. T. (1989). *The formation and use of conceptual structures in problem-solving domains.* Technical Report for the Air Force Office of Scientific Research, University of Idaho.

HERGENHAHN, B. R. (1988). *An introduction to theories of learning* (3rd ed.). Englewood Cliffs, NJ: Prentice Hall.

JONASSEN, D. H. (1988). Integrating learning strategies into courseware to facilitate deeper processing. In D. H. Jonassen (ed.), *Instructional designs for microcomputer courseware* (pp. 151–181). Hillsdale, NJ: Lawrence Erlbaum Associates.

KELLER, J. M. (1984). The use of the ARCS model of motivation in teacher training. In K. E. Shaw (ed.), *Aspects of educational technology, Volume XVII: Staff development and career updating.* London: Kogan Page.

KELLER, J. M., & KOPP, T. W. (1987). An application of the ARCS model of motivational design. In C. M. Reigeluth (ed.), *Instructional theories in action: Lessons illustrating selected theories and models* (pp. 289–320). Hillsdale, NJ: Lawrence Erlbaum Associates.

KELLER. J. M., & SUZUKI, K. (1988). Use of the ARCS motivation model in courseware design. In D. H. Jonassen (ed.), *Instructional designs for microcomputer courseware* (pp. 401–434). Hillsdale, NJ: Lawrence Erlbaum Associates.

KOPP, T. W. (1982). Designing boredom out of instruction. *NSPI Journal,* May, 23–27, 32.

LACHMAN, R., LACHMAN, J. L., & BUTTERFIELD, E. C. (1979). *Cognitive psychology and information processing: An introduction.* Hillsdale, NJ: Lawrence Erlbaum Associates.

LESGOLD, A. M., & CURTIS, M. E. (1981). Learning to read words efficiently. In A. M. Lesgold and C. A. Perfetti (eds.), *Interactive processes in reading.* Hillsdale, NJ: Lawrence Erlbaum Associates.

LESGOLD, A. M., & RESNICK, L. B. (1982). How reading difficulties develop: Perspectives from a longitudinal study. In J. P. Das, R. F. Mulcahy, and A. E. Wall (eds.), *Theory and research in reading disabilities.* New York: Plenum Press.

LINTERN, G., & WICKENS, C. D. (1987). *Multiple resource analysis for part training of air-intercept control* (ARL-87-2/NASA-87-3). Savoy, IL: University of Illinois, Aviation Research Laboratory.

LLANERAS, R. E., SWEZEY, R. W., & PEREZ, R. S. (1989). *Effects of content sequencing on acquisition and transfer of troubleshooting performance* (Tech. Report SAIC-89-04-178). McLean, VA: Science Applications International Corp.

MANE, A. M. (1984). Acquisition of perceptual-motor skills: adaptive and part-whole training. *Proceedings of the Human Factors Society 28th Annual Meeting* (pp. 522–526). Santa Monica: CA: Human Factors Society.

MERRILL, M. D. (1988). Applying component display theory to the design of courseware. In D. H. Jonassen (ed.), *Instructional designs for microcomputer courseware* (pp. 61–102). Hillsdale, NJ: Lawrence Erlbaum Associates.

MERRILL, P. F. (1980). Analysis of a procedural task. *NSPI Journal, 19,* 11–15.

MILLER, G. A. (1956). The magical number seven, plus or minus 2: Some limits on our capacity for processing information. *Psychological Review, 63,* 81–97.

NIEDERMEYER, F. C. (1968). The relevance of frame sequence in programmed instruction: An addition to the dialogue. *Teaching Machines and Programmed Instruction, 16,* 301–317.

PAYNE, D. A., KRATHWOHL, D. R., & GORDON, J. (1967). The effect of sequence on programmed instruction. *American Educational Research Journal, 4,* 125–132.

PERFETTO, B. A., BRANSFORD, J. D., & FRANKS, J. J. (1983). Constraints on access in a problem solving context. *Memory and Cognition, 11,* 24–31.

PYATTE, J. A. (1969). Some effects of unit structure on achievement and transfer. *American Educational Research Journal, 6,* 241–261.

REIGELUTH, C. M. (1987). Lesson blueprints based on the elaboration theory of instruction. In C. M. Reigeluth (ed.), *Instructional theories in action: Lessons illustrating selected theories and models* (pp. 245–288). Hillsdale, NJ: Lawrence Erlbaum Associates.

ROSS, B. H. (1989). Remindings in learning and instruction. In S. Vosniadou and A. Ortony (eds.), *Similarity and analogical reasoning* (pp. 438–469). Cambridge: Cambridge University Press.

SANDERS, M. S., & MCCORMICK, E. J. (1987). *Human factors in engineering and design* (6th ed.). New York: McGraw-Hill.

SCHNEIDER, W. (1985). Training high-performance skills: Fallacies and guidelines. *Human Factors, 27,* 285–300.

SCHNEIDER, W., & SHIFFRIN, R. M. (1977). Controlled and automatic human information processing: I. Detection, search, and attention. *Psychological Review, 84,* 1–66.

SHIFFRIN, R. M., & SCHNEIDER, W. (1977). Controlled and automatic human information processing: I. Perceptual learning, automatic attending, and a general theory. *Psychological Review, 84,* 127–190.

SILVER, E. A. (1987). Foundations of cognitive theory and research for mathematics problem-solving instruction. In A. H. Schoenfeld (ed.), *Cognitive science and mathematics education* (pp. 33–60). Hillsdale, NJ: Lawrence Erlbaum Associates.

SIMON, H. A. (1980). Problem solving and education. In D. T. Tuma and R. Reif (eds.), *Problem solving and education: Issues in teaching and research* (pp. 81–96). Hillsdale, NJ: Lawrence Erlbaum Associates.

SINGLEY, M. K., & ANDERSON, J. R. (1989). *The transfer of cognitive skill.* Cambridge, MA: Harvard University Press.

SPERLING, G. A. (1960). The information available in brief visual presentation. *Psychological Monographs, 74*, Whole No. 498.

THOMAS, J. W., & ROHWER, W. D., JR. (1993). Proficient autonomous learning: Problems and prospects. In M. Rabinowitz (ed.), *Cognitive science foundations of instruction* (pp. 1–32). Hillsdale, NJ: Lawrence Erlbaum Associates.

THORNDIKE, E. L. (1906). *Principles of teaching.* New York: A. G. Seiler.

UNDERWOOD, B. J. (1969). Attributes of memory. *Psychological Review, 76*, 559–573.

UNDERWOOD, B. J. (1983). *Attributes of memory.* Glenview, IL: Scott, Foresman.

VANPATTEN, J., CHAO, C., & REIGELUTH, C. M. (1986). A review of strategies for sequencing and synthesizing instruction. *Review of Educational Research, 56*, 437–471.

WEINER, B. (1972). *Theories of motivation from mechanism to cognition.* Chicago, IL: Markham.

WEINER, B. (1980). *Human motivation.* New York: Holt, Rinehart, and Winston.

WEINER, B. (1982). The emotional consequences of causal ascriptions. In M. S. Clark and S. T. Fiske (eds.), *Affect and cognition: The 17th annual Carnegie symposium on cognition* (pp. 185–208). Hillsdale, NJ: Lawrence Erlbaum Associates.

WHITEHEAD, A. N. (1929). *The aims of education.* New York: MacMillan.

WICKENS, C. D. (1980). The structure of attentional resources. In R. S. Nickerson (ed.), *Attention and performance VIII.* Hillsdale, NJ: Lawrence Erlbaum Associates.

WICKENS, C. D. (1992). *Engineering psychology and human performance* (2nd ed.). New York: HarperCollins.

WICKENS, C. D., & KESSEL, C. (1980). Processing resource demands of failure detection in dynamic systems. *Journal of Experimental Psychology: Human Perception and Performance, 6*, 564–577.

WICKENS, C. D., MOUNTFORD, S. J., & SCHRIENER, W. (1981). Multiple resources, task-hemispheric integrity and individual differences in time sharing. *Human Factors, 23*, 211–229.

WIGHTMAN, D. C., & LINTERN, G. (1985). Part-task training for tracking and manual control. *Human Factors, 27(3)*, 267–283.

ADDITIONAL RESOURCES

ANDERSON. J. R. (ed.) (1981). *Cognitive skills and their acquisition.* Hillsdale, NJ: Lawrence Erlbaum Associates.

KLAHR, D. & KOTOVSKY, K. (1989). *Complex information processing: The impact of Herbert A. Simon.* Hillsdale, NJ: Lawrence Erlbaum Associates.

RABINOWITZ, M. (ed.) (1993). *Cognitive science foundations of instruction.* Hillsdale, NJ: Lawrence Erlbaum Associates.

SCHIMMEL, B. J. (1988). Providing meaningful feedback in courseware. In D. H. Jonassen (Ed.), *Instructional designs for microcomputer courseware* (pp. 183–195). Hillsdale, NJ: Lawrence Erlbaum Associates.

SCHOENFELD. A H. (Ed.), (1987). *Cognitive science and mathematics education.* Hillsdale, NJ: Lawrence Erlbaum Associates.

7

Procedure: Writing Functional Specifications

At this point in the design cycle, the training analyst will have several types of data available to support subsequent stages of development. These can be classified as (a) results of the organizational, task, and trainee analyses; and (b) knowledge of instructional strategies to use for various types of training tasks. Armed with this knowledge, it is typical for the designer to feel ready to consider and choose among design options for the training program. However, before doing so, there is one last step that should be completed before actual design begins. This step consists of putting certain information down on paper; the information then acts as a boundary around the design process. Following conventional practice in most engineering domains, we will term the document containing this information the **functional design specifications** (specs).

BACKGROUND

Some readers may be familiar with the fact that, historically, training analysts have conducted a front-end analysis, sometimes termed "needs assessment," and then written specific **instructional goals and objectives** (e.g., Goldstein, 1986). Because I am expanding on this tradition and giving the final written document a different name, I will briefly describe the previous approach and explain the rationale for its expansion. Then we will move on to procedures for writing the functional specifications.

Instructional Goals and Objectives

In 1962, Robert Mager published a small but very influential book titled *Preparing Objectives for Programmed Instruction*. Because Mager's methods came to be used for other types of training programs in addition to programmed instruction, the title was changed in subsequent publications to *Preparing Instructional Objectives* (Mager, 1975, 1984). For three decades, development of instructional objectives has been a methodology considered essential for any training program, large or small.

The rationale behind instructional objectives is as follows: If you want a student or trainee to perform in a certain manner as the result of participating in an instructional program, you must first identify the nature of that performance. After the analyst conducts the initial analysis and determines that instruction or training is the appropriate problem solution, he or she writes instructional objectives that specify what will be accomplished by the training program. An objective is "a description of a performance you want learners to be able to exhibit before you consider them competent. An objective describes an intended *result* of instruction, rather than the *process* of instruction itself" (Mager 1984, 3).

Instructional objectives are often called behavioral objectives, because they specify the overt *performance* to be exhibited by the trainee. Ideally, they should also specify the *conditions* under which the trainee performs the behavior, as well as *criteria* for acceptable performance. An example of an instructional objective is: "On the 25-yard range, be able to draw your service revolver and fire five rounds from the hip within three seconds. At 25 yards, all rounds must hit the standard silhouette target" (from Mager 1984, 128).

Writing instructional objectives is beneficial because it keeps the training analyst focused. At any point in the design process, one can ask oneself, is this instructional sequence going to result in the objective that is written for this particular topic? In addition to acting as reminders, instructional objectives have another, probably greater role. That role is to support a feedback mechanism in a design loop. Figure 7.1 shows the rationale behind this approach. The instructional objectives act as input to the instructional design system (left side). The design system is represented as a box, where the designer uses whatever instructional strategies are available and seemingly appropriate to develop the instructional system. The system is used to train learners, who are then tested in an evaluation phase using criteria based upon the instructional objectives. Based on the results of the evaluative tests, the

Figure 7.1   Feedback loop based on instructional objectives.

instructional system is modified and tested again. By iterating through the feedback loop in this fashion, instruction is changed to bring learners successively closer to the desired goal *as specified by* the instructional objectives.

Such a feedback loop design rationale is effective and defensible only as long as it is properly used. However, the method has frequently failed the instructional design community. The reason is that many instructional designers do not follow through with the necessary loops. Development of a full-scale training program takes a great deal of effort and is very time-consuming, and post-training evaluation is difficult and time-consuming. Performing these two phases from start to finish is enough work when done once, and is something people often don't wish to repeat at all, much less several times.

It can be seen that if iteration of the design cycle is not performed, instructional objectives do not serve their second purpose, feedback to the instructional design process. Without such iteration, the design philosophy described above is useless. In the design model presented in this text, I have included several steps that reduce this problem. One is the expansion of performance objectives into a functional specification document that explicitly describes instructional strategies to optimize learning of specific *types of knowledge*. The second is the addition of extensive user testing within the design and development phase itself (covered in a later chapter).

Liability Issues

Before prescribing contents for the functional specifications, I will mention a second reason for writing this document. Chapter 16 presents a relatively detailed discussion of liability issues related to post-training performance. The relevance to this chapter is that if you are ever in a position of having to defend the adequacy of your training program design, one important tool is any design documentation you might have. If you are like most analysts, much of the rationale for what you do is in your head. In designing and developing training programs or job aids, you will probably not be very compulsive about writing things down. Unfortunately, if you need to defend your actions, you will wish you had made a very explicit record of your design decisions and the rationale behind them, including references to relevant literature. Thus, writing design documents, including the functional specification, is one of the best procedural defenses against future litigation problems.

FUNCTIONAL SPECIFICATIONS

In our instructional design model, functional specifications (specs) are statements about *what* your system will do (goals and objectives), *how* it will functionally do it, and what things will *constrain* how it accomplishes the identified goals. I have divided the information to be written in the specs into four categories;

- Program goal
- Post-training goals and objectives
- System performance requirements
- System constraints

The **program goal** specifies the overall goal of the training program, what it is supposed to accomplish in a general sense. The **post-training goals and objectives** are much more specific; they identify what trainees should be able to do when they are finished. This category essentially contains the goals and behavioral objectives that were described earlier. These goals and objectives should be directly tied to the results of the task analysis. The **system performance requirements** section contains prescriptions for the instructional functioning of your program (Bailey, 1982). These prescriptions describe basic *instructional strategies* the program should include to accomplish the desired objectives, and can be taken from the instructional strategies described in Chapter 6 or other literature. Finally, the system constraints section specifies any circumstances that constrain the design of the training program. This is usually based on characteristics of the job organization, training organization, and trainees themselves.

To determine the information that should be contained in the last three sections, the analyst can simply answer the questions contained in Table 7.1. However, each training project has unique characteristics, and an analyst should determine whether there is other critical information in addition to what is called for by the questions in Table 7.1. Despite what some researchers may suggest, training program design is still more art than science.

**TABLE 7.1   Questions Suggesting Information That Should
be Contained in the Functional Specifications Document**

POST-TRAINING GOALS AND OBJECTIVES

What are the goals and specific behavioral objectives with regard to trainee performance after training?

OPTIONAL: What are the performance criteria for evaluating training program effectiveness?

SYSTEM PERFORMANCE REQUIREMENTS

What performance characteristics are essential for the training program (for interface design and instructional strategies)?

What performance characteristics would be desirable for the training program?

SYSTEM CONSTRAINTS

1. With respect to training program development:
   What financial resources are available?
   What human resources are available?
   What equipment and other tools are available?
   What are the time limits?

2. With respect to training program delivery:
   When do the trainees or students need to be trained?
   How many trainees need to be trained?
   Where do trainees or students need to be trained?
   What is the frequency of tasks being trained?
   What is the time span from training to actual job performance?
   Are there any other constraints on method of delivery?
   Will instructional content change over time or is it stable?
   What is the estimated lifespan of the training program?
   What financial resources are available for delivery?
   What human resources are available for delivery or distribution?
   What equipment is available for delivery? (including computers, simulators,
      actual job equipment)
   What are the needs or biases of management or other personnel involved?

3. With respect to trainees:
   Who are the trainees and what are their characteristics?
   Is there anything in trainees' background or attitudes that would constrain the
      program?
   What is the relevant knowledge and skills of the trainees?

Program Goal

The first section, program goal, provides a general overview of what the program is all about. What are you training the learner to do? Usually, the goal is stated in terms of a job, or in terms of some subset of a job. An example would be:

> The purpose of the training program is to train all new waiters/waitresses for the Merriweather restaurants to wait tables in the main dining room, and also all current waiters/waitresses referred to the program by management.

In industrial settings, it is frequently the case that when the task analysis has been finished, the analyst determines that an entire job contains too many different tasks for them all to be included in one training program. In this case, the analyst will have to evaluate the data provided by the task, trainee, and organizational analyses, in order to identify subtasks to be contained in the program. Certain questions should be considered, such as:

- What subtasks are performed most poorly?
- What subtasks contain the most potential for accident or injury?
- What are the resources available for training program development?

By answering questions such as these and working with management and workers, the training analyst can identify one or a few subtasks that seem most in need of training. The program goal then specifies these particular areas or subtasks as the focus of the program. In certain cases, the program goal might contain a statement to the effect that the program is to be the first training program in a series of modules, with the program to serve as a design model for subsequent efforts.

Often, the training program is part of some larger "system" with many people working on different components, such as the manufacture of a complex piece of equipment. In this case, the program goal can state the general purpose of the training program as well as the context within which it occurs. An example of a more complex goal statement with the context specified would be:

> The purpose of the training program is to train new owners of the FLY HIGH bicycle to adjust, ride, and maintain their bicycle, and to do so in as safe and responsible a manner as possible. Primary responsibility for repair is assumed to reside with the retailer rather than with the owner. It is also assumed that the retailer will be a knowledgeable expert, and available to act as a resource for the owner.

Regardless of the unique aspects of the particular design case, the program goal should give an outsider a clear, concise, and precise view of what the training program is about, and the context within which it resides.

Post-training Goals and Objectives

While the general program goal is necessary for giving an overview of the program, it is also important to specify the exact training goals and objectives. The source for this information is the task analysis in which you will have identified the tasks performed for a particular job. Out of this set of tasks, there will be some subset that is to be covered by the training program. For this subset, the analyst describes the specific goals and objectives for the trainee as a result of participating in the training program.

An instructional goal is a statement of what is to be learned. Examples given by Hannafin and Peck (1988) include "Learning to drive safely," "Learning to budget," or "Learning to power up a computer." Instructional goals are more specific than the overall goal of the training program, but not sufficiently specific to guide program design and evaluation. Once written, the goals should be translated into specific behavioral objectives. These objectives are designed to guide the instructional designer, but are sometimes also used as objectives explicitly given to students before the instructional sections. As an example, in each chapter of their text on designing instructional software, Hannafin and Peck (1988) provide objectives such as:

After completing this chapter you will be able to

1. List the four steps in analyzing learning tasks.
2. List and describe the activities performed during each of the four steps.
   (p. 91)

Notice the emphasis on measurable activities (*listing* in this case). Objectives should specify an overt behavior to be exhibited (Mager 1962). Some verbs are clues that the objective is too vague. These include *understand, know, appreciate, feel,* and *have a positive attitude.* Better words are verbs that denote some observable behavior or action. Table 7.2 lists some potentially useful verbs for writing instructional objectives.

In summary, the designer should write an instructional goal for each topic or subtask to be trained. These should then be further specified by writing behavioral objectives. As an example, for our general goal of training mountain bike owners to adjust, ride, and maintain their bicycles, we might have a list of goals that includes many items, some of which pertain to safety:

- Be knowledgeable about safe bike riding on roads and streets.
- Be knowledgeable about safe bike riding in off-road areas.
- Be knowledgeable about the use of safety equipment.
- Know how to check the bike for damage.

TABLE 7.2   Sample Verbs for Use in Instructional Objectives

adjust	describe	operate
assemble	determine	perform
build	develop	program
calculate	diagnose	recall
categorize	discriminate	recite
choose	explain	recognize
classify	evaluate	repair
compare	generate	select
count	identify	state
create	list	summarize
demonstrate	match	write

This last goal might fall under both safety and maintenance categories. How would we know that the owner knows how to check the bike for safety? We would have to ask him or her to perform some behavior. Example of a *behavioral objective:*

> On a diagram of the FLYING HIGH bicycle, with no documents or other materials available for consultation, circle the areas on the bicycle where fractures are most likely to show up. The learner must circle at least four of the five fracture areas correctly, and circle no more than one area that is not a fracture point.

Notice that this objective specifies more than just the performance to be exhibited. It also specifies conditions under which it is to be performed (using a diagram and no other documents), and the criteria to be used for successful performance (last sentence). This is the type of instructional objective suggested by Mager (1962) for guiding the analyst, because it includes *performance, conditions,* and *criteria.* However, it should be pointed out that an objective which includes all three factors is seldom given to learners as a "guiding" instructional objective. (For example, it would be rather ineffective to tell students ahead of time that "In response to the question 'Is the platypus a mammal or a reptile?,' you will identify the platypus as a mammal.")

Recall from our model presented in Chapter 1 that at some point, the trainer needs to develop performance criteria for the evaluation phase. These criteria are used to determine whether the training program has accomplished its identified goals. These performance criteria can follow directly from behavioral objectives (sometimes there is no difference between the two). Because of the tie between instructional objectives and evaluation criteria, it is often convenient to include the evaluation criteria in the system specifications document.

Sometimes evaluation criteria are not so directly related to the instructional objectives. Instructional objectives typically specify behaviors to be exhibited during or immediately after the training program. However, some evaluation criteria extend to a wider time period and/or range of variables. For example. one performance criterion might be "a salesperson's increase in the number of sales over the year following the training program, as compared to the year prior to the training program." Such broad job-oriented criteria might be specified in the functional specifications document, or might be postponed until a later point.

System Performance Requirements

The performance requirements state the general characteristics that the system should have from an instructional and human factors point of view, describing the characteristics that the program should have in order to train or teach the knowledge and tasks that have been specified under goals and objectives. If all of the information is declarative conceptual knowledge, the analyst identifies *strategies* appropriate for that type of knowledge (e.g., see Chapter 5) and lists the strategies that should be used in the program. If the objective is for the trainee to have a high level of expertise in some skill, the requirement might specify that the learner have a large number of opportunities to practice the skill in a real or simulated environment.

The designer has some latitude in writing the requirements. For example, he or she may wish to list some methods as essential, such as "the program must list the parts of _____ physical device, and other methods as desirable, such as providing a dynamic animation or video of the physical device in action. Sometimes there is little data on which to base the decision of whether an attribute is essential or simply desirable. In this case, the characteristic can be left unspecified as essential or desirable, and its importance tested later during the design phase. As an example, let's say that we don't know whether providing color photos of actual fractures on a mountain bike is necessary to teach a person to recognize fractures on their own bike. This is something that can be tested during prototyping by giving some people photos and other people alternatives such as black and white drawings. Data is collected and if subjects can successfully identify actual fractures after seeing only black and white drawings, then we have determined that color photos are desirable but not essential.

When the instructional program will be training or teaching very homogeneous types of knowledge or skills, the instructional and motivational factors may be listed in a general fashion for the program as a whole. When the program includes different *types* of knowledge, the requirements may need to be specified for different topics and tasks to be trained. For example, it may be the case that only a certain set of subtasks for a complex job are to be trained to automation. One item in the performance requirements might specify that "tasks X, Y, and Z must be highly practiced by the learner under realistic task conditions until testing reveals them to be automated" (with the tests specified under objectives).

In addition to the instructional and motivational design strategies, this section can contain any "interface" requirements such as ease of use of the system, or ease of learning. For example, a requirement might specify one of the following characteristics:

1. If the program implements computer-based training, the interface of the training program must be designed in accordance with standard human factors guidelines.
2. The program must be easy to initiate and use for the entire range of learners; it must not require any external documents or instructions in order for a learner to successfully engage the system.
3. The training program must not require more than 5 minutes of explanation from an instructor regarding how to use the program.

In short, anything that describes how the system will operate to meet the goals and objectives should be explicitly stated in this section. In any case where future liability might be an issue, not only should the design requirement be listed, but also references to the principles, guidelines, research data, theory, or other sources that led to that particular specification.

System Constraints

In the last section, information will be detailed that puts some boundaries on the training program. These boundaries may be light, or they may be quite severe. However, there will always be boundaries of some type. It can be seen from Table 7.1 that constraint information usually falls into one of three categories:

- Constraints relevant to training program development
- Constraints relevant to training program delivery
- Constraints imposed because of certain characteristics of the trainees

For the first category, the designer should identify any factors that will impact development of the training program. These factors include financial and equipment resources, the number and types of people available to develop the program (e.g., are any programmers available for the project, etc.), people available to act as subject matter experts, and so forth.

For the second category, the analyst must evaluate information relevant to methods of delivery. This task will be easier once you have become more familiar with the training program options (as discussed in the next few chapters). Then you will be more familiar with the resources and needs of the different types of instructional programs. For the moment, you can simply answer the questions listed in Table 7.1, such as "Will the instructional content change over time or is it stable?" If the information to be taught is going to become outdated very rapidly, that will impact the choice of delivery method for the training program.

Finally, the nature of the system as well as the content of the system will be constrained by characteristics of the trainees. Briefly describe the range and mean or median of central characteristics of the trainees. These may include age, gender, occupation and brief occupational history, educational level, primary language, and level of literacy. In addition, it is often relevant to determine trainee computer literacy, as well as other factors such as attitude toward their job and the tasks being trained in the program. The 1990 American Disabilities Act requires that environments and "systems" accommodate people with disabilities, such as the sight impaired. This will begin to have a strong impact on the design of training programs, and should be evaluated as part of the system constraints.

The trainee analysis should have included assessment of how well employees (or others) can currently perform the tasks identified in the goals and objectives section, and also assessment of knowledge and skills that are prerequisite to successful participation in the training program. If some trainees are deficient in prerequisite knowledge or skills, the functional specification document must state how that problem will be addressed. If some trainees have knowledge or skills contained in the training program, the performance requirements section should specify how the program will address this as a potential motivational problem.

FINAL COMMENTS

Specifying what the training program is to accomplish (goals and objectives), the general strategies that should be taken to accomplish those goals, and the constraints imposed by the design environment are all critical to successful training program design. If the steps up to and including writing of the system specifications have been rigorously carried out, the design and development process will more easily fall out as a direct result from that work.

The analyst may have realized during writing of the specifications that the objectives and requirements sections are in direct conflict with the constraints section. For example, the type of skill being taught might require extensive training in a complex real or simulated environment. However, a constraint written in the next section might specify that the program must be completed within two weeks and be deliverable without any technical equipment. It will be relatively common for the designer to find at least some conflict which exists between what the program should do and the resources that are available. The analyst must then work to find the best trade-off. This work might require some prototyping to see what is actually required, and some discussion with personnel in the job and training environments to modify critical constraints.

A training analyst may find him or herself in a position where, due to some factor such as resource constraints, they are simply instructed to go ahead and create a certain type of training program, even when it cannot utilize the recommended instructional requirements (strategies). For example, an analyst or a design team might be asked to make an instructional videotape, even when the task to be taught is a skill that should have hands-on practice. Again, the designer is seeing a disparity between the requirements and the constraints. When this happens, the analyst should evaluate whether there is any serious consequence, such as accident or injury, that may result from deficiencies in the training program. If the analyst determines that there is such a possibility, he/she should strongly consider writing a document to appropriate personnel stating these concerns. This should be done even if the analyst expects that nothing will be changed by virtue of writing the document. The document will still serve two functions: first, it may make people more aware of the seriousness of the design decisions being made. Second, it can protect the analyst from personal liability in the case of future litigation.

SUMMARY

In summary, the analyst writes the functional specifications to guide the system design process and also to provide documentation for any subsequent design or litigation needs. The specs are written via the following steps:

1. Based on the analyses performed to this point and discussions with appropriate personnel, identify the task(s) to be trained and the *type* of knowledge required by each task or subtask.

2. Write the overall program goal.

3. Write specific instructional goals and objectives for each subtask.

4. Write the functional requirements—desirable and essential instructional strategies for each topic and/or major type of knowledge.

5. Write system constraints with respect to development, delivery, and the trainee population.

REFERENCES

BAILEY, R. W. (1982). *Human performance engineering: A guide for system designers.* Englewood Cliffs, NJ: Prentice Hall.

GOLDSTEIN, I. L. (1986). *Training in organizations: Needs assessment, development, and evaluation* (2nd ed.). Monterey, CA: Brooks/Cole.

HANNAFIN, M. J., & PECK, K. L. (1938). *The design, development, and evaluation of instructional software.* New York: Macmillan.

MAGER, R. F. (1962). *Preparing objectives for programmed instruction.* Belmont, CA: David S. Lake.

MAGER, R. F. (1975). *Preparing instructional objectives.* Belmont, CA: David S. Lake.

MAGER, R. F. (1984). *Preparing instructional objectives* (revised 2nd ed.). Belmont, CA: David S. Lake.

8

Background:
Training Systems

Once the functional design specifications have been written, design and development of the training program begins. It would be helpful if we had a design guideline that specified the most appropriate instructional techniques and media given certain design objectives and constraints. However, the current state of the instructional design science has not yet reached that level, and at any rate, technology is advancing at too rapid a rate to make such a static guideline feasible.

While it is true that researchers are currently working on expert systems to support training designers in making instructional design decisions, these are still in the prototype stage. Therefore, it is up to the analyst to evaluate the requirements contained in the functional specifications document and identify instructional techniques and media that will satisfy those requirements. In order to do this, a designer must be familiar with the various training and performance support techniques, and their respective strengths and weaknesses. The goal of this chapter is to present basic information about the most commonly used training techniques. Because it has become a very large field in and of itself, computer-based training will be reviewed more extensively in Chapter 9.

TAXONOMIES FOR INSTRUCTIONAL TECHNIQUES

It is common to talk about various training and instructional methods as
though they are mutually exclusive and clear-cut techniques. While these
taxonomies are necessary for structuring discussions and reviews, they tend to
break down once we begin to analyze the characteristics of each method. For
example, when we evaluate "lecture," we must ask: What does this concept
actually mean? Are we talking only about live lecture in front of a class? What
is a videotape of that lecture? Does the method include homework problems
or working problems in class? Compounding this problem of creating mutually exclusive categories is the fact that simulations have changed from being
predominantly created through hardware or paper-and-pencil methods to
being predominantly controlled by and delivered through computers. It is no
longer possible to keep the category of simulation completely separate from
computer-assisted instruction.

Because taxonomies are helpful in discussing the characteristics,
strengths, and weaknesses of various methods, we will use the taxonomy presented in Table 8.1. While it is possible to distinguish between these methods
to some degree, there will still be times when the differences break down. In
addition to the traditional media categories, I will briefly discuss leadership,
interpersonal skills, and team training because they tend to be special types of
training programs.

The different instructional technologies listed in Table 8.1 vary on many
dimensions. However, there are two dimensions that are critical in determin-

TABLE 8.1   Taxonomy of Instructional Techniques

- Text
- Audiovisual Techniques
- Lecture
- Inquiry Learning
- Tutoring
- Traditional Programmed Instruction
- Computer-Assisted Instruction (CAI)
- Intelligent Computer-Assisted Instruction (ICAI) Simulations
- Simulations
- On-the-Job Training
- Leadership and Interpersonal Skills
- Team Training

ing system effectiveness in promoting acquisition of knowledge and skills. The first is whether the system allows only one-way communication, some interaction, or strongly supports two-way interaction and communication. The second dimension is whether the content is oriented more toward abstract concepts and principles or toward concrete, realistic situations. Figure 8.1 shows these two dimensions, and also suggests the relative placement of the first ten of our technologies on the two continuums. It can be seen that the older technologies such as text and lecture tend to be more one-way and abstract, and the newer instructional methods such as simulation tend to lie more on the concrete, interactive area. In discussing each of these ten techniques, we will move from the top of Figure 8.1 (one-way communication methods) down to the bottom.

In the following sections, I will define all of the techniques listed above and discuss the strengths and weaknesses of each. In addition, in Chapter 9 we will review in much greater depth the various types of computer-assisted instruction (CAI), including simulations, and discuss some of the tools and techniques for developing computer-based instructional programs.

As a final note, leadership, interpersonal skills, and team training are not

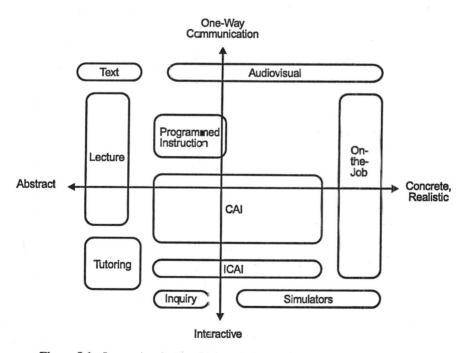

Figure 8.1 Instructional technologies relative to abstractness and type of interaction.

shown in the figure because they are implemented via a great variety of techniques. These categories will be discussed later in the chapter following review of the media techniques.

REVIEW OF INSTRUCTIONAL TECHNIQUES

Text

One of the oldest and most widely used instructional methods is written text. Even today, it is hard to overestimate our reliance on text for instructional programs. Present-day methods tend to include two types of text—hardcopy text on paper, and electronic text. The first category, hardcopy text, includes books, instruction manuals, owners manuals, help and reference documents, etc. The content can be written text, graphics, or some combination.

For the purposes of this review, if a document is stored on a computer and can be accessed by traditional "booklike" search mechanisms, it is considered to fall into the *text* category, although technically it would be referred to as **electronic text.** This includes some systems which are marketed as "interactive" systems. To explain this, we can look at how text in hardcopy books is typically accessed. The material is found or "retrieved" in one of several ways: (a) it is read from beginning to end; (b) a topic is first found in the table of contents, then retrieved using a page number; (c) a topic is first found in an index at the back and then retrieved using a page number; or (d) the reader thumbs through the book, and based on various cues, finds the topic or material for which he or she is searching.

Now consider a computer-based text system such as an "on-line" encyclopedia. The search mechanisms are really very similar to those for hardcopy. The knowledge is contained in linear chunks of text organized according to the same structure as a hardcopy encyclopedia. Functionally, the topics are found the same way. The user can look through a list of topics analogous to a textbook table of contents, search alphabetically, or use an index. In addition, since computers are inefficient at presenting index lists, the user is often able to simply type in the name of the topic being requested. The point here is that the medium provides several mechanisms for searching through the knowledge base to find a particular piece of information. However, the nature of the knowledge and its access is not functionally very different from hardcopy text. Therefore, all text stored electronically and accessed through these traditional types of interchanges is being considered functionally equivalent to traditional linear, noninteractive text.

Strengths of text.   There are many strengths of written documents that are easy to overlook in these days of high-technology training methods. The major strengths are listed below.

1. Text has a very low development cost.
2. Text has a low delivery cost, not counting the cost of the hardware itself for electronic text.
3. Text is convenient for delivery to trainees. There doesn't have to be an instructor at the same location as the learners, and the paper or computer disc can be easily transported.
4. Learners are familiar with the medium, so it requires no special training (recall that we are restricting discussion of electronic text to standard access protocols).
5. Learners can control the learning pace, when they learn, for how long, etc. They can stop, start, go back, and review as much as needed.
6. Text is relatively easy to modify, although not as much so as lecture, tutoring, or on-the-job training.
7. Text can be designed to act as a future reference or job aid.
8. Text can provide graphic information.
9. It is possible to have practice with "feedback" in terms of telling the learner the right answer to questions and practice problems (although there is no way to force the learner to try the problem before looking up the answer).

Weaknesses of text.   The major weaknesses of linear text are related to the fact that it lies at the top of the area shown in Figure 8.1. That is, it is one-way communication that is neither actively interactive nor adaptive. To overcome this problem, a writer would have to foresee the needs and background knowledge of all potential readers and design a document that provided the right information for all of them. This is virtually an impossible task, and in addition, inclusion of all possibly needed information usually makes texts overly cumbersome.

1. Linear text is not interactive. It cannot respond to the needs of the reader by asking questions or providing additional examples or problems.
2. Linear text cannot adapt to the level or individual needs of the reader.
3. Text is not capable of diagnosing student misconceptions or errors.

4. Text is more impersonal than most other forms of instruction, including lecture, videotape, tutoring, and on-the-job training.

5. Text requires certain reading and studying skills on the part of the learner.

6. Text doesn't provide dynamic, visual information that shows how a system changes over time.

7. Text doesn't make use of the auditory channel.

8. Text is more difficult to modify than lecture or tutoring.

Despite the shortcomings of linear text, it is still a very efficient medium to develop and deliver. For these reasons, we can expect text, either hardcopy or on-line, to continue to be a frequently used method of instruction, even if it is only to augment other formats.

Audiovisual Techniques

By definition, audiovisual techniques refer to any audio, video, or audiovisual methods, including training programs in the form of cassette tapes. However, because of the decreasing use of audio methods, this section will focus on audiovisual techniques such as TV and videotape. In addition, this section deliberately discludes discussion of computer-based, interactive, multimedia technologies.

Like text, TV/videotape is a very familiar medium to most people, and is capable of presenting a wide variety of verbal and visual information. Videotape goes beyond the capabilities of text in many important ways. Text is restricted to static text and graphics, and is usually restricted to black and white materials with few photographs. Videotape can present a wide range of information, from abstract information to concrete real-world events in a vivid array of colors. It is not uncommon to see training videotapes that incorporate text, audio, graphics, full motion video, and advanced animations.

Videotapes are used in a wide variety of training domains, from interpersonal skills and leadership training, to dentistry, driver education, and literacy training (Goldstein, 1986). One of the most familiar and successful audiovisual training programs is *Sesame Street,* developed in the early 1970s as an experimental training program to teach children basic skills (Ball & Bogatz, 1970). Sesame Street was an overwhelming success, due mostly to the fact that it was carefully designed by a team of experts using basic principles of instructional design. For example, instructional objectives were delineated in the areas of symbolic representation, cognitive processes, physical environment, and social environment (Ball & Bogatz, 1970). Obviously, one reason the program has been so successful is that children simply enjoy watching it, therefore motivation and attention is inherently high. This is an important lesson for all of us; even adults prefer to engage in enjoyable activities over boring or tedious ones.

Many studies have been conducted over the years in an effort to evaluate the effectiveness of audiovisuals as an instructional medium. A great many of these studies were performed some time ago, and did not use proper control procedures (see reviews in Chu & Schramm, 1967; Schramm, 1962) Most of these studies found that audiovisuals resulted in "few significant" differences in achievement (Goldstein, 1986). A more important finding is that audiovisual methods were significantly more effective when they promoted active processing in the learner, as opposed to simply passive viewing (e.g., Lumsdaine, May, & Hadsell, 1958). This is consistent with our analyses in previous chapters of this text.

Strengths of audiovisual techniques. The following list provides the major strengths of audiovisual techniques.

1. Videotaped information tends to be visually and auditorially interesting. This is especially true for present-day systems given recent advances in animation and other computer-mediated design methods.
2. Videotape tends to be more motivating than many other methods such as lecture or text.
3. Audiovisual training programs are easy, inexpensive, and simple to *deliver*. They can be used for widespread distribution to environments where it is difficult or impossible to provide an instructor. They can be used by individuals at their convenience, or in a variety of settings with large groups.
4. Videotapes are user-friendly in that almost everyone knows how to play them.
5. The content of videotape can be broader than static media such as text. They can present the dynamic unfolding of events, even events that couldn't be observed or experienced in real life. Similarly, a videotape can present specific skills and behaviors being modeled by experts, a phenomenon that is often difficult or impossible to present in a typical lecture or text format.
6. When being used by an individual, learners can control the pace of information presentation. By pausing or reviewing the material, the learner can adapt the instruction to their own rate of learning.

7. Audiovisual materials rely on reading abilities to a much lesser extent than text.

8. Videotapes can be more efficient than either lecture or text because they make simultaneous use of both audio and visual cognitive processing channels. This increases the efficiency and ease of learning for trainees.

Weaknesses of audiovisual techniques. The weaknesses or disadvantages listed below are based strictly on the audiovisual medium. The weaknesses are less applicable to the extent that additional techniques are used, such as formal exercises with feedback, human tutoring, skill practice, opportunities for questions and answers, etc.

1. Despite increases in technology, videotapes tend to be more difficult and costly to produce than most of the other methods discussed here.

2. Videotapes are more costly and difficult to deliver than hardcopy text. However, they are more cost-effective than lecture.

3. Videotape only provides one-way communication; it is neither adaptive to the needs of individual learners nor is it interactive, allowing question and answering processes.

4. It usually promotes passive learning on the part of the viewer (although exceptions such as Sesame Street show that this doesn't have to be the case).

5. It doesn't allow for any testing of declarative knowledge.

6. There is no opportunity for skill practice with feedback.

7. The learner may not be able to control the pace if the videotape is being presented over TV, cable, or to an entire group of people.

Lecture

Lecture usually refers to the process of a subject matter expert presenting material on a particular topic to a group of people. They typically verbalize and elaborate conceptual declarative knowledge, show examples of objects, concepts, or systems, verbalize procedural declarative knowledge, and model skills. They may encourage group discussion and answer student questions. These last two activities tend to vary widely across individual instructors. Lecture traditionally takes place in a relatively one-way communication pattern, especially if the class size is large. There is also little time allotted to student skill practice.[1]

---

[1] *Workshops* are usually a combination of lecture and skills practice.

Strengths of lecture. Most of the strengths associated with lecture revolve around the ease with which a training program can be developed, relative to most other methods. A second major strength is the ease with which it can be modified, either before or during implementation.

1. Lecture is relatively easy and inexpensive to develop.
2. Lecture material is easy to modify at any time before presentation, making it good for information that changes frequently.
3. Lecture can be somewhat adaptive (more than text or videotape). It can be modified according to the needs of the class; however, this can still only be done for the *whole* class and not for individuals to any great extent.
4. Lecture has the potential for being interactive, students can ask questions, and be asked by the instructor to generate ideas or answers. However, in reality, it tends to be less interactive than some methods such as ICAI, OTJ training, and simulations.
5. Lecture can be relatively motivating, certainly more so than reading text.
6. The instructor has control over the material to which the learner is exposed. This "captive audience" factor can be important for certain training areas such as safety.

Weaknesses of lecture. As with the other methods, the weaknesses described are based on what is *typically* done in a lecture, not what is possible to do by augmenting the method.

1. Students for a course must be able to congregate in one location.
2. An instructor must be available at the site where students are to take the course or program.
3. Development may be inexpensive, but delivery is relatively costly. This is especially true if instructors must be transported to various parts of the country to implement the programs.
4. Lecture does not adapt to individual students. By definition, it must be aimed at one point or one knowledge and skill level within the class. This can be a major weakness depending on the heterogeneity of the class.
5. In reality, there is usually little opportunity for questions and answers.
6. At least for traditional lecturing methods, the learning tends to be passive, where the instructor presents information and learners listen and try to remember. Only rarely does an instructor attempt to use methods

where students actively generate knowledge, such as in the inquiry learning method.

7. Relative to some methods such as tutoring and CAI, there is little opportunity to test learners' declarative knowledge "on the fly," identifying holes or inconsistencies in student knowledge, and remediating accordingly.

8. Trainees have relatively little control over the pace of instruction.

9. Lecture may be less motivating than some other methods, such as CAI, videotape, or simulations.

Naturally, the degree of these strengths and weaknesses depends strongly on the skills of a given instructor, and the make-up of the class (including class size and heterogeneity).

Special cases of lecture. Sometimes lecture is delivered using methods other than live presentation within a classroom. The two predominant methods are satellite uplink and videotape. In the case of satellite transmission, the method has the same strengths and weaknesses as traditional lecture with one exception—there is typically less interactivity between students and instructor.

The other condition is where the lecture is videotaped and sent to learners. In this case, the method will have essentially the same strengths and weaknesses as videotape, except that there will be less features such as animation, computer-generated text, movies, etc. That is, the content will be that of lecture, while the delivery mode is that of videotape.

Inquiry Learning

Recent trends in instructional methods have emphasized active involvement of the learners (Brown, 1990; Esler & Sciortino, 1991). Some researchers are using simulated microworlds for students to explore (Jonassen, 1988). However, even abstract areas such as mathematics can be taught using principles of active learning. One such approach is that of **inquiry learning.** Some researchers also call this discovery learning, while others reserve the term discovery learning to refer to learning within the context of a simulation.

Inquiry learning is a method that has been developed to foster high-level thinking skills in students. They are given some task to perform which requires organizing information, applying rules, analyzing information, and inferring and generalizing from data or qualitative information. According to Esler and Sciortino (1991), inquiry learning can be promoted through a wide variety of activities such as those where:

Students manipulate materials and equipment, participate in problem solving discussion groups, respond to open ended questions posed by the teacher, and collect data from the library or from direct observations. (Esler and Sciortino, 1991, p. 113)

While there is no evidence that inquiry learning promotes domain-general thinking skills (skills that would transfer to a new domain), there is some evidence that this type of instruction does promote inferential and other reasoning processes *within* the domain. In addition, there is evidence that this type of concrete, experiential learning enhances acquisition of declarative domain knowledge, and results in better long-term memory for the information. Readers wishing to know more about this method of instruction and study specific examples are referred to Esler & Sciortino (1991).

Strengths of inquiry learning. The strengths of inquiry learning mostly derive from the fact that it is a method that promotes active, generative knowledge in the learners.

1. The method relies on concrete, experiential activities which tend to promote better retention of declarative knowledge.
2. The method is highly adaptive to the level of the learner and is interactive (even more so than tutoring). Also, learners usually *do* more than in tutoring or other methods.
3. Inquiry learning tends to be highly motivating. An important aspect of this factor is that the motivation tends to be *internal* rather than external (see Chapter 5).
4. Learners tend to acquire better attitudes toward the subject matter and the instructor.
5. This method tends to increase learners' self-concepts.
6. Procedural skills can be practiced and feedback received. Declarative knowledge can be added where appropriate.

Weaknesses of inquiry learning. Weaknesses of the method mostly derive from the fact that it is relatively resource intensive with respect to time and personnel.

1. Inquiry or discovery methods take a fair amount of time and work to develop. Instructors must identify and plan activities to promote inquiry learning. Perhaps it is because of past experience, but instructors find it much easier to develop a traditional *didactic* lecture than to develop inquiry or discovery types of projects.

2. Inquiry learning is expensive in terms of delivery. It requires an instructor, a low instructor to student ratio, and more time to teach the same concepts than more traditional didactic lecture methods. In this sense it might be considered somewhat inefficient.

3. Learners or trainees must be congregated in one area, and an instructor must be available.

4. The method is appropriate for learning declarative conceptual knowledge, and higher order thinking skills, but is highly inefficient and not particularly effective for more specific procedural skills. Imagine teaching a technician to interpret blood tests using only this method.

5. There are few guidelines for developing inquiry learning instructional programs.

Tutoring

Tutoring is defined as a one-to-one interaction between a human instructor and the learner. This section explicitly does not include computer-based tutoring, which is in a category of its own. Although tutoring is generally considered one of the most effective methods for imparting knowledge (Bloom, 1984; Cohen, Kulik, & Kulik, 1982; Merrill, Reiser, Ranney, & Trafton, 1992), it is usually not seriously considered by a training analyst because of the high costs associated with its implementation. This cost and rareness of expertise is why so much research effort is going into the design of intelligent tutoring systems, in the hopes that such systems can be used to augment or replace human tutors (e.g., Merrill et al., 1992).

Strengths of tutoring methods. The strengths of tutoring derive mostly from the fact that no other technology can adapt to an individual learner to the extent that a (good) human tutor can.

1. Because of the extensive involvement of the student, the method tends to be more motivating than many others.

2. This method is both adaptive to the level of the student and highly interactive. It is the most adaptive of any discussed in this chapter.

3. With tutoring, the student is almost always involved in active, generative learning rather than passive reception of information.

4. There is good opportunity to recall declarative knowledge and get immediate feedback.

5. There is opportunity to practice skills and get immediate feedback, both

on the correctness of the response, and specific knowledge to guide production rule formation.

6. Like lecture, the content can change easily either before or during instruction.

Weaknesses of tutoring methods.   The major weaknesses of tutoring derive from the great demands in terms of instructor time.

1. The method incurs a very high cost, and takes a great deal of time.
2. Tutoring can wear on instructors because they have to repeatedly go over the same concepts.
3. Instructors have to be available at the same location as students. They also have to be numerous enough to provide the necessary low instructor-to-student ratio.
4. Some students may be intimidated, not wanting to look stupid in front of the instructor. One thing can be said for textbooks and computers—they aren't judgmental.
5. Tutoring tends to include enough abstract declarative knowledge, but not enough concrete, realistic knowledge and skill practice.

This method is instructionally one of the best, given that the instructor knows how to teach well. This is because people are still the best at adapting to the needs of the individual student, to their particular knowledge and skills. While the method is currently not sufficiently oriented toward concrete, real-world experiences, we are seeing this change as instructors realize the value of grounding instruction in "real" and meaningful experiences. Unfortunately, nothing can be done about the high *cost* of providing individual human tutoring.

Programmed Instruction

In the 1950s, educators and psychologists began to develop systematically designed instructional programs based on certain principles in psychology. These principles came from the *Behaviorists,* who said that learning would be optimized if instruction allowed the learner to respond frequently and be immediately reinforced for a correct response. This led to the two principle characteristics of programmed instruction:

- It teaches in small, incremental steps.
- It elicits responses and follows with immediate feedback.

Programmed instruction consists of very stringently designed presentations of small segments of information. Interleaved in with this information is a set of questions to which the learner responds. The system gives immediate feedback as to whether the learner was correct or not. Then depending upon the type of programmed instruction, the system may provide remedial information.

The use of programmed instruction was popular for many years. According to a report from the U.S. Civil Service in 1970, there were at least 2,300 programmed instruction systems in use at that time, just in the government alone. And the method has been used by training analysts in industry for topics ranging from training of sales personnel to the use of life insurance mortality tables (Goldstein, 1986).

Programmed instruction has historically had two forms, linear programmed instruction and branching programmed instruction. Each of these will be described below, along with their associated strengths and weaknesses. This is a good point to note that many "interactive" CAI systems that are currently being marketed are actually no more than modern-day versions of branching programmed instruction. These implementations will obviously have the same strengths and weakness as the programmed learning systems that are delivered via hardcopy text. In addition, they will have the strengths and weaknesses of "lower-end" computer-aided instruction programs.

Linear programmed instruction.   The goal of linear programmed instruction is to get the learner or trainee to take small steps in a successful fashion. Each step can then result in positive reinforcement (such as "Correct" or "Good, that is correct"). This is based on the view that making correct responses and receiving reinforcements was viewed as the optimal method of instruction.

Linear programmed instruction systems thus consist of small instructional segments, or frames as they are called, where the learner must make a response at the end of each frame. The next frame tells the learner the correct response to the previous question. Figure 8.2 shows an example of two successive linear programmed instruction frames. One important characteristic of linear programmed instruction is that it is specifically and explicitly designed so that the majority of learners (at least 90%) will perform each step successfully (Goldstein, 1986). Notice that this requires the system to take *very small* steps between instructional units so that no one is left behind. Also, there must be very small steps from the information presented to the question, so that most learners can make correct responses.

1

1. It is important to wear a safety helmet when riding a bicycle. Being hit by a car when not wearing a safety helmet could result in serious injury or death. When riding a bicycle, one should always wear a _____.

2

helmet   2. When riding at night, the use of reflective tape or reflective clothing can make the rider easier to see. Whenever you ride at night, you will enhance your safety by wearing _____.

Figure 8.2 Example of frames in a linear programmed instruction sequence.

Strengths of linear programmed instruction. This method essentially has the same strengths as linear text, with the following additions:

1. Linear programmed instruction is more interactive than text because it elicits responses and provides feedback.
2. It potentially boosts self-esteem and motivation through reinforcement.

Weaknesses of linear programmed instruction. Linear programmed instruction has not proven to be a particularly successful method for training and instruction. While it has the advantages of text, it has the same weaknesses and many more. Primarily, it is tedious for learners to go through the small steps prescribed by the program. It not only is not adaptive to different levels of learners, but it is targeted to almost the lowest learning level of the learner population. Because there are really no ways to reduce the weaknesses, this method is *not recommended*.

1. The technique takes a relatively large amount of development effort and time.
2. The technique requires people with specialized development knowledge.

3. If the program is computer-based, it requires equipment.

4. Linear programmed instruction is not adaptive to the level of the individual learner.

5. The method is not flexible, and is difficult to modify if the instructional content needs to be changed.

6. The method promotes passive reception of information rather than active, generative learning.

7. Testing of declarative knowledge occurs immediately after learning each piece of information. This method of testing has poor validity, that is, it is completely different from how learners will have to retrieve and use knowledge in the real world.

8. People don't like linear programmed instruction because of the small steps.

9. The realism is poor, therefore the transfer of skills is very low.

10. The system is not interactive.

11. There is no opportunity for exploratory learning.

12. Linear programmed learning systems are very impersonal and rigid in style.

13. Linear programmed instruction is not good for presenting complex or dynamic information.

14. Finally, like other text formats, it does not make use of the audio channel.

Branching programmed instruction.   Realizing the insurmountable difficulties associated with linear programmed instruction, many designers moved to a more reasonable and adaptive approach, programmed instruction with branching depending on the response of the learner. Branching programmed instruction can be delivered on paper, such as a textbook, or on a computer. The computer-based versions are one of many types of CAI. The method is included in this section because the computer-based method is functionally the same as a paper-based system. Therefore, their strengths and weaknesses will be similar. However, as we discuss CAI in a later section, keep in mind that many CAI systems are no more than sophisticated implementations of branching programmed instruction.

An example of branching programmed instruction is provided in Figure 8.3. Note that the content still consists of chunks of information followed by items requiring learner response. These items are almost always multiple choice in format because this limited "menu" of responses provides the basis for subsequent branching. When the response to a question is correct, the program will provide feedback that the learner is correct and move to the next

---

Page 19

If you reject the null hypothesis when it is really true, you have made a Type 1 error. Although the independent variable had little effect, you are saying that its effect was significant. The chance of making a Type 1 error is the same as the significance level one has set for rejecting the null hypothesis.

Let's apply this idea to some examples.
Which of the following is more likely to lead to a Type 1 error?

Dr. G. is studying the effects of a laser treatment on cancer. He sets his significance level at .001 . . . . turn to page 20.

Dr. S. is studying the effects of study groups on exam scores. She sets her significance level at .05 . . . . . turn to page 21.

Page 20

This answer was incorrect.

The study by Dr. G. has a very small significance level, so the chances of a *p* value falling below it are very small.

The probability of making a Type 1 error is exactly the same as the value chosen for the critical value of *p*. Therefore choosing a significance value of .05 will result in 5/100 Type 1 errors. Choosing a significance value of .001 will result in 1/1000 Type 1 errors. Comparing the two, a study with a significance value of .05 will result in more Type 1 errors.

Now return to page 19 and select the other answer.

Figure 8.3 Example of branching programmed instruction.

instructional unit. If the trainee chooses a response that is not correct, the program will branch to a chunk of information that remediates the learner.

Strengths of branching programmed instruction.   Branching programmed instruction can be a relatively powerful technique. It eliminates the problems of linear programmed instruction by allowing the program to adapt to the level of the individual. This strength plus others are listed below.

1. It is easier to develop this type of computer-based program than many of the more powerful, "intelligent" systems. (Also see discussions under CAI.)
2. Programmed instruction systems are a bit more difficult to develop than text, lecture, or videotape, but that is only because they force the instructional designer to carefully analyze the material. This forced analysis can be seen as a benefit from an instructional design point of view.
3. The programs are cheap and easy to *deliver.*
4. Students learn as much with programmed instruction as they do with lecture, and there is evidence that learning time is significantly reduced (Eberts & Brock, 1988; Goldstein, 1986).
5. The method is often more motivating than other more traditional techniques such as lecture or text.
6. The method provides immediate feedback to the learner, and can be designed to provide remedial instruction.
7. The method is moderately interactive; it does allow testing and practice with feedback. However, it is not as interactive as many of the newer CAI methods we will be discussing.
8. Programmed instruction is moderately adaptive to the individual; more so than text, lecture, or videotape, less so than newer and more powerful programs in CAI, tutoring, simulations, or OTJ training.
9. Branching programmed instruction systems are typically very easy to use, especially on a computer.

Weaknesses of branching programmed instruction.   The weaknesses of branching programmed instruction are difficult to present because it depends a *great deal* on how the method is carried out. I was recently at a conference where a salesman was proudly presenting his company's computer-based tutorial on concepts and principles of introductory psychology. I watched for a brief period of time and saw frames much like the three shown in Figure 8.4. I asked the salesman to explain how his system was functionally

Once a subject has learned to give a conditioned response in the presence of a conditioned stimulus, it is possible to build on that learning. By pairing the old stimulus with a new stimulus, the new stimulus comes to elicit the response. This higher order conditioning is the basis for new learning.

ILLUSTRATION:

After Pavlov's dogs had learned to salivate when they heard a bell, Pavlov was able to use the bell (without food) to teach the dogs to salivate at the sight of a black square

| CONTINUE WITH LESSON | TEST |

Reacting to a stimulus that is similar to the one to which you have learned to react is called _____.

a. stimulus generalization
b. response generalization
c. higher order conditioning
d. modeling

YOUR ANSWER WAS...... stimulus generalization

        THIS ANSWER IS INCORRECT.

Stimulus generalization occurs when a stimulus elicits the conditioned response because it is similar to the conditioned stimulus.

Figure 8.4   Tutorial screens in Psychology CAI.

any different from the "text/lecture/test method" that has been used to teach Psychology 100 for the last 50 or 60 years. He assured me that the color and "interactive" capabilities of the computer made it much more interesting and motivating for students. I assured him that taking the massive amount of detail from an introductory psychology text and displaying it on a color monitor was probably not going to make any earth-shattering differences in the degree to which students learned the material.

The point of this story is that lower-end CAI branching programmed instruction certainly allows the student to be "interactive" with the course material in some broad sense of the term (i.e., pressing buttons). However, it is not interactive in the sense of letting the student interact with the domain content. It is still based on the old paradigm of "present and test" information. It is *very* easy to assume that presenting material in this way, on a computer with lots of dazzling color and graphics, will cause the learner to miraculously learn the instructional material. This problem is reflected in many of the weaknesses listed below.

1. Branching programmed instruction is somewhat more time-consuming and expensive to develop than text, and definitely more time-consuming than lecture or OTJ training. It can be difficult and complex to develop all of the branching mechanisms, depending on the nature and extent of the program.

2. Computer-based systems require equipment and instructional spaces for delivery.

3. When implemented in hardcopy text, the information is difficult for a reader to access in ways that we typically use in accessing material in a text (index, etc.). This can make it problematic if the information is used as reference material after the training program.

4. Branching programmed instruction usually focuses on piecemeal information in an isolated fashion.

5. The method is based on the old instructional view of *present a chunk of information* then test knowledge of the information. This places the learner in a very passive role. In addition, the systems too often test for knowledge after too short a period of time. This results in poor assessment of how much the learner will retain after the training program. It also does not test the learner under conditions similar to the post-training environment.

6. Because of the passive role of the learner, the system is usually not very motivating.

7. The method has traditionally not been used much for skill training, although this is changing (see CAI).

8. The method is not as adaptive to the learner as tutoring, more powerful versions of CAI, or intelligent CAI. If the learner has a question about some concept or topic, it is not possible to query the system.

Computer-Assisted Instruction

CAI is one of those techniques that is so varied, it is hard to describe. Computer-assisted instruction, or computer-based instruction,[2] can be widely defined as any instructional activity that takes place through use of a computer. There are many varieties of computer-based instructional activities, and a few go by special names. The most commonly used computer-based systems are described in Table 8.2.

In this section, we will briefly review CAI methods. We will not be concerned with computer-managed instruction, and the ICAI and simulation methods will be discussed under separate headings. When we speak of CAI, we are referring to a broad spectrum of computer-based systems ranging from knowledge bases used to support the instructor to completely stand-alone tutoring systems. Typical implementations of CAI will be described in Chapter 9. The normative system consists of presenting information to trainees, testing trainees' declarative knowledge, demonstration of relevant skills, opportunity for practice of skills, and individualized feedback during and after skill performance. Learners can almost always control the pace of presentation, and usually also the order of topic study. The computer may go through complex branches and loops in adapting to each individual learner's needs, and some systems have a mechanism for the student to ask questions. Some computers assess student knowledge and other characteristics such as learning style at the beginning of the session to be more adaptive.

Strengths of CAI. There have been hundreds of studies conducted to determine whether there are benefits to using CAI. Most of these have compared CAI with traditional lecture methods. A number of researchers have performed reviews and meta-analyses of this literature (e.g., Kulik & Kulik, 1991; Kulik, Kulik, & Shwalb, 1986; Niemiec & Walberg, 1987; Thomas, 1979). The conclusion published by Kulik and Kulik (1991) is that, based on 254 controlled evaluation studies, CAI appears to enhance learning by 0.30

---

[2]While some researchers discriminate between the two, the terms *computer-assisted instruction* and *computer-based instruction* will be used synonymously.

TABLE 8.2   Types of Computer-based Instruction

Hypertext	*Hypertext* refers to programs where text and graphics are stored electronically as chunks. The learner can travel from one chunk to another via electronic links, usually implemented as places on the screen that one clicks a mouse.
Hypermedia, Multimedia	*Hypermedia* is hypertext with the addition of information that is audio and/or dynamic video in nature. *Multimedia* is a more general term for the computer-based access of media beyond simple text and graphics (e.g., video and sound).
CBT	When used for training in industry or government, CAI systems are often referred to as *computer-based training.*
CMI	*Computer-managed instruction* (CMI) refers to systems where the computer is used for administrative tasks such as course registration, grade keeping, and test scoring.
ICAI	*Intelligent computer-assisted instruction* refers to instances where CAI systems exhibit some of the major characteristics of a human tutor. With present technology, these systems often have an expert system for the tutoring component. Some systems model the student's knowledge and use that for adaptive tutoring.
ITS	Sometimes intelligent computer-assisted systems are referred to as *intelligent tutoring systems* (ITS). Some researchers insist that ITSs are one step up from ICAI systems.
Simulations	*Simulations* provide anywhere from a very abstract to a very realistic portrayal of the task in its setting. The learner actually performs subtasks within that simulated environment.

standard deviations; that is, "the typical student in an average CBI class would perform at the 62nd percentile on an achievement exam, whereas the typical student in a conventionally taught class would perform at the 50th percentile on the same exam" (p. 80). Thus, while the effects of CAI were moderate, and not as large as one might hope, they were consistent. In addition, in 29 of the 32 studies that reported instructional time, the CAI learners acquired the material more quickly, taking only two-thirds the time required for lecture covering the same material.

Hannafin and Peck (1988) describe a number of advantages in using CAI. These are listed below (items 2 through 9) along with some additional strengths.

1. Improving ease of development. With the advent of high-end authoring systems, CAI is now relatively easy to program (relative to previous years and relative to simulations and ICAI). It has also advanced to the point of being more cost-effective than lecture over the time of development and delivery (Blackwell, Niemiec, & Walberg, 1986; Eberts & Brock, 1988).

2. Potential for high level of interactivity. The major strength of CAI is the increased level of interactivity over more conventional methods such as lecture and text.

3. Potential for adaptivity. CAI is very adaptive to the individual learner. This "individualization" factor (Eberts & Brock, 1988; Hannafin and Peck, 1988) is responsible for making CAI very efficient as an instructional system.

4. Convenience. Because computers are so commonly available, CAI is more convenient for instructional delivery than lecture or tutoring (but less convenient than hardcopy text). CAI lessons tend to be very inexpensive to reproduce and distribute. They can be used when the trainee has the time, and in a variety of places. Many require only a basic PC to run.

5. Motivation. Learners tend to be more motivated with CAI than with traditional methods such as lecture or text. Trainees generally like CAI, citing reasons such as the low threat posed by the machine compared to an instructor (Brophy, 1981), and the high degree of control that they have over the pace and order of instruction (Hannafin, 1984). In addition, many CAI programs have the "bells and whistles" that lead to visual and auditory interest.

6. Speed of response and feedback. CAI is able to provide responses, simple or complex, to input virtually immediately.

7. Supports course management. CAI provides a method for automatic record keeping. This can be used to keep track of student progress for grading purposes, but can also be used to monitor effectiveness of the instructional system.

8. Potential for insuring topic coverage. Like videotape, CAI can be used to make sure that a certain set of topics is covered. With lecture, the instructor might get sidetracked and forget to cover something. In addition, CAI has the means to keep records to make sure that learners did cover the material, and students can be asked to show proficiency on a topic before moving on.

9. Efficient. Training time has been found to be significantly shorter for CAI as compared to traditional lecture (Eberts & Brock, 1988; Kulik & Kulik, 1991). This is attributed to two factors. First, the pace of instruction is under the control of the learner for CAI but not lecture, and lecture is often driven by the slowest learners. Second, the individualization of CAI allows the pace to be optimized for each student, resulting in more efficient and therefore less time-consuming training.

10. Low cost relative to on-the-job training. Computer equipment can be much less costly than actual equipment used in on-the-job training. In addition, it avoids problems of danger in certain job circumstances.

11. Potential for practice. Extensive practice with feedback is feasible.

12. Potential for testing. Extensive testing of knowledge and skills is more feasible with CAI than most other methods.

13. Learner control. Trainees usually have a great deal of control with CAI. They can decide the order in which to progress through instructional units, whether they are ready for a test, how many problems they wish to attempt, when to review material, whether they want hints, etc.

Weaknesses of CAI. The weaknesses of CAI are not only dependent on the particular system, but are also shifting over time. It used to be only a few years ago that CAI was considered extremely time-consuming to develop (e.g., see Eberts & Brock, 1988). Now there are very sophisticated authoring systems that make the task easier and faster. This makes CAI more accessible to a wider population, although it still takes a substantial effort to develop a reasonably complex instructional program.

1. Work and time intensive. Even considering recent advances in the science, CAI development tends to be more work and time intensive than most other methods, including lecture, tutoring, text, videotape, and on-the-job training. While many instructional designers realize the value of CAI, it is still viewed as too challenging and time consuming to utilize.

2. Requires specialized knowledge. To be effective and "user-friendly," CAI must be designed according to principles for sound interface design. Some designers do not have this training, and bringing in an interface designer adds to the development costs.

3. Unnatural language. Another drawback revolves around the nature by which the learner and computer communicate. The easiest way for learners to communicate is through natural language, particularly *spoken* natural language. Computers do not yet have the ability to communicate in this manner.

4. Requires predicting learner needs. Another drawback is that the designer must anticipate every possible response of every learner. This places a burden on the designer, and is still not a realistic expectation.

5. Trainees cannot practice and test certain skills on a computer. Some researchers (e.g., Eberts & Brock, 1988; Goldstein, 1986) suggest that CAI is more appropriate for areas where the content to be learned is factual. This implies that it is not appropriate for skill acquisition. This is not really true anymore, especially with the increasing emphasis on multimedia and video capabilities. Many skills, such as speaking a foreign language or taking a photograph, can be modeled and simulated on the computer. However, there are still two remaining weaknesses with current CAI systems. First, it is difficult to test the trainee on certain skills using a computer. For example, if we are training car salespersons to interact with clients, it is not possible to *directly* test their ability via computer. That is, the computer has no way of judging their performance. This weakness goes back to the current deficiencies in natural language interfaces. A second situation where CAI is not appropriate is teaching of psychomotor skills, such as how to use a lathe or how to hit a tennis ball (although virtual reality designs may be changing this).

Intelligent Computer-Assisted Instruction (ICAI)

Many educators, cognitive scientists, and computer scientists are attempting to imbue CAI with intelligence using technologies currently being developed in artificial intelligence (AI). These researchers seem to agree on a relatively well-defined set of characteristics that should be exhibited by a CAI system in order to be considered *intelligent*. The following list of **defining characteristics** is taken from Seidel, Park, and Perez (1988, 239–41).

1. Generative instructional presentation. In conventional CBI, the system's capability to adapt instructional processes to the individual student's learning needs is very limited because every instructional event needs to be specified in advance in programming algorithms. To overcome this limitation, the notion of generative CBI was proposed in the late 1960s (see Uttal, Rogers, Hieronymous, & Pasich, 1969). Generative CAI has the capability to generate new presentations from the combinations of different elements in a large data base (see Koffman & Blunt, 1975).

2. Mixed initiative function of instruction. In conventional CBI, every possible interaction is specified in advance in programming algorithms. Consequently, the system initiates every instructional activity and the student is just passively involved in the decision process. However, AI methods and programming techniques have been applied to develop ICAI systems in which the student is allowed to interrupt the instructional process, ask questions, and provide comments and other inputs.

[NOTE that it isn't a difference of *who can initiate what,* as much as it is a difference in input capabilities. The ICAI systems have something closer to natural language capability than older CAI systems.]

3. Modeling function of the student learning process. The main purpose of modeling the student's learning process is to assess the current state of his/her knowledge and skills on a given task, to identify his/her misconceptions, learning problems and needs, and to provide the most reliable diagnostic information for the prescription of instructional treatments.

4. Qualitative decision-making function of instruction. The decisions made in ICAI are made on the basis of qualitative information (Clancey, 1986) obtained in the process of the system's inference about the student's current learning state and misconceptions/learning problems.

5. Function of Inference. When the ICAI system interprets the student's inputs, diagnoses his/her misconceptions and learning needs, and generates instructional presentations, it draws inferences on the basis of available information at that time which is usually insufficient, inconclusive, and incomplete.

6. Self-improving function. Another important intelligent characteristic of ICAI is the capability to monitor, evaluate, and improve its own teaching performance.

These are not only defining characteristics of ICAI, but can also be used to distinguish it from standard CAI. That is, ICAI has "inferential reasoning" abilities that allow it to choose instructional strategies, develop an internal model of the individual student and tutor accordingly, accept natural language input, and generate responses to the student by searching through a domain knowledge base and compile completely new material for presentation (Eberts & Brock, 1988; Seidel et al., 1988).

Strengths and weaknesses of ICAI.   Clearly this is a very powerful technique for teaching difficult or complex material. The major problem is that no system exists with even the majority of the characteristics listed above

(Frasson & Gauthier, 1990; Eberts & Brock, 1988; Siedel et al., 1988). Many researchers are currently working on prototypes with intelligent reasoning capabilities (e.g., see Frasson & Gauthier, 1990; Polson & Richardson, 1988; Psotka, Massey, & Mutter, 1988). So one major weakness from the trainer's point of view is that not only are ICAI systems extremely difficult to develop, the technical capability to create such a system barely exists.

Simulations

Simulations are artificial recreations of some real-world environment. They can be carried out by complex hardware (such as flight simulators), computers, or even just groups of people interacting. Some simulations are used to train managerial or other interpersonal skills; these will be described in the next section. Simulations differ from the environment being mirrored along various dimensions and to different degrees. It used to be the case that the highest priority in developing complex physical simulators was that they have "high fidelity" (Flexman & Stark, 1987). This meant that they should have the look and feel of the real system to the greatest extent possible. Unfortunately, this led to an extremely high price for most physical simulators. More recently, researchers are trying to identify the similarities that are important rather than simply trying to make a simulator as realistic as possible in all ways.

Simulators vary in the types of system that deliver them and in the types of task that they train. We can break them down into three categories for the sake of convenience. First, there are physical systems that must be specially manufactured, such as driving simulators, flight simulators, space simulators for training astronauts, and diesel locomotive simulators for training locomotive crews.

The second type of simulator is a computer-based simulator. This category includes task simulations that are carried out using only standard computer hardware configurations. For example, power plant control can be simulated using colored displays and controls on multiple computer screens. Instead of manipulating actual controls, the trainee clicks on buttons and moves slide switches on the screen. A few computer-based simulations are now using virtual reality as well (discussed in the next chapter).

The third type of simulation requires little to no special equipment, and is used to teach leadership and interpersonal skills in organizations. This type of simulation depends on paper and pencil materials, or simple role playing. Because this third type is very different from the first two, it will be included in the following section on leadership and interpersonal skills training.

Strengths of simulations.   According to Flexman and Stark (1987), the value of a simulator derives from its unique ability to provide a realistic context for specific skill acquisition. The simulated context is often a more desirable training environment than the real world for a number of reasons, including cost, safety, and efficiency/speed of practicing only certain job components. This and other strengths of simulators are listed below.

1. Simulators are infinitely patient. They are less intimidating than a human trainer.
2. Obviously, one of the critical strengths is a simulator's ability to provide good transfer to the actual job environment.
3. Simulators can provide massive opportunity for skill practice with feedback.
4. Simulators invoke generative, "active" learning in the student.
5. There are several advantages over OTJ training. As compared with the "real" environment, simulation can be cheaper, safer, and provide a much broader array of tasks.
6. Simulations can be used to delete irrelevant stimuli from the training environment and focus learning on critical variables.
7. Feedback can be very realistic and detailed.
8. Simulations are often stand-alone systems, meaning that they can be used without another person there to act as trainer.
9. Simulations are highly motivating.
10. They make use of audio and visual channels.

Weaknesses of simulations.   While simulators have few weaknesses, the ones they have are important.

1. The major weakness is the high cost of developing a simulation. However, there are certain types of computer-based simulations that are becoming relatively inexpensive to develop.
2. Physical simulators are extremely costly to implement, and some simulators requiring high-end computer systems are also relatively costly.
3. Some simulators may be difficult to modify when the real-world system changes. Obviously, this will depend on the type of simulation.
4. Depending on the hardware base, delivery of simulator training systems will generally be more difficult and costly than other methods.
5. It is hard to assess the likelihood of this event because it is not frequently

documented, but the increasing ease of developing simulations may cause designers to bypass good instructional design, because instead they can simply *recreate* reality. Developing a simulator still requires instructional design to determine its properties and use.

6. High-fidelity (realistic) simulations may actually interfere with instruction by providing too many cues in too complex an environment (Rouse, 1991).

7. Simulators basically teach procedural/skill knowledge, and unless combined with some other method, do not teach declarative knowledge.

8. Similar to #7, simulators do not provide any specific guidance during the training, or instructional documents afterward. Thus, to be used effectively, they must be combined with some other type of instruction. This can be overcome by having a computer-based instructional system embedded in the simulation.

On-the-Job (OTJ) Training

On-the-job (OTJ) training is usually an informal procedure where someone either tells the worker what he/she is expected to do, or the trainee watches a more experienced worker for some indeterminate period of time. There are rarely guidelines for the trainer to follow during the instructional period. Rather, they simply demonstrate skills for tasks that happen to come up, and any other knowledge or skills of which they happen to think during the training period. A great number of workers receive no other kind of training beyond that which occurs "on-the-job."

As an example, consider the normative training for a restaurant food waitress. Typically, the trainee is asked to follow an experienced waitress for one or two days. Often, the worker/trainer is busy and simply performs the job while the trainee watches. Less than 50% of the time, the trainer will verbalize what she is able to in terms of declarative knowledge. The trainee may or may not then be asked to spend some period of time performing the job while the worker/trainer observes.

The following analysis of OTJ strengths and weaknesses will make it seem that OTJ training is a very poor method. However, it is important to point out that most of the weaknesses come from the fact that there is usually not a well-designed training program to serve as the foundation for OTJ training. The entire instructional program is the responsibility of someone who may or may not know how to perform the job well, and probably does not have expertise in instructional design. The weaknesses described below are based upon the assumption that the OTJ training program is carried out in this manner. To the degree that proper instructional design is carried out, the weaknesses may be mitigated. However, we should note that *effective* OTJ training is usually difficult to implement, mostly because the trainer's first responsibility is to his/her job.

Strengths of OTJ training. Even with little instructional design, OTJ tends to have the following strengths.

1. No special equipment is needed.
2. No professional instructor is needed.
3. OTJ training has a very low cost. People are doing what they normally would do anyway. There are no extra operating costs, except to pay the trainee, which would be done in any case.
4. OTJ training is easy to implement, in fact, it takes no advance efforts or activities at all.
5. The company usually has a worker available to act as trainer.
6. The content is flexible, and automatically changes as jobs change.
7. OTJ provides good transfer of training for skill acquisition because it is, in fact, the same as the ultimate job environment.
8. There is a good opportunity to test declarative knowledge access and use under *realistic* circumstances.
9. There is greater potential for active learning and skill practice than most other techniques, including lecture, video, and text.
10. OTJ automatically makes use of both audio and visual channels, for efficient information processing.

Weaknesses of OTJ training. As many advantages as there are for OTJ training, they are usually outweighed by the weaknesses of the method, as normally implemented. As stated before, these generally derive from a lack of adequate instructional design. This is stated first, as the major weakness of the method.

1. OTJ programs are usually not systematically designed and implemented using proper instructional design principles.
2. The worker/trainer usually doesn't have instructional design skills.
3. Related to #1, the trainer usually doesn't spend any extra time developing a program or thinking about what to train or how to train it.
4. The worker/trainer doing the training is usually expected to continue

doing his/her regular job at the same pace. As a result, he/she may not have much time to spend on training the other person.

5. The worker will usually not cover declarative knowledge in sufficient breadth, but will only verbalize those things of which he/she is spontaneously reminded.

6. The worker/trainer will usually not cover sufficient breadth of skills and task performance. He/she only covers what happens to come up, or what he/she happens to access from memory.

7. The worker may resent his/her role as trainer.

8. Usually the trainee has a rather passive role, and may not actually perform the job to a great degree.

9. There is usually too little practice of the various subtasks. And the trainee may get inadequate feedback from the worker/trainer.

10. What feedback there is may be less than optimal. For example, an expert may not be able to explain the reasoning behind why subtasks are performed in a particular fashion.

In summary, OTJ training can be extremely effective, but only if it is implemented in a manner consistent with good principles of instructional design.

COMBINING METHODS

Each of the methods described above has certain strengths and weaknesses. Documents may make very good reference manuals but poor tutorials. CAI systems may provide good training but not enough realistic skill practice. And OTJ training may provided critical real-world cues, but lack the necessary structure or breadth of knowledge and skills. These trade-offs can be handled in at least two ways. The analyst can perform a systematic trade-off analysis and choose the most appropriate training or performance support methodology. One can also combine instructional methods (or instructional methods with performance support systems). It is apparent from the lists of methodology strengths and weaknesses that with careful analysis, we should be able to combine different methods and come up with a more complete and instructionally efficient training system. This is, in fact, what many researchers are currently doing.

In Chapter 9, we will look at several tutoring systems that are combined systems integrated through computers. In Chapter 10, which covers the topic of developing the design concept, we will evaluate the various ways that different instructional methods can be combined.

TRAINING FOR LEADERSHIP
AND INTERPERSONAL SKILLS

A large part of the training business involves the use of programs designed to enhance managerial effectiveness, and also to increase interpersonal skills in general. These activities require very little in terms of physical environments or special equipment, but rather are activities structured and led by professional trainers. We do not have room to delve into all of the varieties of training programs that have been created for enhancing interpersonal skills. However, we will briefly look at a few of the methods that have received widespread use over the years.

Simulations and Business Games

Simulations are training programs that recreate some part of a real-world business setting so that trainees can practice making decisions and interacting with others. Simulations for this purpose are also sometimes called **business games** (Friedman & Yarbrough, 1985). Trainees participate in groups and are given large amounts of information about a fictitious business (such as financial, sales, marketing, and production data). They are asked to carry out tasks that they would normally perform in their position.

A technique that is similar to business games is called **role-playing,** and some people consider this one type of simulation (e.g., Goldstein, 1986). However, consistent with Latham (1988), we are keeping the two categories distinct. According to Friedman & Yarbrough (1985):

> They [business games] differ from role playing in that simulations usually exaggerate reality more, run for a much longer period of time, involve more complex interactions among people, and yield learning outcomes that are less under the control of the trainer.

Business games were originally designed in the 1950s and 1960s to train basic business skills. More recently, they have focused on leadership and other interpersonal skills. Once of the most well known business games is Looking Glass, Inc. (Kaplan, Lombardo, & Mazique, 1983; McCall & Lombardo, 1982). This game is a simulation of a glass manufacturing company. Participants are situated in an office-like setting and given company annual reports, financial data, and so forth. The game lasts approximately one day, and trainees start out with an in-basket full of typical items. They spend the day dealing with the items and interacting. The purpose of the training program is to enhance management skills, both as individuals and as team members.

Other examples include competitive games such as the UCLA Executive Decision Game, where trainees interact in a scenario of competing firms. A different type of game that increases people's awareness of and sensitivity towards others is the Starpower game, a simulation where some participants are given much more power than others.

All of these business games are similar in trying to get trainees to expand their viewpoints and try out new behaviors. According to Friedman & Yarbrough (1985) they have the following characteristics:

- They tend to be motivating and enjoyable for most trainees.
- While they are relatively time-consuming, they can also be used to fit a large number of differing activities into a small period of time.
- A simulation can focus solely on certain elements of the job environment, eliminating many of the distractions from the ordinary workplace.
- Simulations are good for getting trainees to expand their perspectives and try new behaviors. Sometimes they are deliberately designed to get trainees to see that certain tasks performed by others may not be as easy as they had thought.
- Simulations can increase self-awareness and sensitivity to the situations and needs of others.

Simulations are conducted by a trainer experienced in this type of program. Much of the success of a simulation depends on the ability of the trainer, even when the program is acquired as a previously developed package. If a trainer is planning on designing and conducting a simulation from scratch, he or she should carefully proceed through a design sequence such as that presented in this book, and also seek guidance on leading such interactions (e.g., Friedman & Yarbrough, 1985).

Role Playing and Behavior Modeling

As one would predict from our discussions of skill acquisition, the best way to train interpersonal skills is to demonstrate the skill and ask the trainee to perform it with subsequent feedback. In some programs, trainees are simply asked to *role play* some particular type of interaction with another person who is also role playing (or, alternatively, is the trainer). Trainees are subsequently given feedback on their performance. Often this is accomplished with the help of a videotape of the behavior. Role playing is generally more specific, shorter in duration, and done with fewer people than business games.

More often than not, role playing is performed after the trainer or other expert has modeled the behavior as it should be performed. This combination is often referred to as **behavioral role modeling** or **behavioral modeling**. And behavioral modeling usually follows a didactic lecture on the topic or a group discussion. As can be imagined, the situation provides opportunity for a variety of activities. Trainers can model numerous positive examples of how interactions should be handled. They can also show examples of poor behavior. Trainees can participate in role playing and can also watch others. They are given feedback and guidance during and/or after the behavior. And the trainer can provide positive reinforcement for behavior as it comes closer to the form of the desired behavior (Goldstein & Sorcher, 1974).

Several studies have shown that behavioral modeling and role playing can be an effective training method (e.g., Davis & Mount, 1984; Latham & Saari, 1979; Meyer & Raich, 1983). Davis and Mount (1984) found that CAI with behavior modeling was significantly more effective than CAI alone. And in a meta-analysis of 70 studies, Burke and Day (1986) showed that behavioral modeling was one of the most effective of training methods.

As with business games, the role of the trainer is critical for the success of role playing programs. Anyone who will be developing and leading a role-playing session should seek advice from appropriate sources.

Human Relations Training: Group Interaction

In this section, we will review one type of training that takes place in groups. However, the context is not a business simulation or specific roles to be enacted. Rather, it is a group discussion focused on open unstructured discussion and sharing of feelings (both positive and negative). According to Campbell, Dunnette, Lawler, and Weick (1970), the goals of the group interaction are (a) to provide insight into the trainee's behavior toward others, (b) to provide insight into and tolerance for other people's behavior, (c) to teach people how to listen to others, (d) to provide insight into group dynamics, and (e) to provide a nonthreatening forum in which a trainee can express feelings and try out new behaviors (with feedback).

There have been numerous types of group interaction methods developed over the years. Some of these "laboratory training" methods, as they are often referred to, have special names such as L (learning) groups, encounter groups, action groups, and T groups (Blumberg & Golembiewski, 1976; Buchanan, 1971; Goldstein, 1986). All of these techniques consist of bringing people together, allowing them to interact in a relatively unstructured manner, and encouraging people to drop their "masks" and say whatever they feel. Some methods bring together employees from within the same company, others like the T group specifically focus on bringing together strangers to encourage freedom of expression.

The methods often result in trainees expressing both positive and negative feelings about themselves and others. In fact, this is considered to be one of the goals of the method. The freedom to express opinion often engenders anxiety and hard feelings in participants. While some researchers feel that this effect is beneficial, others have criticized the method for causing too many negative effects on the trainees afterwards. For example, one study showed that at least one out of ten participants returned from the training program liking themselves less and feeling unsure as to what they should do about it (Klaw, 1965). In addition, other research has shown that there are both positive and negative outcomes of the method (Goldstein, 1986). For example, managers were more effective in some ways, but also were less effective in others. Apparently the trainees were venting emotions in very inappropriate ways. This can be seen as a serious transfer problem from the training setting to the job setting.

In summary, there seem to be two major problems with group interaction methods. First, the method can be too anxiety provoking and cause a loss in trainee self-esteem. Second, there are no mechanisms to provide trainees with specific skills to improve their interactions with others. In addition, there is not enough conclusive evidence that the benefits of the method outweigh the risks to trainee mental health. For these reasons, the method should not be used without a great deal of caution (Jaffe and Scherl,1969).

TEAM TRAINING

Our final category is another type of group interaction that specifically focuses on enhancing team performance. Sometimes people working together experience difficulties such as determining goals, deciding who does what, sharing equally in responsibilities, or simply feeling uneasy about the situation. As Friedman and Yarbrough (1985) state, a team that works well together has a common knowledge of their goals, who is doing what to accomplish the goals, and how they are to work together to accomplish the goals.

Some team training focuses on development of general "working together" skills. Team members are brought together to come to open and explicit agreement as to what constitute the team goals, who will be responsible for what tasks (termed role negotiation), and how the tasks will get accomplished. Trainees leave armed with knowledge of what they must do to enhance group performance, including: "(1) things to do myself, (2) things to do with others on this team, and (3) things to do with others and the environment outside this team" (Friedman & Yarbrough, 1985). The advantage of this type of team training is that implicit assumptions and problems are brought out into the open and addressed. After the training program, all members should be in agreement, arrived at through consensus, as to group goals and individual responsibilities.

Some jobs require team members to work closely in very difficult and complex environments. As an example, jet fighter pilots almost always function as a highly integrated team under strong time constraints. Researchers are searching for ways to train such teams to act more effectively. According to most theories, team performance is a function of **taskwork skills,** which pertain to the execution of autonomous individual tasks by a member, and **teamwork skills,** which involve the behavioral interaction among the team members (Denson, 1981; Glickman et al., 1987). Most team training programs have focused predominantly on taskwork skills (Denson, 1981; Swezey, 1993). However, effective team performance requires both types of skills and there is a growing focus on how to train teamwork skills. In the past, teamwork skills were often taught to individuals. For example, members would be taught appropriate communication skills, coordination strategies, etc. With the growing sophistication of simulators, it is now possible to model such skills as the team actively engages in realistic tasks. This allows the members to practice and receive feedback.

One difficulty associated with team training is measurement of team performance per se (beyond measuring individual performance). Measurement of team performance at both the macro and micro level is important both for front-end needs analysis and also for training program evaluation. Various methods for measuring team performance have recently been suggested (e.g., Dyer, 1992; Prince, Brannick, Prince, & Salas, 1992). Readers who will be responsible for training highly complex and interdependent teamwork are referred to the text by Swezey and Salas (1992).

REFERENCES

BALL, S., & BOGATZ, G. A. (1970). *The first year of Sesame Street: An evaluation.* Princeton, NJ: Educational Testing Service.

BLACKWELL, M., NIEMIEC, R., & WALBERG, H. (1986). CAI can be doubly effective. *Phi Delta Kappan, 67,* 750–751.

BLOOM, B. S. (1984). The 2 sigma problem: The search for methods of group instruction as effective as one-to-one tutoring. *Educational Researcher, 13,* 4–16.

BLUMBERG, A., & GOLEMBIEWSKI, R. (1976). *Learning and change in groups.* Clinton, MA: Colonial Press.

BROPHY, J. (1981). Teacher praise: Functional analysis. *Review of Educational Research, 51,* 5–32.

BROWN, J. S. (1990). Toward a new epistemology of learning. In C. Frasson and G. Gauthier (eds.), *Intelligent tutoring systems: At the crossroads of artificial intelligence and education* (pp. 266–282). Norwood, NJ: Ablex.

BUCHANAN, P. C. (1971). Sensitivity, or laboratory, training in industry. *Sociological Inquiry, 41,* 217–225.

BURKE, M. J., & DAY, R. R. (1986). A cumulative study of the effectiveness of managerial training. *Journal of Applied Psychology, 71,* 232–246.

CAMPBELL, J. P., DUNNETTE, M. D., LAWLER, E. E., & WEICK, K. E. (1970). *Managerial behavior, performance, and effectiveness.* New York: McGraw-Hill.

CHU, G. C., & SCHRAMM, W. (1967). *Learning from television: What the research says.* Washington, DC: National Association of Educational Broadcasters.

CLANCEY, W. J. (1986). Qualitative student models. In J. F. Traub, B. J. Grofz, D. W. Lampson, & N. J. Nilsson (eds.), *Annual review of computer science: Volume 1* (pp. 381–450). Palo Alto, CA: Annual Review Inc.

COHEN, P. A., KULIK, J. A., & KULIK, C.-L. C. (1982). Educational outcomes of tutoring: A meta-analysis of findings. *American Educational Research Journal, 19,* 237–248.

DAVIS, B. L., & MOUNT, M. K. (1984). Effectiveness of performance appraisal training using computer-assisted instruction and behavioral modeling. *Personnel Psychology, 37,* 439–452.

DENSON, R. W. (1981). Team training: Literature review and annotated bibliography (AFHRL-TR-80-40,A9-A099994). Wright Patterson AFB, OH: Logistics and Technical Training Division, Air Force Human Resources Laboratory.

DYER, D. J. (1992). An index for measuring naval team performance. *Proceedings of the 36th Human Factors Society Annual Meeting* (pp. 1356–1360). Santa Monica, CA: Human Factors Society.

EBERTS, R. E., & BROCK, J. F. (1988). Computer-based instruction. In M. Helander (ed.), *Handbook of human-computer interaction* (pp. 599–627). Amsterdam: Elsevier.

ESLER, W. K., & SCIORTINO, P. (1991). *Methods for teaching: An overview of current practices.* Raleigh, NC: Contemporary Publishing.

FLEXMAN, R. E., & STARK, E. A. (1987). Training simulators. In G. Salvendy (ed.), *Handbook of human factors* (pp. 1012–1038). New York: John Wiley & Sons.

FRASSON, C., & GAUTHIER, G. (eds.) (1990). *Intelligent tutoring systems: At the crossroads of artificial intelligence and education.* Norwood, NJ: Ablex.

FRIEDMAN, P. G., & YARBROUGH, E. A. (1985). *Training strategies from start to finish.* Englewood Cliffs, NJ: Prentice Hall.

GLICKMAN, A., ZIMMER, S., MONTERO, R., GUERETTE, P., CAMPBELL, W., MORGAN, B., & SALAS, E. (1987). The evaluation of teamwork skills: An empirical assessment with implications for training (Tech. Rep. No. 87–016). Orlando, FL: U.S. Naval Training Systems Center.

GOLDSTEIN, A. P., & SORCHER, M. (1974). *Changing supervisor behavior.* New York: Pergamon Press.

GOLDSTEIN, I. L. (1986). *Training in organizations: Needs assessment, development, and evaluation* (2nd ed.). Monterey, CA: Brooks/Cole.

HANNAFIN, M. J. (1984). Guidelines for using locus of instructional control in the design of computer-assisted instruction. *Journal of Instructional Development, 7(3),* 6–10.

HANNAFIN, M. J., & PECK, K. L. (1988). *The design, development, and evaluation of instructional software.* New York: Macmillan.

JAFFE, S. L., & SCHERL, D. J. (1969). Acute psychosis precipitated by T-group experiences. *Archives of General Psychiatry, 21,* 443–448.

JONASSEN, D. H. (1988). *Instructional designs for microcomputer courseware.* Hillsdale, NJ: Lawrence Erlbaum Associates.

KAPLAN, R. E., LOMBARDO, M. M., & MAZIQUE, M. S. (1983). *A mirror for managers: Using simulation to develop management teams* (Technical Report #13). Greensboro, NC: Center for Creative Leadership.

KLAW, S. (1965). Inside a T-group. *Think, 31,* 26–30.

KOFFMAN, E. B., & BLUNT, S. E. (1975). Artificial intelligence and automatic programming in CAI. *Artificial Intelligence, 6,* 215–234.

KULIK, C.-L. C., & KULIK, J. A. (1991). Effectiveness of computer-based instruction: An updated analysis. *Computers in Human Behavior, 7,* 75–94.

KULIK, C.-L. C., KULIK, J. A., & SHWALB, B. J. (1986). The effectiveness of computer-based adult education: A meta-analysis. *Journal of Educational Computing Research, 2,* 235–252.

LATHAM, G. P. (1988). Human resource training and development. *Annual review of psychology, 39,* 545–582.

LATHAM, G. P., & SAARI, L. M. (1979). The application of social learning theory to training supervisors through behavior modeling. *Journal of Applied Psychology, 64,* 239–246.

LUMSDAINE, A. A., MAY, M. A., & HADSELL, R. S. (1958). Questions spliced into a film for motivation and pupil participation. In M. A. May & A. A. Lumsdaine (eds.), *Learning from films.* New Haven, CT: Yale University Press.

MAGER, R. F. (1984). *Preparing instructional objectives* (revised 2nd ed.). Belmont, CA: David S. Lake.

MCCALL, JR., M. W., & LOMBARDO, M. M. (1982). Using simulation for leadership and management research: Through the Looking Glass. *Management Science, 28,* 533–549.

MERRILL, D. C., REISER, B. J., RANNEY, M., & TRAFTON, J. G. (1992). Effective tutoring techniques: A comparison of human tutors and intelligent tutoring systems. *The Journal of the Learning Sciences, 2,* 277–305.

MEYER, H. H., & RAICH, M. S. (1983). An objective evaluation of a behavioral modeling training program. *Personnel Psychology, 36,* 755–762.

NIEMIEC, R., & WALBERG, H. J. (1987). Comparative effects of computer-assisted instruction: A synthesis of reviews. *Journal of Educational Computing Research, 3,* 19–37.

POLSON, M. C., & RICHARDSON, J. J. (Eds.) (1988). *Foundations of intelligent tutoring systems.* Hillsdale, NJ: Lawrence Erlbaum Associates.

PRINCE, A., BRANNICK, M., PRINCE, C., & SALAS, E. (1992). Team process measurement and implications for training. *Proceedings of the Human Factors Society 36th Annual Meeting* (pp. 1351–1355). Santa Monica, CA: Human Factors Society.

PSOTKA, J., MASSEY, L. D., & MUTTER, S. A. (eds.) (1988). *Intelligent tutoring systems.* Hillsdale, NJ: Lawrence Erlbaum Associates.

SCHRAMM, W. (1962). Learning from instructional television. *Review of Educational Research, 32,* 156–167.

SEIDEL, R. J., PARK, O., & PEREZ, R. S. (1988). Expertise of ICAI: Development of requirements. *Computers in Human Behavior, 4,* 235–256.

SWEZEY, R. W. (1993). Considerations in the application of Complexity Theory—based measures of individual performance to team and organizational tasks. *Proceedings of the Human Factors and Ergonomics Society 37th Annual Meeting* (pp. 1176–1180). Santa Monica, CA: Human Factors Society.

SWEZEY, R. W. & SALAS, E. (1992). *Teams: Their training and performance.* Norwood, NJ: Ablex.

THOMAS, D. (1979). The effectiveness of computer-assisted instruction in secondary schools. *AEDS Journal, 12,* 103–116.

UTTAL, W. R., ROGERS, M., HIERONYMOUS, R., & PASICH, T. (1969). *Generative computer-assisted instruction in analytic geometry.* Newburyport, MA: Entelek.

Background:
Computer-Assisted
Instruction

Because there is an increasing emphasis on the use of computer-based instruction in both industrial and governmental training programs, we will look at this particular method in more depth than the others. Therefore, in this chapter, I will review the major trends in the CAI business and some examples of typical CAI programs.

BACKGROUND: TRENDS IN CAI

When computers first entered the instructional scene back in the late 1950s and early 1960s, researchers had high expectations that CAI would revolutionize learning. All that held us back was the state of hardware technology, and future advances would allow the computer to completely individualize instruction. In the 1960s, several large-scale development efforts were initiated. One of these was PLATO (Programmed Logic for Automated Teaching Operations), developed at the University of Illinois (Lyman, 1981). The PLATO system has received about $900 million in funding over the years (Eberts & Brock, 1988), and by 1980 had been applied to over 150 subject areas (Lyman, 1981). Two other major projects initiated in 1960 were TICCIT, developed by the MITRE Corporation, and PLANIT, developed by the Systems Development Corporation.

Even with the progress being made on the PLATO, TICCIT, and PLANIT systems, computers did not make substantial headway into the classroom. By 1970, only 13% of public schools used computers in instruction (Price, 1991), and much of this was use of computers for mathematics and science (not CAI per se). Clearly, CAI had limited impact.

In the 1970s, several factors affected the educational computing business. First, computers did continue to grow in computational power and diminish in size, making them more feasible for distribution. The advent of the personal computer finally caught the public eye. The renewed interest in instructional computing led to a rush of software production, and by the end of the decade the "first generation" of instructional software had been marketed (Price, 1991).

While this flurry of (often poor) product development took place in commercial companies, researchers working on major university-based projects continued to make progress. In particular, the Minnesota Educational Computing Consortium (MECC) produced high-quality software that was created by *teachers,* and systematically tested and produced according to sound design principles. Because of their quality and effectiveness, the MECC products remain successful.

The 1970s and 1980s also saw an increasing emphasis on the use of computers in industry and government training programs. Because these institutions have traditionally had the time, money, and organizational infrastructure to follow systematic instructional design methodologies, these CAI programs were generally more instructionally sound and therefore successful than a majority of the educational applications (Roblyer, 1988).

At this point in time, then, we can characterize the state of CAI in the mid 1990s as being a technique that is heavily utilized in industry and government, and becoming even more so. It continues to hold promise for education, but some people have doubts about the reality of CAI as a really effective learning medium. Part of the reason for those doubts is the fact that most CAI has been based on the "accumulation" model of instruction (Brown, 1990; Soloway, 1991). This model is now being viewed as an undesirable one. As Elliot Soloway (1991) aptly described it:

> By and large, computers in education have been used to implement the accumulation model of learning: with technology we can transfer "stuff" to students faster; with technology we can deliver instruction more effectively—as if ideas were just bags of potatoes in need of transportation. *That* model has not worked—and it will not work, even if we add multimedia. (p. 30)

He goes on to point out what other researchers and educators are telling us— learning is not a passive process, but requires a person to be active. People learn by "building artifacts—whether they be formal reports, private scribblings, Lego choo-choo trains, or computer programs, the path to learning is strewn with *things*."

Clearly, if there has been one lesson learned, it is that a particular medium is not a panacea. Just as learning to use special carpenter's tools does not guarantee a beautiful piece of furniture, so learning to use the latest tools (read *media*) in training design does not guarantee a successful instructional program.

MORE TECHNOLOGICAL ADVANCES: MULTIMEDIA

Things seem to occur roughly in decades. With the entrance of the 1990s, we are once again seeing a new technology that is being held up to the world as an answer to the problem of too many people learning too little. This new wonder is termed multimedia. According to an article in PC WORLD, "Multimedia can teach, entertain, sell products, or simply make business presentations sizzle" (Holsinger, 1991). The bells and whistles of multimedia make old computer-based applications look absolutely ho-hum. For what it's worth, most people hyping multimedia are in the computer business. Knowledgeable educators and trainers realize that the technology has much promise as a tool, but only if its potential is realized in a way that is consistent with our emerging knowledge of how people learn (Montague, 1988; Sorohan, 1993).

Multimedia in a Nutshell

Just what is multimedia? Whereas traditional computer systems presented text and graphic images, multimedia gives computer users random access to a variety of information and sensory experiences such as stereo sound, animation, full color video, combinations of graphics and video, and all manner of cinematic effects such as wipes and dissolves. It is extremely difficult to describe multimedia in a way that conveys its power as an entertaining medium. In fact, its closeness to the entertainment medium of *film* is evidenced by the fact that George Lucas, a giant in the movie industry with a talent for dazzling special effects, has recently developed a multimillion dollar complex in California where, among other things, "entertaining" instructional programs will be developed.

Multimedia allows us to come much closer to the full world of all our senses. We can touch a computer screen and instantly see a lion running through the jungle after its prey. This is accompanied by all of the sounds you would hear at the scene. We can watch as a doctor performs operations, stop the scene, ask questions, and replay any part we wish. By clicking on a button, animation allows us to travel into the bloodstream and experience the micro-world within the body.

Multimedia has the capability to bring information to us in a form much closer to that of the real world, including a wide variety of task simulations (Todd, 1991). And as we saw in an earlier chapter, there is good reason to believe that this capability might enhance learning. (Computerized "virtual realities" push this concept one step further, and the reader is encouraged to discover more about this new and engaging technology, e.g., Heeter, 1993.)

The Technologies Behind Multimedia

Multimedia instructional programs are made possible by an integrated system of hardware and software components. Up until recently, computers could only display digitally stored information on the computer monitor. Advances in technology have allowed computers to display images that are either digi-tally stored (as on a hard disk) or are in analog form (as on a VCR). Likewise, digitizing capabilities allow sound and video from input sources such as micro-phones, cassette tapes, and videotapes to be stored on the computer and played out again through headphones, speakers, and a high resolution monitor.

The computer is used to manage information from a variety of sources. All of this can be done easily through graphical user interfaces. Before dis-cussing these interfaces, let's briefly consider the hardware components of a multimedia system. The components, other than the computer itself, can be divided into input and output devices. Input devices include microphones, VCR or cassette players, MIDI synthesizers, fax or other telecommunications devices, scanners, digital storage (such as an external hard drive), magnetic tape, digitizing camera, and laserdisc (see Figure 9.1). Output devices include computer monitor, TV, speakers, headphones, laser printer, magnetic storage, writable CD-ROM, etc. (see Anderson & Veljkov, 1990, for a description of the various hardware components).

Current standards are moving toward use of a computer, microphone, speakers, and CD-ROM drive as the prototypical multimedia workstation. Both Macintosh and IBM PC desktop computers can be easily configured for multimedia using CD-ROM. In addition, some workstations also use laserdisc because at this time, it is the storage medium of choice for full-motion video of any length. Because of storage requirements, digital video is still restricted to short segments. General descriptions of the major hardware components are given below.

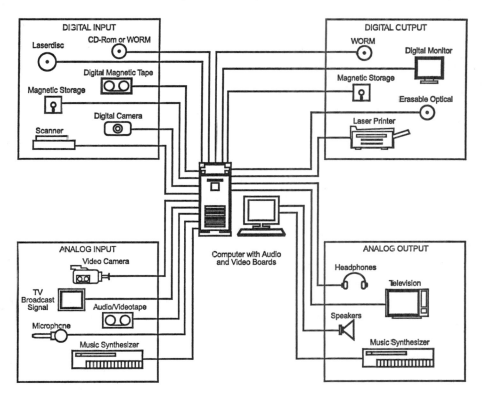

Figure 9.1 Components of multimedia computer workstation.

External hard drives. Many of the tutorials being developed for personal computers require large amounts of memory. A typical tutorial may require anywhere between 10 and 50 megabytes of memory. While the hard drives on most computers being marketed these days can easily handle that amount, it becomes hard to "deliver" such tutorials because a standard 3.5" disc holds less than 1 megabyte (or around 800K). Thus, if you want to be able to distribute the tutorial or move it to another system, it can be difficult. One solution to this problem is to add external drives to the computers. These drives use special cartridges that can currently store anywhere from 35 to 150 megabytes of information (the smaller size drives are relatively inexpensive, the larger size is about $700). The tutorial is stored on a cartridge which can then be carried away and used on any machine with one of the disc drives. One disadvantage to this system is that most computers are not configured with such external drives. Another solution that is gaining in popularity is networking.

Optical memory devices. Computer hard drives, floppy drives, and external hard drives all store information magnetically. That is, they read and write information using magnetic recording heads. There is an increasing number of hardware components that read and write information using a finely focused laser beam. These are optical storage devices. Instances of optical storage devices include CD-ROM (compact disc read only memory), CD-I (compact disc interactive), DVI (digital video interactive), read only laser videodisc and WORM (write once and read many) laserdiscs.

The reason that optical storage devices are becoming so popular is that their storage capacities are *vastly* greater than that of a computer. A single CD-ROM disc currently holds over 600 megabytes, or the equivalent of about 700 floppy discs. Conservatively, they can hold over a quarter of a million pages of text (Van Horn, 1991). I am hesitant to give the capabilities of other hardware components because technology is currently moving at a rapid pace in the optical memory industry.

It is important that we have storage devices with large memory capacity because people like to see images and movies that look real. This type of information, as well as sound, takes up a great deal of space when it is converted from its analog format to a digital format on the computer. Creating such rich tutorial environments takes a great deal of storage capacity.

At the time of this writing, manufacturers are just starting to market both laserdisc and CD-ROM machines that have the capability to write as well as play back information. These advances will significantly ease the task of developing instructional applications. Other technologies will also be making multimedia production easier and more efficient. For example, DVI technology currently allows a CD-ROM disc to hold up to 72 minutes of VCR-quality full-motion video. Unfortunately, the field is so new and changing so rapidly, that it is difficult to predict future technology and standards. Readers should investigate the potentials for peripheral information storage devices available at the time of training program development.

Other input devices. There are other devices that can be used to create multimedia applications. Color photos can be digitized using a scanner. Digital cameras are relatively inexpensive that allow you to take pictures and store them as digital files using a special piece of equipment that comes with the camera. This allows the training designer to take still single pictures for the application, or take a series to play back as dynamic motion. In addition, film can be sent in to be put on a photo-CD, videoboards and software allow trainers to digitize video from videocassette tapes, and audio boards allow trainers to digitize music from cassettes, CDs, or a MIDI music synthesizer.

One of the most useful input devices is a standard microphone. We discussed the usefulness of using both video and audio channels for reducing trainee cognitive workload. The text, graphics, and video can be developed for a multimedia application, and much of the instruction, explanation, and tutoring can be done by using a microphone and sound board.

Output devices. All of the information we just discussed can be played back on a computer monitor and speakers. Thus, the designer can blend video movies, animation, text, and complex graphics. For groups of trainees or students, the information can be displayed on a large (e.g., 35″) TV monitor, or alternatively on a projection screen using a special panel laid on top of an overhead projector. These panels are now capable of displaying color video at the full speed of 30 frames per second. The second important output device is speakers (or headphones) for presentation of audio information.

In summary, there are many new hardware components that are making it easier to develop very highly integrated multimedia training and educational programs. On the other side of the coin, the technological advances in this field will probably continue for some period of time, making it difficult to know what hardware configurations to purchase for current applications. Readers are advised to attend trade shows and stay in touch with journals and magazines on multimedia advances.

Implications for Training

Even a few years ago, it was relatively difficult to bring together the various "media" in multimedia. However, new software that is object-oriented and has a graphical interface, such as Compel™ or Authorware™, makes integration of the various media relatively simple. At the end of this chapter, I will briefly discuss three types or "levels" of software for creating CAI and ICAI applications, in multimedia or otherwise.

What does the availability of multimedia mean for training and instructional design? Probably the most straightforward and conservative answer is that it gives us one more tool in our bag of tricks, a very powerful tool, and one that no one has had ample opportunity to evaluate. No one really knows how to best take advantage of its unique abilities. In this text, therefore, we must just assume that the instructional design should progress in a "business as usual" fashion. Keeping in mind what we know about humans and the learning process, we can try to create ways to take advantage of multimedia's unique characteristics *when the need is there and the situation warrants it.* And sound methodologies such as prototype testing and final system evaluation should be performed as with any other instructional product.

In the next section, I review the basic functions that can be served by CAI. These functions generally fall into a small number of categories. All of the "types" of CAI systems we will discuss can be implemented within a multimedia environment as well as the more traditional text plus graphics environment.

CAI DESIGNS

In describing the many roles of CAI, researchers have often chosen to classify instructional programs as falling into one of four basic categories: **drill and practice, tutorials, simulations,** and **instructional games** (e.g., Alessi & Trollip, 1985; Hannafin & Peck, 1988; Price, 1991). Because of an increasing emphasis on hypertext and hypermedia (e.g., see Berk & Devlin, 1991), I am also adding a fifth category, learnercontrolled **information access and generation.** Information access refers to instances where the learner accesses chunks of text or multimedia information through menu or other hypertext linking mechanisms. Information generation refers to the use of software tools such as hypertext and hypermedia by the learner to create his/her own knowledge bases.

Drill and Practice

As we saw in Chapter 6, the learning of many skills requires extensive practice to promote proficient performance. Providing the learner with the opportunity to practice skills and skill components is a fundamental requirement in many training environments. Providing practice alone is usually not enough. Psychological research has shown that feedback closely following each practice trial is needed for efficient learning. CAI systems can be used to provide the opportunity for a learner to practice newly acquired skills. Traditionally, such systems have been referred to as **drill and practice.** They typically present problems to a learner one at a time (such as math problems), and the learner must try to respond correctly. The computer analyzes the learner's response and provides appropriate feedback. A basic feature of drill and practice programs is that they are used for practice of skills where the basic information has already been taught using some other type of instructional system (either computer or human).

Drill and practice programs have numerous advantages and disadvantages. Many of these are dependent on the way the method is implemented. Like many other instructional designs, they can be created in such a fashion to be interesting, colorful, and fun to do (even for adults). Or they may be dull and uninspiring, nothing more than computerized flashcards (Goldstein, 1986; Hannafin & Peck, 1988; Price, 1991). Consult guidelines on instructional design of drill and practice if you are going to develop this type of system (e.g., Hannafin & Peck, 1988; Salisbury, 1988). In general, drill and practice programs should be designed so that students have a clear idea of what is expected of them, the means of inputting responses, and the type of feedback they will receive. In addition, designers should try to *push* the systems toward a job simulation to the extent possible.

Tutorials

In addition to providing a context for skill practice, CAI can be used to teach declarative and procedural/skill knowledge at the outset. Such systems are referred to as tutorials. Tutorials are often designed to be self-contained (hence the name), and according to Hannafin and Peck (1988), "Effective tutorials include lesson orientation information, learner guidance during the lesson, appropriate feedback and remediation, and strategies for making the instruction more meaningful to the learners."

Most tutorials are designed such that information is presented to the learner, then they are questioned on that information (as in programmed instruction) or they are allowed to practice associated skills. As described in Chapter 8, feedback is provided through branching mechanisms. Because the learner is required to answer questions or otherwise perform various tasks, the tutorial is considered to be an interactive method.

While this type of instruction may sound dry (and sometimes is), the instructional program can actually be very interesting. For example, by using random access video in a multimedia system, tutorials can teach people skills that are difficult to teach through lecture or text. As an example, a medical doctor might be learning the proper procedures for performing triage in an emergency. Triage is the process of quickly looking over multiple patients and identifying what to do next. A tutorial might begin by giving some information and showing video clips of representative sets of patients as well as good examples and poor examples of doctor behavior. At this point, a *simulation* can require the learner to practice by seeing new scenarios (with sound) and being asked to state what he or she would do next. Based on the learner's answer, the computer could simply demonstrate via video the results that would likely occur from the doctor's particular response. In this way, he or she could see more concretely whether the response was or was not the best.

In summary, the best tutorials incorporate the various learning strategies we discussed earlier. While tutorials are becoming implemented for an ever wider array of tasks, we should continue to ask critical questions before automatically embarking on the long and sometimes difficult task of producing CAI. The major one is: Does this method offer something that can't be obtained using a less costly technique? If the answer is "yes, it leads to the most efficient learning and the highest proficiency rates," it may well be worth the extra work and cost. It is dangerous to assume that CAI will be better than lecture. We recently completed a study showing that lecture with multimedia support resulted in greater learning than a stand-alone multimedia tutorial, even though both systems were well-designed and *liked* by trainees (Brown, 1993).

Information Access and Generation

Sometimes instructional systems present new information but don't really provide opportunity for learners to perform any tasks (such as answering questions or performing procedural skills). Such systems can still be effective learning aids because they may provide a more efficient means for learners to access the necessary information than traditional means such as text or lecture. These *learner-accessed information systems* usually consist of chunks of information stored on a computer or other peripheral device such as CD-ROM or laserdisc player. The various chunks can be text, graphics, sounds, animation sequences, or video clips. The chunks are usually accessed by way of a graphical interface where the learner points to some button on the computer screen, or clicks on some area of the computer screen with a mouse, etc. When the user is accessing text, this is termed **hypertext.** When the information is multimedia, it is termed **hypermedia.**

According to Jaffe and Lynch (1989), "the term 'hypermedia' should not be confused with 'multimedia' which is merely a mix of audio-visual techniques. 'Hypermedia,' by contrast, connotes a highly integrated electronic environment allowing a user to interactively peruse a very large assembly of electronically linked information consisting of real-time moving color video images, sound, text, and electronically searchable data" (Jaffe & Lynch, 1989). This definition notwithstanding, most people are beginning to use the term "interactive multimedia," or simply "multimedia," when referring to a hypermedia system.

Hypertext and hypermedia systems can be very simple or very complex. If you have used an on-line help system, you are familiar with one form of hypertext. In addition to varying in type and complexity, hypermedia systems vary in both purpose and how they are used. They are rapidly gaining popularity for use as information kiosks, explaining all manner of things to people visiting sites for the first time. Examples include systems that provide information to people visiting museums, art galleries, universities, highway rest stations, and business conferences. Retailers are even beginning to see the advantages of a self-access information booth showing products in various configurations.

Hypermedia is also being used for a variety of training and instructional programs. Two examples are:

1. *The Perseus Project.* The Perseus Project is based at Harvard University but involves collaborators at many other institutions (Bannon, 1991). It is a database that contains Greek literature and various aspects of Greek culture, in the broad sense of the word. By clicking on buttons, students can see specific pieces of Greek literature, essays on the development of that literature, information about Greek history, buildings including drawings of how a theatre used to look and photos of how it looks today, architectural plans at different periods in time, a Greek-English dictionary, etc.

2. *Grapevine.* Grapevine is a multimedia application that contains information about America in the 1930s (Campbell & Hanlon, 1988). It runs on a desktop computer with a videodisc player. It contains a large text base that is linked and cross-indexed. It includes music, interviews, photographs, art, video from television and film, statistics, letters, and quotes. It is controlled (accessed) through HyperCard™. According to Campbell and Hanlon (1988), the subject matter is "the major social, economic, political, and cultural issues of the United States in the 1930s as approached from the perspective of John Steinbeck's *The Grapes of Wrath.*"

Both of these examples are typical of the types of hypermedia knowledge bases being used for training and education. There are currently many applications of hypermedia and the number is growing by hundreds, perhaps even thousands. For the sake of simplicity, we have first divided hypertext and hypermedia applications into two categories, information access and information generation. In information access, the learner uses the computer to access different chunks of information to support the learning process. In information generation, the learner uses the hypermedia system to actually compose some piece of work. We will briefly consider each of these alternatives.

Information access. Even within the category of information access, there are a variety of uses of hypermedia. First, it can be used by an instructor as a visual aid in the classroom. By having a text, graphic, and video database, the instructor can click on the computer screen and have instant access to thousands of visual references. For example, an architect could take students on a tour through a neighborhood and show various architectural features of buildings. These could be seamlessly combined with other examples of similar (or contrasting) architecture, blueprints, or any other instructional material. This provides an infinitely more flexible medium than slides or video.

The second major use is where students sit at a computer workstation and browse through various topics and types of information, learning individually or in groups. This method of instruction may be somewhat similar to the beginning phase of a tutorial, although in hypermedia students generally have more control over what they see and hear, and also the sequence of material. In fact, a critical characteristic of hypermedia is that the learner has control of what information is accessed. The learner traverses from one chunk or frame of information to another in a nonlinear fashion.

There are critical design decisions that must be made in creating a hypertext/hypermedia system. In this chapter, we only review the basics of hypermedia. In Chapter 11, we will consider a few design guidelines based on the small amount of research literature available at this time.

Accessing information via links.   Information in hypermedia is usually accessed via electronic links. Part of the computer screen is selected with a mouse or other input method and that part of the screen is linked to other information. Items to be selected on the screen can be menu options, words in a sentence, parts of a diagram, and so forth. Figure 9.2a shows an example of text presentation with numerous buttons embedded in the text, and Figure 9.2b shows a hypertext screen with both menu buttons and buttons embedded in text.

Similarly, graphical images can have hidden buttons called *hot spots* located anywhere on the image. Clicking on or pointing to the location of a hot spot activates the associated link. For example, clicking on the digital camera drawing shown in Figure 9.1 might lead to a new screen showing a close-up of a digital camera and an explanation (verbal or visual) of its key features. It might also show a video of a person using the camera and transferring the photos.

Information generation.   Hypermedia has a second use that, while currently less common, has a great deal of potential. This is where the learner uses the system as a database to create some artifact. He/she may write expository or narrative text in hypertext chunks, add special graphics or video, or integrate subsets of information that already exist in a system. The Grapevine application described above is used in this fashion. The idea behind the approach is that students learn better by creating some product (Brown, 1990); the process forces them to think about the meaning of each piece of information and its relationship to other information. In addition, hypermedia gives them the tools to bring in their own ideas and experiences. This idea can be

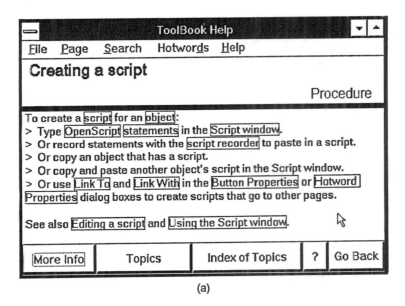

(a)

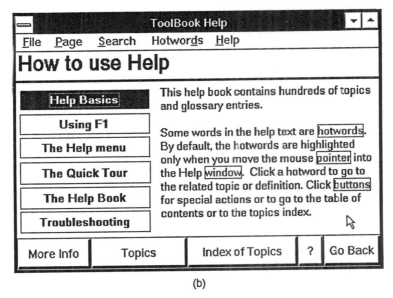

(b)

Figure 9.2 *ToolBook*^{*TM*} screens showing (a) buttons embedded in instructional text; and (b) stand-alone and embedded buttons. (Software by Microsoft.)

applied in industry as the idea of *interdisciplinary teams* working together increases in importance.

In summary, hypertext and hypermedia can do at least two things that are important for training and instructional design. First, they can help us to move from simply *telling* learners about things to *showing* them with vivid imagery and sound. And we can show them not only the information of central concern, but we can allow them access to a great variety of related information, giving them the freedom to choose the material in which they are most interested. Second, with creativity it can allow us to draw learners into the domain and have them *actively* try things out, bringing together declarative ideas and practicing skills. However, this last capability will require more than just an information base. It will require elements of tutorials and simulations. This is the direction in which hypermedia is currently moving.

Simulations

As we saw earlier, simulations are an approximation of some real task or setting. They are used for a variety of reasons including the high cost of real performance, the danger of real performance, the lack of feasibility of real performance (e.g. surgery), or simply the complexity of the real environment (Alessi & Trollip, 1985; Ellinger & Brown, 1979; Hannafin & Peck, 1988). They are also frequently used because a trainee can practice many more times on the simulator than in the real world—certain procedures take too much time in real life. As Price (1991) notes, "Simulations constitute one of the most powerful and potentially valuable applications of CAI. CAI simulation programs are designed to represent the essential elements of some real-life or imaginary event or phenomenon without its attendant hazards and inconveniences." An example of a simulation display for a steam power plant is shown in Figure 9.3.

From a learning perspective, simulations are the next best thing to high-quality on-the-job training. However, certain difficulties with simulations remain. The primary one is that responses from the learner are often far removed from those that would be performed in real life. As an example, consider a trainee learning how to perform CPR. This is a psychomotor skill, and no matter how well the computer presents the simulated scenario, there is no input mechanism for giving mouth-to-mouth resuscitation to a computer (and how would it judge your performance?). Some of these difficulties of input can be overcome by providing special input devices. For example, an instructional system manufactured by Actronics Inc. and described by Van Horn (1991) consists of an interactive videodisc system interfaced with a

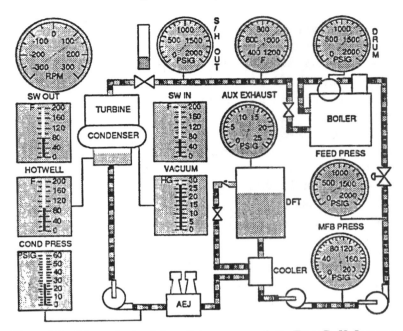

Figure 9.3   Schematic simulation of steam power plant. (From D. H. Jonassen (ed.), *Instructional designs for microcomputer courseware,* 1988, Lawrence Erlbaum Associates. Reprinted with permission.)

CPR mannequin. Students practice CPR on the mannequin and receive feedback from both the computer and mannequin. That is, depending on the correctness of the CPR, the computer will adjust the mannequin's vital signs (dilate pupils, etc.). Another input/output technology that will be able to help in this respect is virtual reality. Finally, we can't overlook the usually high cost of developing computer-based simulations.

Instructional Games

Instructional games are another type of CAI, but are used primarily for teaching basic educational skills such as mathematics. In this method, learners are allowed to play a game as the major instructional activity. It is usually an environment in which there are objects that have to be manipulated, constructed, changed, moved, or some other similar action. They are usually competitive, either against oneself, the machine, or another person.

Instructional games can be both entertaining and educational if designed properly. Games are usually relatively easy to use; however, Hannafin and Peck (1988) have also provided some general guidelines in the construction of instructional games. To be most effective they should (a) attract and retain the interest and enthusiasm of the learner, (b) provide the information necessary for the learner to understand the components of the game (goals, necessary actions, consequences of actions, etc.), (c) provide immediate feedback after learner actions, and (d) reinforce correct behaviors more than incorrect behaviors. This last point was based on the fact that some designers created interesting and elaborate animations that were initiated when a learner made an incorrect response. Clearly, it is not desirable to reinforce the wrong behaviors.

Instructional games have a few problems or weaknesses that should be mentioned. First, it is often difficult to come up with a game that is interesting and also teaches what we need taught. Some also criticize the method because the ratio of learning activity to "fun" activity is sometimes low, hence the method is viewed as inefficient. Probably their biggest drawback from a standard training point of view is that they traditionally have been designed more for younger trainees than for adults. A related difficulty is that they may be limited in terms of the types of knowledge that can be taught. However, it also seems that this approach does have applications for adults if we just think of them. Adults enjoy challenges of skill. As an example, when Army tank gunnerymen were given instructional games requiring them to fire at simulated enemy tanks, they showed high motivation and were extremely competitive in trying to outperform one another. This example shows that as CAI technology advances, we may see an increasingly blurry distinction between instructional games and instructional simulations.

Hybrid Designs

Each type of CAI has its own unique advantages and disadvantages. There are certain circumstances where using one method by itself will fulfill all of the requirements specified in the system design specifications document. For example, a person may need to automate some particular task, and so the training program would implement a simulation of that task with a wide variety of training instances. However, it is often the case where learners need to acquire declarative knowledge and also practice skill acquisition. In these instances, none of the methods described will be as appropriate as a combination or **hybrid** design. One example of a hybrid design is given below.

Case Study: Antibody Identification
for Blood Transfusion

When blood is to be transfused into patients, one task that must be performed is identification of the antibodies that exist in the blood. The choice of blood to be used for transfusion is based on this analysis. Technicians are trained to perform the analysis, but it is complex, difficult, and prone to inefficient or incorrect decision making processes.

A tutorial has been developed at Ohio State University to train technicians to perform the antibody identification task (Miller et al., 1991; Smith et al., 1991a; Smith et al. 1991b). The trainees have already had some instruction on the antibody identification task; the TMT (transfusion medicine tutor) program is designed to augment and extend that training. The designers explicitly ruled out the use of two common CAI instructional strategies. They stated that:

> They could have presented declarative factual information and then tested trainees on that information.

> They could have developed a hypermedia program that would give students freedom to explore a knowledge base (Miller et al., 1991).

The designers ruled out both of these because the focus would be too much on declarative knowledge rather than skill development. Instead, the design concept chosen for the project was a microworld with coaching system (see Chapter 10 for a description of the design concept in this case).

The basic concept of TMT is a simulation or *microworld* that supports the trainee in performing the task. The computer generates simulated cases on which the students practice, receiving coaching when they have difficulties. The designers created screens that captured the important features of the task. One computer screen provides buttons, each listing a specific blood test that the trainee can request. As trainees request tests, the results are shown in a matrix display (see Figure 9.4). The trainee must then identify the antibodies that he or she decides are likely, possible, or ruled out. At this point, the trainee can click on "show interpretation" which will then provide feedback on what an expert would have concluded for the same task. In addition, the trainee can receive additional (declarative) information that explains the reasoning behind the expert's choices.

It can be seen that this instructional system is a simulation combined with performance feedback and tutoring. Distinctive features of the system concern the use of a matrix for subject response that supports small cognitive

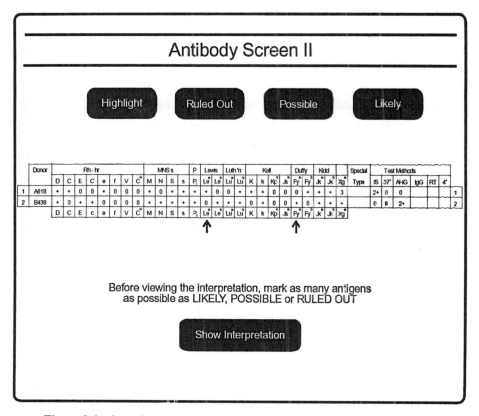

Figure 9.4 One of three screens for TMT; displays results of blood tests and also provides context for problem solving activities of trainee. (Used with permission.)

sub-task level inputs. This provides a mechanism for the tutor to monitor trainee performance at the micro rather than the macro level. This feature then allows for feedback and remediation at the subtask level where it is needed. Such combinations of task simulations with tutoring capabilities are becoming increasingly common.

AUTHORING TOOLS

We have now taken a brief look at the possibilities offered by CAI systems. A reaction that many people express at this point is: OK, it looks good, but exactly how difficult is it to do these things? The purpose of this last section is to briefly describe the various methods for authoring CAI systems, so when you are generating and evaluating design concepts, you will have some idea of the feasibility of different approaches.

Levels of Authoring Tools

There has been a steady progression in the development of authoring tools over the past ten years. Originally, designers had little choice but to use basic programming languages such as C or Pascal to produce instructional systems. However, object-oriented programming and graphical manipulation interfaces have advanced the art of application development. There is now a continuum of tools moving from basic programming languages to icon-driven tools that require no programming. For the sake of convenience, I have divided this continuum into three levels: **base-level** programming tools, **mid-level hybrid** tools, and **high-level graphical interface** tools that do not require programming.

Basic Programming Languages

One method for developing a CAI system is to write the program using one of the standard computer science programming languages. Typical languages include *Basic, Fortran, Pascal, C, C++, Actor Professional,* and *Smalltalk. Basic, Fortran,* and *Pascal* are typical of traditional programming languages where the entire application is developed by writing lines of computer code.

Some tools have been upgraded to make program development easier. For example, *Turbo Pascal for Windows* provides an entire development environment called the Integrated Development Environment (IDE). This includes the programming language along with several tools such as an editor, compiler, linker, debugging tools, and a special toolkit for creating dialog boxes, menus, etc.

A new type of programming language has become increasingly popular because of its greater power and efficiency. It is generically termed **object-oriented** programming, meaning that certain *objects* can be defined within the program, and the attributes defining it are always *attached* to the object and used, no matter where it occurs in a program. Classes of concepts, objects, or ideas can be created, and any instance falling within that class will automatically *inherit* all of the properties and laws that pertain to that class. Object-oriented programming makes "bundles" of information that can be used more efficiently than older types of programming. *C++* is a language that has some elements of object-oriented programming and some elements of traditional "molecular" programming. Other languages such as *Actor Professional* and *Smalltalk* are pure object-oriented languages where everything is an object.

Mid-Level Hybrid Tools

There has been pressure for programming tools that are easier for the layperson to use in developing applications. This has led to a "second generation" of tools that are object-oriented but also are developed via a graphical interface rather than writing lines of code. These programs provide a graphical user interface that allows the designer to create applications by choosing icons or options from menus, and filling in dialogue boxes. To create most applications, these tools require a combination of graphical interface operations, and use of a programming language. One such program is *VisualBASIC* by Microsoft. *VisualBASIC* is a program that allows the user to develop programs using the *Basic* programming language, but implemented via graphic design. The program provides programming design tools such as pull-down menus, scroll bars, dialog boxes, text fields, etc. In other words, the programmer doesn't write "code." There are many instructional design programs that are partially graphical interface and partially programming language, including *Ask Me 2000™*, *Quest™*, and Tencore's *Language Authoring System™*.

There are several programs that were initially developed to create hypertext knowledge bases. However, designers discovered that these tools can be used for a variety of purposes including traditional drill and practice, programmed instruction, some simple simulations, problem solving, creating document generation, etc. The major commercially available programs in this category are *HyperCard™* and *SuperCard™* (for the Macintosh) and *ToolBook™* (for IBMs and compatibles).

HyperCard and ToolBook metaphors. *HyperCard* is a development tool created for Macintosh computers in the mid–1980s. It was welcomed as the first development tool that instructors and trainers could use with little to no computer science background. *HyperCard* is based on a metaphor of "cards." The designer creates a stack of cards, each card containing a relatively small amount of text and/or graphic information. The cards are then electronically linked together with buttons (see earlier discussion under hypertext).

ToolBook was designed by Asymetrix to be an analogous programming tool for IBM compatible systems. The applications are authored in a manner similar to creation of programs using *HyperCard.* The underlying metaphor is that of a book containing pages (rather than a stack of cards). Separate pages are developed, then linked together electronically. Researchers have not tested the efficacy of the two metaphors, cards vs. pages. However, it seems intuitively plausible that cards would be preferable because pages in a book are always bound together in a particular *linear* order, and this property does not transfer to the ToolBook electronic environment.

Authoring in HyperCard (now Supercard) and ToolBook.   Both
programs have two modes of interaction, author mode and reader mode. The mode is chosen by using a pull-down menu at the top of the screen. Figure 9.5 shows the pull-down menu for *ToolBook*. When the user is in author mode, tools for developing applications are displayed (see left side of Figure 9.5). To go to reader mode, the user clicks on the menu option of "Reader." When the program is in reader mode, the user is interfaced with the instructional program without development tools, and depending on the settings of the program, may be locked out from making any changes to the application.

HyperCard, Supercard, and *ToolBook* applications are considered to be object-oriented development tools, and programs are authored by using similar activities. Applications are created by using a variety of input methods, including: Choosing menu options, choosing icons, filling in dialogue boxes, and writing directions in "scripting" language. Figure 9.6 shows examples of (a) choosing icons from a set in HyperCard, (b) a dialogue box for ToolBook buttons with menu options and fill-in forms, and (c) written directions using

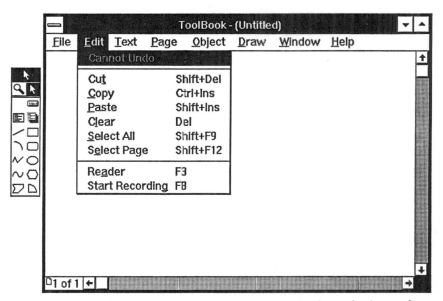

Figure 9.5   Pull-down menu for choosing *Author* or *Reader* mode, Asymetrix *ToolBook*.

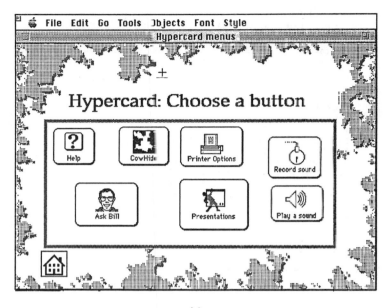

(a)

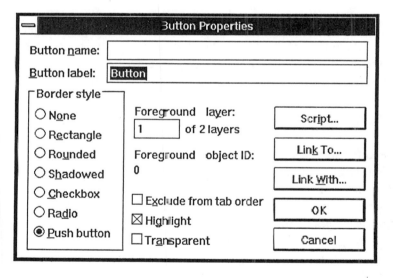

(b)

Figure 9.6   Input methods for using *HyperCard* and *ToolBook*.

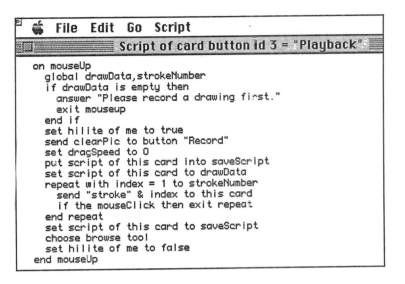

(c)

Figure 9.6 Continued

HyperTalk, the HyperCard "scripting" language. It can be seen that much of the programming work required for using a traditional language has been replaced with easier menu selection and form filling functions.

High-End Graphical Interface Tools

The need for writing "script" (i.e., programming) has been viewed by many as a negative characteristic of the mid-level languages such as *HyperCard* and *ToolBook*. As a result, several authoring tools have been developed that are self-contained, graphical, object-oriented environments. Some of these are easy-to-use programs for creating multimedia presentations, such as Compel™ by Asymmetrix. Others are more powerful instructional design programs such as *Icon Author*™ (for IBM), and *Authorware Professional*™ (for either Macintosh or IBM systems). All of these tools allow the designer to integrate text, graphics, color photos, animation, sound, and full-motion video using an object-oriented graphical interface and no programming whatsoever.

As an example, let's consider Authorware Professional in more detail. It is object-oriented through a graphical interface that eliminates the need for a programming or "scripting" language. The interface for Authorware is based on a flow chart metaphor. The development environment is shown in Figure 9.7. The program begins with a single line representing the "blank" flow chart

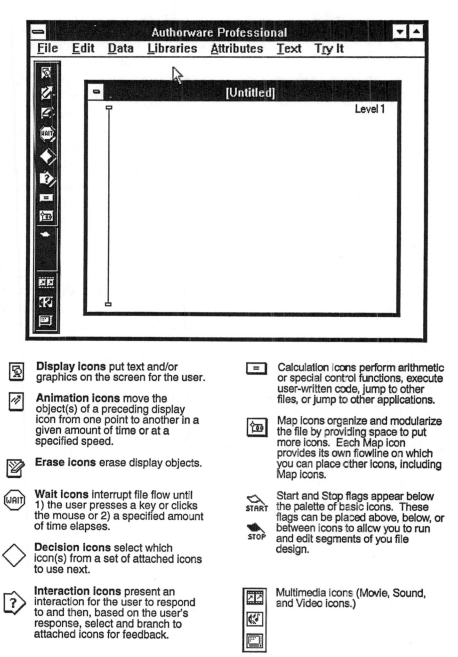

Display icons put text and/or graphics on the screen for the user.

Animation icons move the object(s) of a preceding display icon from one point to another in a given amount of time or at a specified speed.

Erase icons erase display objects.

Wait icons interrupt file flow until 1) the user presses a key or clicks the mouse or 2) a specified amount of time elapses.

Decision icons select which icon(s) from a set of attached icons to use next.

Interaction icons present an interaction for the user to respond to and then, based on the user's response, select and branch to attached icons for feedback.

Calculation icons perform arithmetic or special control functions, execute user-written code, jump to other files, or jump to other applications.

Map icons organize and modularize the file by providing space to put more icons. Each Map icon provides its own flowline on which you can place other icons, including Map icons.

Start and Stop flags appear below the palette of basic icons. These flags can be placed above, below, or between icons to allow you to run and edit segments of you file design.

Multimedia icons (Movie, Sound, and Video icons.)

Figure 9.7 Initial development screen for *Authorware Professional*, with icons defined below.

for a new application. The designer literally builds the application by dragging the icons from the left onto the flow chart line. The icons (defined in the lower part of the figure) allow the designer to directly add in video segments, animation sequences, presentation screens, interactive screens responding to trainee inputs, etc. The program allows building screens with multiple layers, making it easier to develop complex animation sequences, and provides support for integrating still and motion video segments with graphical overlays. These options allow the designer to seamlessly blend real-life images or video sequences with instructional graphics. An example would be, for a physics tutorial, showing a car slowing in a turn and overlaying the changing forces and moments right onto the system.

Comparison of Authoring Tools

Each of the types of authoring tools described above have certain advantages and disadvantages. There is not space in this chapter for an in-depth review and comparison. As a result, I will only present some of the salient differences. Readers can consult Pinheiro and Pet (1991) and other comparative reviews for a more extensive treatment.

Table 9.1 lists the three types of authoring tools in the order that we have discussed them. Generally speaking, we can say that the time and difficulty of using the tools decreases from the basic programming languages to the high-end tools. However, it is also the case that, *generally speaking*, the tools also decrease in flexibility. Thus, a designer wanting to create a simulation such as the TMT tutorial described earlier would have little choice but to use a basic programming language. Such languages are almost always necessary for development of complex ICAI programs. They are also necessary for some simulations. However, we should also note that many simulations can be very adequately created with the more flexible high-end systems. One way to develop effective simulations is by using modeling/rendering/animation packages such as Autodesk's *3-D Studio*™ to create animation segments. These segments are then imported into the instructional application using the authoring package.

COMMENTS

CAI (including ICAI) is still a very young field. While hundreds of studies have been conducted to evaluate the effectiveness and efficiency of CAI methods, we have not yet amassed enough knowledge to determine which types of system are best for the myriad of different training and instructional

TABLE 9.1 Comparison of Authoring Tools

Level	Examples	Characteristics	Uses
Basic	Basic Fortran Pascal C, C++ Actor Prof. Smalltalk	Rely solely on programming Most time-consuming Most flexible Usually least expensive	Simulations ICAI Special purposes

Mid-Level Hybrid	VisualBASIC
	AskMe 1000
	Quest
	LAS
	Hypercard
	ToolBook
		Rely on graphical
		  interface plus
		  programming or
		  scripting
		Moderately flexible
		Moderate time to
		  develop
			Unstructured
			  hypertext
			Static information
			Moderate levels
			  of CAI

High-End Graphical Interface	Compel
	IconAuthor
	Authorware
		Rely solely on graphical
		  interface for design
			Good for most CAI
			  applications

situations. In addition, recent advances in application development tools mean that CAI programs can be relatively easily created that are vastly different from those created only ten years ago. These may be important technological advances because they allow the learner to become immersed in experiential learning that once could take place only in the real world. An example is the new category of instructional programs that teach language by creating interactive video microworlds. Given that there are no set answers or solutions to training needs, it is more important than ever that instructional designers use systematic design and analysis methods such as those described in this book, but also combine them with creativity and vision for "what could be" and not "what has been."

REFERENCES

ALESSI, S. M., & TROLLIP, S. R. (1985). *Computer-based instruction: Methods and development.* Englewood Cliffs, NJ: Prentice Hall.

ANDERSON, C. J., & VELIJKOV, M. D. (1990). *Creating interactive multimedia: A practical guide.* Glenview, IL: Scott, Foresman.

BANNON, C. J. (1991). The Perseus project. In E. Berk & J. Devlin (eds.), *Hypertext/ Hypermedia handbook.* New York: McGraw-Hill.

BERK, E., & DEVLIN, J. (eds.) (1991). *Hypertext/Hypermedia handbook.* New York: McGraw-Hill.

BROWN, B. (1993). *Comparison of instructional strategies for the classroom.* Unpublished master's thesis, University of Idaho, Moscow, ID.

BROWN, J. S. (1990). Toward a new epistemology for learning. In C. Frasson & G. Gauthier (eds.), *Intelligent tutoring systems* (pp. 266–282). Norwood, NJ: Ablex.

CAMPBELL, R., & HANLON, P. (1988). Grapevine: A high-tech voyage through the "thirties." *American Educator,* Winter.

EBERTS, R. E., & BROCK, J. F. (1988). Computer-based instruction. In M. Helander (ed.), *Handbook of human-computer interaction* (pp. 599–627). Amsterdam: North-Holland.

ELLINGER, R. S., & BROWN, B. R. (1979). The whens and hows of computer-based instructional simulations. *AEDS Journal, 12,* 51–62.

GOLDSTEIN, I. L. (1986). *Training in organizations: Needs assessment, development, and evaluation* (2nd ed). Monterey, CA: Brooks/Cole.

HANNAFIN, M. J., & PECK, K. L. (1988). *The design, development, and evaluation of instructional software.* New York: Macmillan.

HEETER, C. (1993). The thin line: Hypermedia meets virtual reality. *Ed Tech Review,* Spring/Summer, 37–46.

HOLSINGER, E. (1991). How to build your own multimedia presentation. *PC World,* November, 250–255.

JAFFE, C. C., & LYNCH, P. J. (1989). Hypermedia for education in the life sciences. *Academic Computing,* September.

JONASSEN, D. H. (ed.) (1988). *Instructional designs for microcomputer courseware.* Hillsdale, NJ: Lawrence Erlbaum Associates.

LYMAN, E. R. (1981). *PLATO highlights.* Urbana, IL: University of Illinois Computer-Based Education Research Laboratory.

MILLER, T., SMITH, P. J., GROSS, S., GUERLAIN, S., RUDMANN, S., STROHM, P., SMITH, J., & SVIRBELY, J. (1991). *Problem-based learning in clinical laboratory science: The role of computerized learning environments.* Technical Report #CSEL 1991–5, The Ohio State University.

MONTAGUE, W. E. (1988). Promoting cognitive processing and learning by designing the learning environment. In D. H. Jonassen (ed.), *Instructional designs for microcomputer courseware* (pp. 125–149). Hillsdale, NJ: Lawrence Erlbaum Associates.

PINHEIRO, E. J., & PET, W. J. A. (1991). *Developing applications for Windows.* The Institute for Academic Technology, University of North Carolina, Chapel Hill.

PRICE, R. V. (1991). *Computer-aided instruction: A guide for authors.* Pacific Grove, CA: Brooks/Cole.

ROBLYER, M. D. (1988). Fundamental problems and principles of designing effective courseware. In D. H. Jonassen (ed.), *Instructional designs for microcomputer courseware* (pp. 7–33). Hillsdale, NJ: Lawrence Erlbaum Associates.

SALISBURY, D. F. (1988). Effective drill and practice strategies. In D. Jonassen (ed.), *Instructional designs for microcomputer courseware* (pp. 103–124). Hillsdale, NJ: Lawrence Erlbaum Associates.

SMITH, P. J., MILLER, T. E., GROSS, S., GUERLAIN, S., SMITH, J. W., SVIRBELY, J., RUDMANN, S., & STROHM, P. (1991a). The transfusion medicine tutor: Methods and results from the development of an interactive learning environment for teaching problem-solving skills. *Proceedings of the Human Factors Society 35th Annual Meeting* (pp. 1408–1411). Santa Monica, CA: Human Factors Society.

SMITH, P. J., SMITH, J. W., SVIRBELY, J. R., KRAWCZAK, D., FRASER, J. M., RUDMANN, S., MILLER, T. E., & BLAZINA, J. (1991b). Coping with complexities of multiple-solution problems: A case study. *International Journal of Man-Machine Studies, 35*, 429–453.

SOLOWAY, E. (1991). Quick, where do the computers go? *Communications of the ACM, 34* (Feb), 29–33.

SOROHAN, E. G. (1993). We do: Therefore, we learn. *Training & Development*, October, pp. 47–55.

TODD, D. (1991). HyperActive: The story of The HyperMedia Group. *New Media, 6*, 60–62.

VAN HORN, R. (1991). *Advanced technology in education.* Monterey, CA: Brooks/Cole.

ADDITIONAL RESOURCES

CAI and ICAI

DEBLOOIS, M., MAKI, K. C., & HALL, A. F. (1984). *Effectiveness of interactive videodisc training: A comprehensive review.* Falls Church, VA: Future Systems Inc.

CLARK, R. E., & SALOMON, G. (1985). Media in teaching. In M. Wittrock (ed.), *Handbook of research on teaching* (3rd ed). Chicago: Rand McNally.

CONKRIGHT, T. D. (1984). Linear, branching, and complex: A taxonomy of simulations. *Training News, 6(3)*, 6–7.

JOHNSTON, J. (1987). *Electronic learning: From audiotape to videodisc.* Hillsdale, NJ: Lawrence Erlbaum Associates.

KELLER, A. (1987). *When machines teach: Designing computer courseware.* New York: Harper and Row.

SALOMON, G. (1979). *Interaction of media, cognition and learning.* San Francisco, CA: Jossey-Bass.

SHLECHTER, T. M. (Ed.) (1991). *Problems and promises of computer-based training.* Norwood, NJ: Ablex.

WALKER, D. F., & HESS, R. D. (Eds.) (1984). *Instructional software: Principles and perspectives for design and use.* Belmont, CA: Wadsworth.

Hypermedia and Multimedia

AMBRON, S., & HOOPER, K. (eds.)(1990). *Learning with interactive multimedia: Developing and using multimedia tools in education.* Redmond, WA: Microsoft Press.

ARWADY, J., & GAYESKI, D. (1989). *Using video: Interactive & Linear Design.* Englewood Cliffs, NJ: Educational Technology Publications.

BARRETT, E. (1992). *Sociomedia: Multimedia, hypermedia, and the social construction of knowledge.* Cambridge, MA: MIT Press.

BEEKMAN, G. (1990). *HyperCard in a hurry.* Belmont, CA: Wadsworth.

BERGMAN, R., & MOORE, T. (1990). *Managing interactive video/multimedia projects.* Englewood Cliffs, NJ: Educational Technology Publications.

FLOYD. S. (1991). *The IBM multimedia handbook.* New York: Brady Publishing.

FRANKLIN, C. (1989). Hypertext defined and applied. *Online,* May, 37–48.

FUTURE SYSTEMS INC. *Multimedia & Videodisc Monitor.* A catalog for books, discs, and other products relevant to interactive multimedia. Falls Church, VA: Future Systems Inc.

GRIFFIN, S. (1991). Multimedia computing in higher education. *IAT Briefing,* Winter, 13–14. Institute for Academic Technology, University of North Carolina, Chapel Hill.

HAYNES, G. R. (1989). *Opening minds: The evolution of videodiscs and interactive learning.* Dubuque, IA: Kendall Hunt.

HYPERMEDIA COMMUNICATIONS INC., *New Media.* A journal focusing on multimedia business and educational products. Hypermedia Communications Inc.

IMKE, S. (1991). *Interactive video management and production.* Englewood Cliffs, NJ: Educational Technology Publications.

IUPPA, N., & ANDERSON, K. (1988). *Advanced interactive video design.* White Plains, NY: Knowledge Industry Publications.

JONASSEN, D., & MANDL, H. (Eds.) (1990). *Designing hypermedia for learning.* New York: Springer Verlag.

KINZIE, M. B., & BERDEL, R. L. (1990). Design and use of hypermedia systems. *Educational Technology Research & Development, 38,* 61–68.

LOCATIS, C., LETOURNEAU, G., & BANVARD, R. (1989). Hypermedia and instruction. *Educational Technology Research & Development, 37,* 65–77.

LUTHER, A. C. (1989). *Digital video in the PC environment.* New York: Intertext Publications and McGraw-Hill.

NIX, D., & SPIRO, R. (eds.) (1990). *Cognition, education, & multimedia.* Hillsdale, NJ: Lawrence Erlbaum Associates.

PERLMUTTER, M. (1991). *Producer's guide to interactive videodiscs.* White Plains, NY: Knowledge Industry Publications.

RICHARDS, T., CHIGNELL, M. H., & LACY, R. M. (1990). Integrated Hypermedia: Bridging the missing links. *Academic Computing,* January, 24–44.

SCHWARTZ, E. (1987). *Educator's handbook to interactive videodisc* (2nd ed.). Washington, DC: Association for Educational Communications and Technology.

VERBUM, *The Verbum Book Series.* A series of how-to books for new media art and design. San Diego, CA: Verbum.

VIDEOSOFT PUBLISHING. *Video Computing: The Journal of Interactive Video Technology.* A journal that acts as a "digest" of other sources. India antic, FL: Videosoft Publishing.

Virtual Reality

HELSEL, S., & ROTH, J. P. (eds.) (1991). *Virtual reality: Theory, practice, and promise.* Westport, CT: Meckler Publishing.

KRUEGER, M. (1991). *Artificial reality II.* Reading, MA: Addison-Wesley.

10

Procedure: Developing the Design Concept

Design concepts can be generated in one of two ways. In the first, the designer works with the objectives, requirements, and constraints all at once. That is, you try to generate ideas that fit within all of the information in the system design specifications. However, this approach has the disadvantage of restricting creativity early in the process, which often means that fewer good ideas are generated. You might have a really good idea that doesn't fit within your constraints, so you wouldn't even consider such an idea. Sometimes what seems to be an unworkable idea in the beginning can be reshaped later to fit within the constraints.

The second approach is to first generate ideas, no matter how impractical or unrealistic, that accomplish what you need. These ideas are listed without criticism. Later, the ones that seem to be really good or hold promise can be evaluated to see if there is any way to modify them to fit within the constraints. Or alternatively, the initial "wild" ideas can act as a springboard for thinking of analogous strategies that *would* fit within the constraints. In this chapter, we consider this second approach.

DESIGN CONCEPT GENERATION

The approach described in this section consists of three steps: generating ideas for the design concept, bringing the ideas into line with constraints, and filling in the concepts.

Step 1: Generate General Idea of Instructional Design

The first step in generating design concepts using this approach is to focus only on *what you are trying to accomplish.* Don't worry about being realistic, simply brainstorm and try to come up with ideas for a system that would be exactly what you need. While brainstorming is usually an activity that is done in groups, it can actually be done individually *or* in groups.

Brainstorming is the process of trying to come up with as many ideas as possible for some system. Each person should try to generate ideas for a program that perfectly meets all of the objectives and requirements. In other words, each person should answer the question: "What would we REALLY like our project to do and look like?" (Anderson & Veljkov, 1990).

One person should list all of the ideas on a board or large floor chart, in a place where everyone can see it. List all ideas, no matter how far-fetched. No one should throw out or criticize any of the ideas; try not to do this even to yourself. Simply identify the best way to do it; try to generate new creative methods. Osborn (1957) gives the following suggestions:

1. Evaluation of any idea is not permitted. All ideas, no matter how unusual, are valid and should be recorded. This rule eliminates any need for personal defense of an idea and promotes group acceptance.
2. The group is encouraged to think of the wildest ideas possible. Any creative idea is easier to tone down than to build up. Rule two also discourages placing value judgments on creative ideas.
3. The quantity of ideas is more important than the quality of ideas. The more ideas recorded, the more likely the group will find the most creative solution to the problem.
4. Team members should build upon, modify, or combine ideas. Synthesis of previously suggested ideas will create new ideas that are often superior.

Although these suggestions are useful, the design team must still exercise caution. Research has shown that even using the techniques suggested by Osborn, group dynamics are simply so powerful that they still inhibit creative problem solving (Diehl & Stroebe, 1987).

Given what we have seen from the research literature, team members should focus on the development of interactive, generative activities that involve learners in *doing things*. This is especially critical for lecture and CAI activities to avoid falling into traditional approaches of "giving knowledge to passive learners." Anderson and Veljkov (1990) have described the need for "imagineering." It means the creative use of technology to go beyond what has been done before. Make sure that design ideas follow the guidelines maximizing learning given in Chapter 6.

After team members have generated what seem to be a number of good ideas that meet the current instructional needs, the team discusses them and chooses one or two that seem to be particularly good. These will become the first choices for evaluating the concepts and modifying them to fit within constraints. Before we discuss the second step, we will address the occasional need to work within certain types of constraints.

Working within set design constraints.   Occasionally, you will be given absolute design constraints. For example, you have been asked to design an owner's manual to accompany a piece of equipment, and management is absolutely set on a paper document. If that is the case, it is most efficient to work within that particular constraint, but try to think of new or different approaches to that design solution rather than just ones with which you are familiar.

Using a decision matrix.   There are many types of instructional techniques, not to mention job aids and performance support systems. Sometimes it is difficult to determine which of a number of alternative design solutions would best meet your criteria (as set out in the functional specifications). A decision matrix can help in identifying the optimal design. To use a decision matrix, one might follow a procedure something like this:

1. First **rule out** any design solutions that are not feasible. For example, you might rule out computer-based systems because of a lack of time and funds (or trainees won't have access to computers), job aids or performance support systems because trainees are dealing with people and would not have the opportunity, OTJ training because the tasks are too hazardous, etc.

2. Make a table with the following headings (or something similar)

- General Method
- Specific Instantiation
- Advantages
- Disadvantages
- Possible changes

3. Write the various techniques that have not yet been ruled out, such as lecture, simulation, etc., down the left side of the table under General Method. In the next column, Specific Instantiation, specify how you might use that technique for the particular project under consideration. This might be done several times for one technique, if it could be implemented in more than one way. In the third and fourth columns, write the advantages and disadvantages of that particular instantiation. In the fifth column, write down modifications to the method that would decrease the disadvantages (if there are any).

4. Look for a method that has relatively few disadvantages. Otherwise, look for combinations of methods that cancel out each other's disadvantages (such as lecture with computer-based simulation).

More complex decision matrices can also be developed. It is possible to list a variety of attributes across the top, such as cost, development effort, delivery constraints, equipment requirements, etc. (e.g., Marrelli, 1993).

Step Two: Adjust the Design Concept(s)

In this step the analyst or design team takes the best one or two design options and tries to modify them to work within the constraints identified in the system specifications. If this effort is not successful, then the team should try to generate an instructional design concept that does meet the design constraints, but that also has the desirable characteristics of the original design.

Anderson and Veljkov (1990) describe the following questions as useful:

- If our technical specifications won't support what we REALLY want to do, how can we find another way to achieve the same results?
- Will our efforts enhance or detract from the content and objectives of our project?
- If some ideas detract, can we alter these ideas to enhance our project?
- If our ideas seem wild to us, would they be inappropriate for our end users?

- Is there other technology that might benefit our project?
- How would this other technology enhance our project?
- Will it be enjoyable?

In summary, the design team identifies a concept that seems to work, and perhaps identifies a backup concept.

Step 3: Fill In Design Concept

The third step is to fill in the design concept, getting ready for the prototyping stage. That is, specify how different parts of the instructional strategy will be carried out. If more than one strategy will be used (i.e., simulation and tutoring), specify what content will be addressed by each, or how the different strategies will function differently. Also specify any job performance aids and other pertinent information. Some general guidelines for the more common forms of instructional designs are given in the following sections.

Documents.   Determine the general nature of the document, for example, is it to be a bound text, booklet, one-page brochure, etc. Specify the type of organization, the *type* of information to be presented, and the *impression* you wish to give to the learner. Identify what type of pictures, illustrations, or figures can be used to convey that impression.

As an example, imagine that you have decided to produce an informational brochure on your company (you do multimedia instructional development). You want to get peoples' interest, so it should be visually interesting and appealing. You will do this with a glossy paper format and color photos. You want the image to be upbeat and dynamic, so you decide to format some of the text and photos on the diagonal. The photos show people using high-tech equipment to produce your products and others interacting with your products. The blocks of information are short, and are designed to show the success of your products and the enjoyment that they bring to people. To show your expertise in graphical design, a pale colored computer-generated graphic runs transparently through the materials. Finally, a contact person is clearly blocked out in an almost luminescent fashion.

Lecture or workshop.   A lecture design concept will be relatively simple, unless other methods are combined with it (which hopefully they will). For this strategy, specify who will do what, where, and when. Determine the students per instructor ratio, and identify any equipment or materials needed. Identify the general structure of the information, and how it will be presented (handouts, chalkboard, etc.). Explicitly determine what methods will be used to convey the information (such as advance organizers or instructional objectives), and also how motivation will be increased. If additional teaching methods are to be used, briefly describe these as well.

As an example of a design concept for the lecture/workshop, imagine that you are asked to give a two-hour workshop on how to use your new instructional design software you have developed. You decide to hold the workshop in a room that holds 40 people. Twenty computers will be set up on the tables so that attendees can work in pairs. You will:

1. Spend 30 minutes lecturing on the basic purpose of the software, followed by a description of the interface metaphor, the basic functions, and how to use them.
2. Spend 20 minutes walking participants through a simple tutorial on developing an application.
3. Have a five-minute break.
4. Have a one-hour open session where participants can work on their own with the software. To assist, there will be five or six assistants roaming through the classroom. There will also be a simple performance aid card listing the most common user goals (determined through pretesting) followed by methods for trainees to achieve those goals.

CAI and simulations. For computer-based programs, a design concept should be a bit more worked out. Start with the basic system, its general structure, what type of components, etc. If the system will provide information, then questions with remedial tutoring, specify how this type of system will engage and interest learners. What methods will be used to support learner goals at various junctures? Are there any metaphors being used? If there is a simulation, how will it be carried out, and what major characteristics of the task must be captured?

As an example, consider the TMT tutorial described in a previous chapter. This system places the trainee in a simple simulation that captures the essence of the task (blood antigen testing) that is to be performed, and also provides tutoring during simulated task performance. In considering various design concepts, the system designers (Miller et al., 1991) explicitly ruled out some of the more traditional approaches for definite reasons:

> At a conceptual level, a critical decision focused on the general approach to tutoring. We could have taken a traditional approach to computer-aided instruction in which declarative (factual) knowledge is first presented (often ac

companied by supporting graphics), and the student is then queried about this knowledge (or alternatively, the student is queried before the correct answer is presented) .... A second possible approach would have been to develop a hypermedia environment which would give the student much more control over the flow of material. The primary focus, however, would still have been on presenting declarative knowledge, rather than on embedding instruction in the practice of actual problem-solving tasks.

A third approach, and the one we chose to explore, is the design of a true problem-based learning environment in which the computer plays an active role in tutoring the student. Specifically, we chose to design a coaching system which:

1. Allows the student to get practice solving cases within a computer-supported microworld;
2. Provides coaching in the form of suggested problem-solving strategies, immediate feedback in response to errors made by students, and a case summary indicating how an expert might solve a particular case. (Miller et al., 1991)

Additional elements in the TMT design concept included the need to provide a microworld which captured the important characteristics of the real-world task, and to develop "a vocabulary for interactions between the student and the computer which both supported the problem-solving activities of the student and provided the computer with data about the student's thought processes" (Miller et al., 1991).

After getting this far in the design concept, the team or analyst should specify how the interface will work within the context. How will the task be represented (what characteristics will be captured by a simulation)? What types of interface mechanisms will be used (e.g., hypertext, menus, icons with buttons, etc.). Some basic design decisions should be made at this point so that prototyping can proceed.

OTJ training.   Design concepts for on-the-job training are not usually developed, but should ideally be created just as in any other program. It may be more important to design and develop OTJ programs because the instruction is typically carried out by a worker without expertise in training. As in lecture, the design concept specifies who, where, when, and what knowledge and tasks are to be taught. Materials should be provided to the trainer, if only a task list of items to include in the program. This list can be developed in conjunction with trainers and management, and should be directly tied to the material in the task analysis. Specify the methods to be used for training, including what type of practice and feedback mechanisms are to be implemented.

As an example, a full-time secretary might be asked to train a new half-time secretary who will be helping him or her. The design concept might specify that:

1. The existing full-time secretary will provide the training.
2. The trainer will be given a task checksheet; all tasks must be performed correctly for the trainee, followed by trainee performance of the task.
3. Training will continue until the trainee can perform all tasks on the checksheet without error or advice; the trainer is to check off each task.
4. The trainee must pass an exit exam to be developed on basis of instructional objectives.

PRODUCT DEVELOPMENT PLAN

Now you are almost ready to develop the program. Even if you have been on a budget and timeline up until this point, you will need to pin things down more firmly. Since you now have an idea of what the instructional program will look like, it will be easier to specify resources, cost, and time for completion. If necessary, develop a project plan that specifies:

- Personnel
- Budget
- Equipment
- Outside services
- Tasks to be accomplished, by whom
- Task timeline

As examples, Table 10.1 shows a sample project budget form, Figure 10.1 shows an example of a task worksheet with personnel allocations, and Figure 10.2 shows part of a task timeline.

OBTAINING PROJECT APPROVAL

Once the design concept has been fairly well worked out along with a development plan, this is a good point to check with management to ensure that you are not wandering down a dead-end path. That is, make sure that what you think is a good design concept is also a good design concept in the eyes of people who asked for the program in the first place. Hopefully, the organizational analysis will have revealed any biases or constraints with respect to the types of programs acceptable to managers and trainees. But occasionally you

TABLE 10.1   Example of a Budget Form for a Multimedia Program

ITEM	$/Hr	# Hours	Total Cost
1. Design Team Salaries Team leader/director Instructional designer Writer Subject matter expert Video producer			
2. Consultant/Specialist Fees Graphic designer Programmers Actors Technicians			
3. Program Design Facilities Phone Computers Other multimedia equipment Travel			
4. Audio/Video Facility Audio equipment Video cameras Studio equipment Editing equipment			
5. Materials Computer software Videodiscs Computer discs or other storage media			
6. Videodisc Production			
7. Documentation Development			
8. Delivery Equipment			
9. Packaging Costs Personnel Materials			
10. Dissemination and Service			

TASKS	PRIMARY PERSONNEL REQUIRED			
	Director/Facilitator	Technical Director	Writer/Designer	Video Producer
Identify Goals	X	O	X	O
Assess Needs	X	O	X	O
Define Objectives	X	O	X	O
Develop Budget	X	O	O	O
Global Visual Overview (GVO)	X	O	X	O
Writing Scripts	O	O	X	X
Video Production	O	X	O	X
Course Authoring	O	X	X	O
Packaging/Dissemination	O	O	X	O

TASKS	SUPPORT PERSONNEL REQUIRED					
	Content Specialist	Instructional Designer	Graphic Designer	Programmers	Actors	Technicians
Identify Goals	X	X				
Assess Needs	X	X				
Define Objectives	X	X				
Develop Budget						
Global Visual Overview (GVO)	O	O				
Writing Scripts						
Video Production					X	X
Course Authoring			X	X		
Packaging/Dissemination			X			

X - Indicates Primary Involvement
O - Indicates Possible Involvement

Figure 10.1 Example of a task list with personnel allocation. (From C. Anderson and M. Veljkov, *Creating interactive multimedia*, 1990, Scott Foresman. Reprinted with permission.)

will have failed to uncover all existing biases, and you may still get some surprises at this time.

When you preview the idea with managers or other administrative personnel, explain the rationale behind your concept, and why other (usually less expensive) methods are less acceptable. If the concept is still rejected, you must go back to the drawing board. But remember, you have a lot of empirical

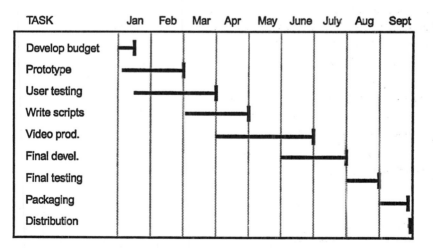

Figure 10.2 Sample of a task timeline.

research upon which you should be able to draw to argue for an instructionally sound program

It is also a good idea to get the reactions of several representative trainees. Ask them for their initial reaction, and ask them to evaluate the idea relative to some alternatives. To get reliable and valid feedback from trainees, you will have to be able to describe or show them a fairly concrete design concept. In some cases, such as interactive video, it may be too difficult to convey the design concept at this point. In this case, simply wait for the user testing phase. Getting initial reactions to your program is informative for two reasons. First, if reactions are strongly negative, you can find out why. There may be ways to overcome this reaction, or you may decide to identify a different design concept. The second reason is that even if you stick with the current concept, you will know the initial reaction of trainees when they first hear of the program or begin training. This will help you anticipate any negative reactions and hopefully design the program to counteract these attitudes.

SYSTEM DESIGN SPECIFICATIONS

At this point, if you are developing a project that will require work from other people, you will probably need to write detailed design specifications. This is a document that builds on your functional specifications. Adding to those items, you will describe all of the components in the training program or job aid. This includes hardware, software, interface characteristics, locations, and any other relevant details of the program.

There was a time when the design specifications were written in great detail before development of a program, especially a computer program, was begun. However, with advances in technology, it is often more advisable to perform iterative prototyping and user testing before lengthy design specifications are written. With this in mind, it is justifiable to write the design specifications in just enough detail that other people can work on the project and get the prototyping started. Alternatively, you may be able to do the prototyping yourself and then write the design specifications after some preliminary design decisions have been made.

A CASE STUDY

In this section, I will briefly describe a partly fictitious example of the design concept generation process. One reason for presenting it is that this particular case is typical of what a designer might go through in overcoming initial biases.

Laura Smith works for a social service agency that provides services for cognitively disabled adults. One of the state mandates is that people with cognitive disabilities become more self-sufficient through training and home service. Laura has been asked by her supervisor to develop some training programs for several adults who have been referred to their agency. These adults range in capabilities, but most live with family members or a person who does the majority of the caregiving. While Laura hopes to develop a wide variety of programs in "living skills," she has decided to focus first on basic cooking. A trainee analysis revealed that almost all of the adults serviced by her agency would like to be able to do more in the kitchen, and would like to learn basic skills in following a recipe.

Laura's supervisor has just acquired a new personal computer with multimedia capability for the office. He makes it clear to Laura that he is very much in favor of the idea of multimedia training at their center. Laura begins development of the training programs with the idea that she will create a multimedia training program for each topic. The first topic, following a simple recipe, will serve as a model for units to follow.

Laura routinely performs the needs assessment tasks: organizational analysis, task analysis, and trainee analysis. During the task analysis, she identifies several cooking tasks that are positively valued by the trainees. One of these tasks is that of making a hamburger. As Laura delves into the task analysis further, she realizes that there are certain dangers associated with using kitchen appliances, and certain potential hazards associated with cooking a hamburger in a skillet. If the contents of the pan were to catch fire, the trainee must know the proper actions to take. Laura decides to teach basic safety skills along with the cooking skills. When she teaches the task of cooking a hamburger, she will also teach safe procedures for dealing with a grease fire on the stove.

Laura now comes to the step where one would normally generate the design concept. She knows that her supervisor would like a computer-based training program using multimedia. However, she also considers the following information:

- The task of dealing with a grease fire is a critical one, with serious consequences of an error.
- Dealing with a grease fire may occur while a person is performing some other task (high cognitive workload), and require swift action.
- The training context for learning the safe procedures should be very similar to the actual task performance environment, in order to maximally cue the behavior later.
- The trainees have lower cognitive capabilities than the average person.

While it would be possible for Laura to develop a multimedia training program on the safety procedures, she thought that there were many problems with the approach. The disadvantages or problems she identified with the use of a multimedia training program were:

- First, the computer context and required actions would be very different from the trainee's home environment and required actions, allowing for little *transfer of training*. A more similar context, providing greater positive transfer of training, would be to provide the instruction in the person's own kitchen, or at least in a real kitchen at the center.
- Second, Laura was worried that the diminished cognitive capabilities of her trainees might prove a problem in multimedia instruction. That is, if a sizable proportion of a trainee's attention and/or cognitive resources were taken up with just the task of interacting with the computer, there might be problems with the person comprehending and especially *storing* the safety information.
- Finally, Laura was worried that the trainees might learn the information but not retain it over a sufficiently long period of time.

For all of these reasons, Laura realized that a multimedia training program was probably not the best answer for the kitchen safety training portion of the program. She chose to develop a workshop to be attended at the center by a small number of trainees and their primary care providers. Laura would provide a realistic context and cues, model the behavior with explanations, and each trainee would practice the proper actions. Laminated job aids would be printed with large letters and given to trainees to hang in their kitchen. The care providers would be asked to review this information with the trainees on a periodic basis, and help them practice the correct procedures.

FINAL COMMENTS

While there are many final considerations that we could review at this point, I will focus on two. First, during the design process, always keep in mind the thing that you identified as your **goal** for trainees or learners. The lure of using computer-based instruction can be strong, and sometimes a person is tempted to use that design solution because it is easiest to sell to management. However, remember that the best design solution is the one that meets the needs of the employees. That may be something high tech, or it may be something very simple.

Don't hesitate to use your prospective trainees as a resource throughout the design process. During the design generation, you can ask, "What
would you think of X?" Or, "Without knowing much about it, do you think
you would rather have X or Y?" Even if you don't go with their opinion,
you will have more information as to what their first gut-level reaction will
be when they are introduced to the program. You can also get a feel for
what will interest or motivate them, and these are critically important
factors.

SUMMARY

Development of the design concept for a training program or performance
support system can be accomplished in a variety of ways. In this chapter, we
reviewed a method whereby the analyst brainstorms some initial concepts
without criticism, and then works with those ideas. The entire sequence of
steps is as follows:

1. **Generate several basic design concepts.** Use a decision matrix or other
   device to identify what seems to be the best design concept or combination.
2. **Adjust the design concept** and check to determine that it meets requirements and constraints in the functional specifications.
3. **Fill in the design concept** with more details.
4. **Write a product development plan,** and include necessary information
   such as timelines, budgets, etc.
5. **Obtain project approval,** and check out with users.
6. **Write detailed specifications if required** for system development.

REFERENCES

ANDERSON, C. J., & VELJKOV, M. D. (1990). *Creating interactive multimedia.* Glenview, IL: Scott, Foresman.

DIEHL, M., & STROEBE, W. (1987). Productivity loss in brainstorming groups: Toward the solution of a riddle. *Journal of Personality and Social Psychology, 53,*
497–509.

MARRELLI, A. F. (1993). Determining costs, benefits, and results. *Technical and
Skills Training,* November/December, 8–14.

MILLER, T. E., SMITH, P. J., GROSS, S., GUERLEIN, S., RUDMANN, S., STROHM, P. L., SMITH, J. W., & SVIRBELY, J. R. (1991). *Problem-based learning in clinical laboratory science: The role of computerized learning environments.* Technical Report #CSEL 1991–5, The Ohio State University.

OSBORN, A. F. (1957). *Applied imagination.* New York: Charles Scribner's Sons.

11

Background: Document and Computer Interface Design

Many instructional programs and performance support systems rely heavily on hardcopy text and/or computer-based informational systems. In this chapter, I provide some basic principles and guidelines for the design of written documents such as manuals and job aids, and for the design of computer interfaces. There have been entire books written on each of these topics, so the coverage here will obviously be incomplete. Readers are referred to several outside resources, especially Weiss (1991) and Gottfredson and Guymon (1989) for user documentation, and Helander (1988) and Mayhew (1992) for computer interface design.

DOCUMENT DESIGN

There is currently a trend toward storing documents on the computer in addition to, or in place of, hardcopy text. Even with this change occurring, we can say that both forms will be used for instructional systems in the foreseeable future. If nothing else, the frequency of receiving warning literature and operating instructions with new equipment is increasing, and the convenience of hardcopy text cannot currently be replaced. On the other hand, there are instances such as maintenance of large and complex equipment, where it is actually more convenient to take one portable computer than several volumes of repair manuals.

Instruction and Job Aid Documents

Some workers still perform relatively simple psychomotor tasks (such as moving a product off of an assembly line), but many are facing the operation and maintenance of increasingly complex systems. Some tasks are performed quite infrequently with these systems. This makes training difficult because trainees tend to forget relevant knowledge and procedures. Instructional designers have at least four choices:

1. Insist that the equipment and tasks be designed so that no instruction is necessary. In such a situation, correct task performance or system operation is apparent and the person can do the task without initial instruction or later help.
2. Design the task so that once a person has learned it they will remember it, even after long periods of time.
3. Provide a job aid that is sufficient for "training."
4. Provide instruction initially, then provide some type of prompt or reference information that can support job performance later.

Many instructional texts are designed in a manner consistent with this last approach. That is, they serve as an instructional document *and* as a job aid. Other instructional programs are developed so that one text serves as the initial instructional document and another serves as a job aid. Computer program documentation often comes packaged this way as a set of components, each of which serves a separate purpose. A prototypical example is a word processing program that comes with:

- A hardcopy tutorial
- An interactive computer-based tutorial
- A hardcopy user's guide (reference manual)
- A quick reference guide (short job aid)
- An on-line help system (usually a short reference manual in hypertext format).

This example illustrates that job aids are increasingly being provided along with instruction of other types, including longer documents such as user's guides. While instructional documents and job aids are different, there are some principles of document design that apply to both. We will briefly review these concepts and then discuss job aid design as a separate issue.

Factors to Consider in Document Design

User biases. Before reviewing some of the basic principles of document design, we should note an important characteristic of people learning a task, or learning to use equipment or software. People often have a "hierarchy" of information source preference. That is, if given a choice of where to go for information, they will generally choose people first. If there is not a person readily available to ask for help or information, they will next go to a written document. If that is not available, they will use computer-based information (unless it has some advantage over the paper document). In general then, we can say that the order of preference is people, paper, computer (Bailey, 1989).

A related finding is that people have a generally negative reaction to instructional documents that are overly long (Conrad, 1962). By rewriting instructions to include only information directly related to the task, Conrad showed that learners had a better attitude toward the material, were better able to learn the information, and as a result were better at performing the task.

The fact that people react negatively to large amounts of documentation and tend to not use it is a fact that computer scientists have learned the hard way. Commercial software for the general public is successful to the extent that a person can sit down and use it; not necessarily all of its power or functions, but at least the basics. Research has consistently shown that people prefer to learn software by trying it out rather than looking things up (Carroll & Mack, 1984; Carroll, Mack, Lewis, Grischkowski, & Robertson, 1985). This finding was the basis of a "minimalist" approach to instruction (Carroll & Carrithers, 1984; Carroll, Smith-Kerker, Ford, & Mazur, 1986). In this approach, the learner is given only the minimal amount of instructions necessary to get started. The system itself is designed to support exploratory learning, where performance errors cannot result in any seriously negative consequences (Catrambone & Carroll, 1987).

Document design process. Training systems in the form of a document may be as short as one paragraph of instruction or as long as a complex textbook. Many of the most commonly marketed "instructional texts" consist of the owner's manuals that accompany new equipment. The development of any of these types of documents should follow the design guidelines we have discussed throughout this text. That is, the training analyst should first conduct a cognitive and behavioral task analysis to identify the material that might be included. This information is then narrowed down to only what is needed for the learner to gain the necessary knowledge and skills. This is a judgment call, and may require user testing during the development phase. When there is a wide range of information need (e.g., some users need minimal knowledge and others need large amounts of detail), it is better to provide both but in a format that segregates the levels of detail. An example of this will be described shortly.

In our experience, the majority of instructional documents are a combination of *declarative conceptual* knowledge, such as how a system works, and *declarative rule* knowledge, the rules and procedures for carrying out a set of tasks. One difficulty in document design is determining how to integrate these two types of knowledge. This is an issue that has not received a great deal of empirical research. However, Bailey (1989) recommends placing basic conceptual information adjacent to the procedures with which it is associated.

In addition to identifying the instructional material through the task analysis, it is important to identify appropriate warning or safety information. This can be done by consulting an expert in the field of safety who is familiar with the research literature as well as ANSI, OSHA, and other standards. (*Training Requirements in OSHA Standards and Training Guidelines* is available from the Superintendent of Documents, U.S. Government Printing Office, Washington, D.C. 20402-9329.)

Next, the analyst must identify the document's intended use and design it accordingly. For example, if it is expected that the trainee will read the document once, such as for the assembly of new equipment, it will be written differently than if it is to serve as a reference document, as in an automobile owner's manual. One way to approach this task is to develop a hierarchy of the goals that the reader might have at any time while interacting with the document. The major groupings and headings can be designed to map onto those goals, making it easier for the user to access needed information. One way to determine how people cognitively cluster ideas and goals is to use the sorting and multidimensional scaling methods described in Chapter 4 (Morrow, Leirer, & Andrassy, 1993).

The following sections provide general guidelines for document design. Whether the document is for initial instruction, a job aid, or both, it should be written in a manner that is consistent with good principles of design and submitted to *user testing* before release.

Principles of Document Design

Before designing the document, consider the users and tasks carefully. In discussing the methods for writing software documentation, Sheppard (1987, p. 1543) suggests asking oneself the following questions:

1. What information is to be documented? Are there some areas of information that should be emphasized more than other areas? It is important to focus on the objectives of the document.

2. What is the nature of the audience? How abstract should the presentation be? Potential users of the manuals should be identified, and the contents should be tailored to their skill levels and requirements.

3. What kinds of tasks does the user need to know how to do? Instructions need to be clear and complete. Errors should be anticipated, and procedures should be provided for actions to amend errors.

4. How will the manual be used? Training workbooks will have different contents and be organized differently from reference books.

5. Will reference materials be readily available? If not, the manual must be self-contained.

6. In what environment will the manual be used? Desks and workstations vary in available surface area, and the size and shape of documents should be varied to accommodate the environment in which the documentation will be used.

7. What is the budget for producing the documentation? How many people can be supported and at what level? What is the time frame for completing the documentation?

8. Who will evaluate the manual for accuracy, completeness, and clarity of presentation?

9. How will the document be revised and maintained? It is important to provide some mechanism for periodic evaluation and for making changes to the document after it has been tested and used.

Having answers to these questions will provide a good starting point for the general concept and outline of your document.

In addition to these considerations, you can write your document by following some standard guidelines. There are numerous sources for principles of document design. These sources include Adams and Halasz (1983), Bailey (1989), Hartley and Burnhill (1977), Macdonald-Ross (1977), Sheppard (1987), Strunk and White (1972), and Wright (1988). Gottfredson and Guymon (1989) offer a thorough and very usable *style guide* for writing software documentation. The following material includes some of these guidelines plus suggestions based on recent findings in education and psychology. The items can be used to guide document development and/or as a postdevelopment checklist.

TYPOGRAPHICAL AND FORMAT FEATURES

- **Type size** should be between 9 and 11 points (Tinker, 1963).

- **Use type styles that are in common usage,** such as courier, times roman, helvetica, etc.

- **Use upper- and lower-case letters;** the use of only capital letters has been shown to slow down reading.

- Boldface can be read as easily as regular type, although readers tend to like the use of **boldface only for emphasis.**

- **Don't use more than two or three typefaces** counting headings and titles, but not counting italics.

- **Use ample margin widths** (e.g., at least one inch), especially on the inside of book pages because curvatures cause reading difficulties.

- **Use black on white** rather than white on black or colors. Black on a color is acceptable if the color has a reflectance value of at least 70%, and the type size is at least 10 points. Embedded in black on white, use **colored ink** on white for selected *search* topics such as primary goals or procedures.

- **Colored ink on colored paper is less acceptable** than black on white (it degrades readability). If colors are to be used, the best colors are dark ink on light paper.

- **Expand the type size if reading conditions will be poor** (such as low ambient lighting) or if the user population is likely to have visual impairments.

- **Use visual graphics** to break the text into meaningful segments. These include dashed lines, solid lines, bullets, boxes, and variations in print size or type font.

- **Use large amounts of white space.** When in doubt, leave more not less.

CONTENT

NOTE: Content for procedural types of information will be covered in a later section.

- **Structure organization** around reader goals and/or the natural structure of the information (Gottfredson & Guymon, 1989; Hannum, 1988; Moore & Gordon, 1988).

- **Use words that are familiar, meaningful,** and that appear frequently in the language. Use words that have relatively few syllables. Use abbreviations sparingly, and define each the first time it is used.

- **Sentences should be complete, well organized, unambiguous, and concise.** Obviously, a sentence should be grammatically correct.

- **Use simple grammatical structures.**

- **Avoid being wordy** and including information just because it is there. For each piece of information ask yourself, does the learner really need this? An example is computer software: You might identify a particular subtask where there are four different methods of accomplishing a goal; is it necessary to tell readers all four?

- Try to use **positive active statements,** which are more easily comprehended than other types (Broadbent; 1977; Slobin, 1966). An example of an active statement is "Press the ON button."

POOR:

To open the door, the handle must be pressed down.

BETTER:

Press the handle down to open the door.

- Place the **main topic of a sentence near the beginning** of the sentence.

- Use **technical terms (jargon) sparingly,** and only when really necessary. Ask yourself: What is the reason for the trainee learning this term? Use it only if it will be used repeatedly in subsequent parts of the training program or in their job. When using any terms which might be new for the reader, make sure they are defined the first time they are introduced.

- Consistently **use the same term for the same idea.** Do not use a wide variety of terms for the same idea, and don't use the same term for different ideas.

- **Avoid ambiguous terms** or phrases.

- **Maximize the coherence of the information** by tying sentences together for readers. The easiest way to do this is to make sure that each sentence has a clear reference back to at least one previous concept or idea (see Britton & Gulgoz, 1991, for a more in-depth explanation).

- **Use examples (and nonexamples) close to the introduction and definition of concepts.**

- **Use words at the reading level of the target population.** Try to target the material to the lower level of trainees, unless this is awkward or inappropriate for the remaining trainees. If you are unable to estimate or determine the reading level of the trainees, use the standard level; most texts are written at about the eighth grade reading level. There are several sources that provide quantitative indices of the reading level of text (e.g., Bailey, 1989, Sheppard, 1987). Sheppard (1987, p. 1544) suggests the Gunning Readability Formula:

  a. Randomly select a sample segment of 100 words from the document.
  b. Count the number of sentences and divide by 100 to get an average sentence length.
  c. Count the words in the sample with three or more syllables.
  d. Add that number to the average sentence length.
  e. Multiply the sum by 0.4. This figure is an estimate of the grade level for the documentation.

- **Do not require readers to convert one unit of measurement to another** (otherwise provide conversion aids).

ORGANIZATION

- **Use a short but informative title.**

- **Use advance organizers** to help the reader develop a mental model or general structure of the information (Jonassen, Beissner, & Yacci, 1993).

- **Provide some type of short introduction and overview** of the material. Make sure that the first part of a section introduces the topic and is not some lower level detail.

- **Write section headings that are clear, meaningful,** and **provide high-level structure** to the information. By high-level, we mean the superordinate nodes in an information hierarchy, not just sidenotes or any comment that might introduce a topic.

- When documents have a large procedural orientation, such as user's guides, owner's manuals, etc., **use task goals and subgoals as section headings** when possible.

- **Use an adequate number of heading levels** (first level, second level,

---

EXAMPLE:
 Initiate the Software Program
 Initiate the Hardware
 Change Software Settings
 Scan Image
 Save Scanned Image
 Exit the Software Program

---

etc.) to organize the information for readers. Usually two or three heading levels are best.

- **If the document is long, provide a table of contents and index** to make the contents accessible. The index should include topics and concepts explicitly stated in the text, and also related or synonymous concepts. A good index is critical to a usable text.

- If the material contains procedural elements (telling *how* to perform tasks), **develop an index that is based on appropriate task goals,** in addition to or instead of an index based on system functions or other types of information. This index can be provided separately or interleaved in with the standard index of topics. Likewise, if you are alphabetizing the document sections themselves, be sure to include alphabetized sections using user goals as headings. Do trainee analysis and user testing to support both of these strategies!

- **Use short paragraphs.** Bailey (1989) suggests that paragraphs should contain between 70 and 200 words. Each should present only one basic idea, principle, or relationship (plus elaborations on that idea).

PROCEDURES

- Under most circumstances, **include the following information in an overview** (from Bailey 1989, p. 373):

1. Describe work activities in general terms.
2. Describe inputs to the work activities or the conditions likely to be present whenever that set of procedures is applicable.
3. Briefly describe work activities to be performed in the set of procedures.
4. Describe the outputs or ensuing conditions that result from successful execution of the work activities.
5. Include admonishments (i.e., dangers, cautions, warnings) relevant to

the set of procedures to alert the users before they start the procedures.

6. Identify, when appropriate, the destination of the outputs or subsequent operations that result from correctly performing the procedures.

- For writing procedural statements; **start sentence with a verb, place object of the verb close to the verb, and use 20 words or less** (Bailey, 1989).

- **Use words consistently** among the steps or procedures. Avoid using different words to mean the same thing.

- **If three or more conditions** exist for a step, **use a decision tree or list** format for conditions.

- **Avoid using *AND* and *OR*** in the same conditional action.

- **Tell the reader what to do** instead of what not to do (if possible and appropriate).

- **Write procedural instructions in an order consistent with task performance.**

POOR:

Before exiting the system, save your document.

BETTER:

Save your document before exiting the system

POOR:

Mix the flour in with the egg. Make sure the egg has been well beaten.

BETTER:

Beat the egg well. Mix the flour in with the egg.

The results from the task analysis should specify the order of procedures. If the task analysis information is represented as a hierarchy, this can be used to structure document sections. Present the highest node level information (goals) first to give reader an idea of what the specific actions will be accomplishing.

- When describing complex procedures to be performed, **use a format that reflects the structure of the task, such as list, hierarchical list, flow chart, or decision tree** rather than narrative prose (Blaiwes, 1974; Kamman, 1975; Miller, 1981). The following is an example of narrative prose transformed into a list format.

POOR:

At the top of the screen you will see several boxes that contain FILE, PROC, PLOT, DISPLAY, PRINTOUT, and PARAMETERS. These boxes contain all the NMR commands which you will use. You access the commands with the mouse. Use the mouse to move the arrow to the FILE box. Place the head of the arrow on the F of file, press the FIRST (left) mouse button.

BETTER:

Display: FILE PROC PLOT DISPLAY PRINTOUT PARAMETERS

Actions to select File:	1. Use mouse to move arrow to FILE box.
	2. Place arrow on F of FILE.
	3. Press left mouse button.

The original for this example was taken from a procedure guide written by an instructional assistant for an engineering course. The example points out the fact that we tend to write the way we talk—in long sequential sentences, that is, in narrative text. This tendency must be overcome in developing procedural materials.

USE OF ILLUSTRATIONS

- When appropriate, **illustrations can be effectively used as a primary form of communication.** One application is when readers may have reading deficiencies, or English is not their primary language, as in the instruction cards placed on airplanes.

- **Illustrations can be effectively used to augment text.** However, when augmenting text, the designer should make sure the illustrations add information *beyond* that given in the text. If the figures are just repeating what is contained in the text, leave them out. When in doubt about whether the illustration helps, perform user testing.

- **Make sure readers adequately understand the illustrations.** Use symbols sparingly, and use symbols consistently across different illustrations (i.e., use one symbol per concept or object).

- **Place illustrations near the referring text,** the closer the better.

- **Use color to highlight important areas,** but implement this principle sparingly. When using colored illustrations, use brighter colors for most critical or salient areas. When color is used for coding purposes, do not use more than eight colors (Bailey, 1989).

In summary, a document should have information with the three characteristics suggested by Bethke and colleagues (Bethke, Dean, Kaiser, Ort, and Pessin, 1981). According to their research, the three most important factors for a manual were that the information be *easy to find, easy to understand,* and *task sufficient.* Factors that made the information easy to find were consistency, signposting, and arrangement. Characteristics that determined ease of understanding were simplicity, concreteness, and naturalness. Factors that affected task sufficiency were information completeness, accuracy, and exclusivity. While training analysts can design documents with these criteria in mind, the ability to successfully carry out such design will depend heavily on appropriate use of task analysis and user testing. In addition to these design methods, there is one other set of tools that can help with document design. These are briefly reviewed in the next section.

Document Evaluation Tools

There are now several automated tools that can be used to evaluate a document once it is written. One is the *Writers' Workbench™*, which is a set of computer programs that provide feedback about document variables such as number of words in sentences, the frequency and location of long sentences, the occurrence of verbiage that is too lengthy (with suggestions for better expressions), and the degree of repetition of certain words or phrases (Coke, 1982; Wright, 1988). A similar tool is *CRES,* the *Computerized Readability Editing System™* (Kieras, 1985). Most researchers working on automated tools hope to use AI in these programs in the near future. This would allow document evaluation to occur at more than just a surface level (Kieras, 1985; Wright, 1988).

DESIGN OF JOB AIDS

As mentioned at the beginning of this chapter, some documents are written as job aids rather than instructional documents. Such job aids might be written as separate texts, or they might be integrated into the instructional manual. There have been a few researchers who have looked at the design of job aids separately from the issue of general document design (Folley, 1961; Joyce, Chenzoff, Mulligan, & Mallory, 1973; Pearlstein, Schumacher, Smith, & Rifkin, 1973; Rifkin & Everhart, 1971; Smillie, 1985; and Swezey, 1987). In addition, there are several automated job aiding techniques under development (see Swezey, 1987, for a review). In this section, I will provide some basic design guidelines and describe several examples.

Job aids have been around for quite some time, and design guidelines were written as far back as 1961 (Folley, 1961). While Folley's procedures have been expanded for specific situations (e.g., Rifkin & Everhart, 1971), they have remained essentially intact, and will be included in the list below. Swezey (1987) has noted that much research is needed to address questions of optimal job aid design. For example, "the most easily understandable information presentation techniques are not necessarily those that optimally facilitate performance, recall, and/or retention of alphanumeric material" (Swezey, 1987). In addition, he notes that:

> The motivational properties of leading individuals by the hand through a series of steps without requiring them to exercise their own judgment and/or logic is typically considered demeaning. This is a particular problem in the job aid domain, since the reliability of job aids is demonstrably higher than is the reliability of typical free-wheeling, seat-of-the-pants maintenance approaches that are presumably based on some unknown level of understanding of underlying theoretical principles. (Swezey 1987, p. 1055)

This points to the problem with going different ways in job aid design. If you design the document (or computer system) to provide very simple, straightforward steps to follow, the worker will feel belittled. If you provide less procedural aid, the worker may rely on vague and incomplete knowledge, performing the job less well than he or she would using more complete job aids. These are issues of task allocation and overall system design that will require further empirical investigation. In the meantime, training analysts can perform at least some minimal user testing to identify the best way to provide instruction and job aid documents for a particular problem.

GENERAL PHYSICAL FORMAT

- **Base all job aids on thorough task and trainee analyses.** Consider the functions that must be met by the job aid, and develop the design accordingly. If it is to be used as a checksheet during task performance, provide a mechanism for keeping the user on task and provide a mechanism for checking off items.

- **Consider the physical environment** in which the job aid will be used. Make sure that the product can stand up to its environment. For example, if a job aid will be used next to a computer, you may want to place it on the wall where it won't get abused or lost. In addition, clear lamination can be used on paper job aids, and holes can be punched to hang them when it is appropriate.

- **Consider the portability requirements** of the job aid.

- **Consider the distribution and maintenance costs** of the job aid.

ORGANIZATION

- Job aids most frequently contain the following types of information:

Conceptual—facts, principles, relations, system components, etc.

Rule, procedural

Troubleshooting

Where to get help

There may be a lot of this type of information, or there may be a minimal amount. These factors will determine how the information is organized. The following guidelines can be used to determine the best organization for the various combinations of information. Not all combinations of information are covered, but the most common variations are discussed.

- For any job aid, **make sure that the information is well-organized, the organization is immediately obvious to the user, and the organization of segments is consistent.**

- **Match the organization to the user's cognitive organization or mental model.** Morrow et al. (1993) used a sorting and ordering task to determine the "natural" organization of information for the end user population.

- If the job aid must cover only a small amount of procedural information, typically one page or less, use a **simple list of procedures.** Many if not most job aids are of this type, providing skeletal procedural information. An example would be a checklist of five steps for using the office photocopier, taped on the machine.

- If the document must contain a relatively large amount of procedural information, with conditional information (IF this, DO that) interspersed in many sections, organize the information into a **procedural document** organized around goal-related sections. In each section, include the conceptual and procedural information relevant to the tasks for accomplishing the section goal. *In addition,* make a **short job aid of procedures** (such as a one-page reference guide or flow chart) that provides just the most basic, normal, straight path through the goal structure.

- If the information to be contained in the job aid includes some conceptual information, and medium to large amounts of procedural information, a similar approach can be taken. First, divide the information into chunks organized around major goals. Place the conceptual and procedural information under each relevant heading. Write a **document using the goals as major headings** and also provide a **short list format job aid.** If needed, add a section at the beginning providing an overview of the system and its use.

EXAMPLE:

Analysts developed a job aid to help students perform a literature search using the *PsychLit*[TM] computer software system. The necessary conceptual and procedural information turned out to be more than a list that would fit onto one page. The design solution was a fold-open, standard-sized laminated document; 8½ × 11 when closed, which opened like a book. The front contained a short-form job aid (see Figure 11.1). The middle two pages gave a lengthier version, and the back contained additional information. While Figure 11.1 is black-and-white, the actual job aid used color to signify major goals.

- When the document must contain conceptual information, relatively large amounts of procedural information, and some amount of trouble-shooting information, the approach described above is still useful. Create a **document with goal-oriented sections** containing conceptual and procedural information. Place the relevant troubleshooting material at the end of each section. Place additional troubleshooting material at the back of the document if necessary. At the beginning, place an overview, introduction, and where to find the help section. Develop a **short job aid** that can be pulled out and used separately as a convenient checklist. Designed only to act as a prompt, this short job aid should contain only the most common goals and actions required to accomplish the goals.

As an example, an analyst decides to write a document for about six office workers who must occasionally scan color and black-and-white graphics into a personal computer using a multipurpose imaging and drawing software package. The job aid requires instructing the employees to perform the tasks that were listed earlier (repeated here); initiate the software, initiate the hardware, change settings if necessary, scan the image, save the image, and exit the software.

User's Guide

Contents:

Quick Steps . . . . Front Cover
Operation in Depth . . . . . Inside
Advanced Features . . . . . . Back Cover

What is PsycLit? - PsycLit is a computer database that contains the references and abstracts of all the journal articles, books, and book chapters listed in *Psychological Abstracts*. Some references listed in these sources may not be available at the University of Idaho library, in which case they may be ordered at the Inter Library Loan desks.

Quick Steps

1. Starting PsycLit:

 If you will be storing the information from a search onto a disk, you must have an IBM compatible formatted disk *before* you begin.

1a. Select PsycLit Database: Make sure that PsycLit is highlighted and press [Enter].

1b. Select Journal Articles Or Books: Use the arrow keys to move to the database (Journals or Chapters), press the [Spacebar], then press [Enter].

2. Starting a Search:

2a. Enter Search Terms: Type in search term after "FIND:" prompt appears at the bottom of screen, then press [Enter].

2b. Display Search Results: Press [F4]

3. Print to Paper or Store Results on Disk:

3a. Mark Records: Move blinking cursor to desired record then press [Enter] to mark it.

3b. Select Print or Download for Saving on Disk: To Print; press [P] for print, then press [Enter]. To save on disk; press [D] for Download, then press [Enter].

4. Start Next Search or Quit:

4a. Start Next Search: To start next search, press [F10], then press [Enter].

4b. Quit: Press [F10], then press [Q] for quit.

5. Determine if the Library Has It:

 Call numbers and information about availability at the U of I library are NOT maintained in the PsycLit system. You must go to one of the other library computers (e.g., IDA or CARL).

Figure 11.1   Front page of job aid for literature search.

While the tasks required for 90–95% of the jobs were few and simple, certain types of images required complex changes in the software settings. To handle this problem, a 32-page procedurally structured document was written. The material was organized into the following sections:

- Overview
- Things You Need to Know
- Table of Contents
- Start the Software from Windows
- Initiate the Howtek Scanner
- Change Settings
- Scan Image
- Save Scanned Image
- Exit Software
- Troubleshooting Quick Reference Guide

The material in the Quick Reference Guide (two pages) was also made into a two-sided laminated short job aid. These two pages are shown in Figures 11.2 and 11.3.[1] Note the page references to the long-form job aid.

CONTENT

- **Use the minimum wording necessary** to support performance.
- **Use diagrams and illustrations** when those formats are more efficient than words.
- **Use adequate white space** so users can readily find information.
- **Use color, font, graphics, and other tools to help the organization** of the job aid "pop out" at users.
- **Try to list each goal or subgoal in a salient manner** so users can find it. Follow the goal with the action required to carry it out.

[1]Job aid designed by Bill Brown at the University of Idaho.

To scan an image using the Howtek Personal Color Scanner

1. **START** the Tempra Pro software   Page 3

    A. Double click on the "Non-Windows Applications" icon with the **LEFT mouse button**

    B. Double click on the "Tempra" icon with the **LEFT mouse button**

2. **INITIATE** the Howtek **SCANNER**   Page 5

    A. Click on the **"INP"** button with the **LEFT mouse button**

    B. Click on the **down arrow** on the scroll bar with the **LEFT mouse button**

    C. Click on **"PCS (MC)"** with the **LEFT mouse button**

3. **CHANGE SETTINGS** for desired effect   Page 7

    Click on **"SCAN MENU"** with the **LEFT mouse button**

    If you want a **color** copy, with **smooth color transition**, that **fits on the screen**, choose the default settings for the "SCANNER SETTINGS MENU"

Default settings are:

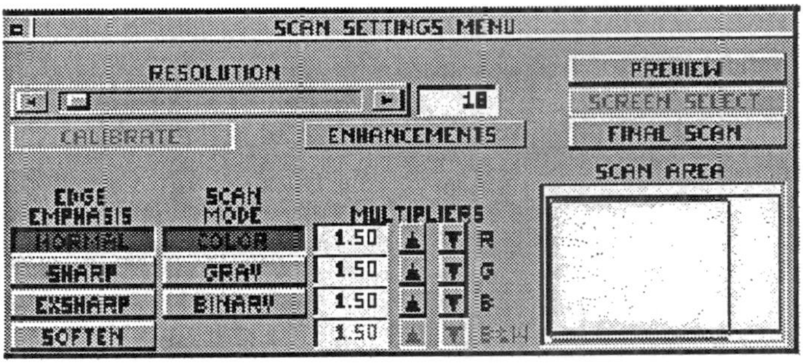

To have all the colors equal, set each of the Multipliers to:
1.50 = R
1.50 = G
1.50 = B

Figure 11.2 Front of short job aid for scanning image into PC. (Used with permission.)

4. SCAN IMAGE   Page 20

    A. PLACE THE IMAGE IN THE SCANNER:
               FACE-DOWN
               TOP OF IMAGE TOWARD YOU
               ANYWHERE ON THE SCANNING SCREEN
               (placing the image next to an edge will ensure perpendicularity)

    B. Click on the **"Preview" button** with the **LEFT mouse button**

    C. Click on the "SCREEN SELECT" button with the **LEFT mouse button**
                       AND
    Draw a box around the desired area by:
        1. Click in the upper left corner of the area you want in the final scan with the LEFT mouse button
        2. Click in the lower right corner of the area you want in the final scan with the LEFT mouse button

    D. Click on the **"Final Scan" button** with the **LEFT mouse button**

5. SAVE Scanned **IMAGE**   Page 24

    A. Click on the **Box above the "DSK" button** with the **LEFT mouse button**

    B. Click on the **Root Directory (Drive) Letter** with the **LEFT mouse button**

    C. Click on the desired **Subdirectory** with the **LEFT mouse button**

    D. Click on the **FORMAT EXTENSION** with the **LEFT mouse button**

    E. Click on the FILENAME box with the **LEFT mouse button**
                       AND
    Type the **FILENAME** (no extension) from the keyboard
                       AND
    Press ENTER

    F. Click on the **"Save" button** with the **LEFT mouse button**

    G. Make sure the file has been saved

6. EXIT Tempra Pro software   Page 29

    A. Click on the **"Exit" button** with the **LEFT mouse button**

    B. Click on the **"Yes" or "No" button** with the **LEFT mouse button**

Figure 11.3   Back side of short job aid for scanning image into PC. (Used with permission.)

POOR:

After you have scanned in your image and checked for accuracy, you are ready to save the image as a file. To save the image as a file, you must first access the root directory drive where you want to save your image. You can do this by clicking on the root directory drive letter.

BETTER:

TO SAVE IMAGE:

Access Root Directory	by	Click on Root Directory Letter
Access Subdirectory	by	Click on Subdirectory Name

While these guidelines and others can give the designer some ideas for job aid organization and content, it will still be more art than science creating a good and useful job aid. The best tool you have is to use prototyping and user testing procedures.

INTERACTIVE COMPUTER INTERFACE DESIGN

Computers are becoming a predominant medium in the training industry. Many of the programs being developed are impressive uses of the new multimedia techniques. For example, an educational program titled "Ulysses" was recently developed for IBM by And Communications Inc. (Holsinger, 1991). Ulysses is a PS/2-based videodisc application that required over a million dollars to produce. And Communications is now developing five new titles using Turbo C++ Objects. It is easy to see that computer-based training is moving into a new era.

Even with this major shift in the nature of CAI, there are some basic principles of design that should be followed. Adherence to these principles will be a major job for the instructional designer because the new multimedia training programs are developed by teams heavily populated by media and graphic designers. This gives the programs their surface "gloss" and sensory appeal. However, these technicians rarely have training with respect to what makes a system usable or good from a learning point of view. The teams also work with instructors who are experts in the subject matter domain, but who also have no background in instructional design. The training analyst or instructional designer must keep this team grounded in sound design, regardless of how simple or complex and glossy the application.

There are entire volumes written on human-computer interface design (e.g., Helander, 1988), and we can't cover all of that material here. I would strongly suggest that those who aspire to CAI production read and take courses in that area. In the meantime, we can review the major or most important goals in designing CAI systems. The one goal that designers should always keep in mind is:

- **Create a system that is easy for the learner to use.**

While this goal is important for software design in general, it is especially critical for training and instructional applications. There are two reasons: (1) in most cases, the student is a novice who will not use the program for a long period of time (as does a person using a word processing system, for example); and (2) in instructional applications, cognitive resources need to be devoted to the content and not to trying to find one's way through the instructional courseware. For these reasons, the instructional designer must focus on developing a system that is simple and straightforward in its use.

What characteristics make things "easy" for a novice user? Among other things, the system should exhibit the following characteristics:

- **At any given point, users should understand what is being presented, what they are required to do or have the option of doing, and how to accomplish their current goal. The system should also display what the system is currently doing, if anything.**

As an example, at any given screen, the user may want to end the session. This is a current goal. The action required to do this should be apparent. Now consider the screen shown in Figure 11.4. It is not entirely apparent what the user must do to end the session. Does one click on the button marked "End" or on the button marked "Quit?" Also note that the screen doesn't meet the requirement of the user knowing what the options are. To know what the options are, the user must know what will happen when any of the icons are activated or clicked on with the mouse. There are several icons where this information is not readily apparent. For example, unbeknownst to the learner, clicking on the star in the upper left corner will cause the screen to show the credits for the instructional package. (I am sad to report that this screen was modeled from a real system.)

Designers will try to sidestep the problem of usability by including icons whose meaning is obscure and then rely on "instruction" regarding how to use the system. This instruction tells users the meaning of the icons, as well as how to get around in the system. However, this is a *particularly* bad design philosophy and should be avoided whenever possible. The approach of using training to teach the learner how to interact with the system places an extra burden on user memory, and takes cognitive resources away from the primary task (not to mention frustrating the user if they fail to remember the information).

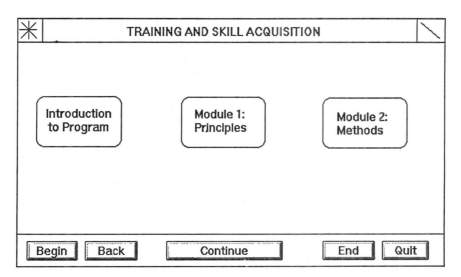

Figure 11.4   Typical screen for interactive computer-based training.

In summary, the primary goal for designing the interface is to make sure that the system allows users to easily understand how to interact with that system (Galitz, 1980). This is done through clear and concise messages, and buttons or other input devices that also have clear and concise meanings. This type of design requires patience, creativity, and most of all, user testing. In the sections below, we will review other more specific design guidelines.

Static Screen Design

Many of the guidelines offered in the document design section will also be relevant to computer screen design. In addition there are several suggestions that pertain to each individual screen. Readers are strongly encouraged to consult Mayhew's 1992 book, *Principles and Guidelines in Software User Interface Design,* for specific and useful guidelines.

- **Don't try to place too much information on the screen.** As a general rule of thumb, Tullis (1983) suggests that, based on research, the alphanumeric characters shouldn't take up more than 25–30% of the total screen space. This is equivalent to leaving plenty of white space in written documents. Include *only* information that is really necessary.

- Even when trying to keep the screens uncluttered, be sure to **include *all* information that *is* necessary.**

- **Group characters** so that they are moderately (but not extremely) packed (Cakir, Hart, & Stewart, 1980).

- **The items on a screen should be logically grouped** so that people perceive the screen as consisting of blocks of related pieces of information. Groupings of items often consist of *functional* groupings. An example is where all buttons to move from screen to screen are grouped together, all buttons to edit are grouped together, etc.

- **Leave adequate space around each group** (Heines, 1984; Jones, 1978; Treisman, 1982).

- In addition, **a smaller number of groups is better** than a larger number of individual items or a large number of groups (Tullis, 1983).

- **Minimize the complexity of the layout.**

- **Maximize visual predictability.** Put another way, "based on a knowledge of the location of some items on the screen, one should be able to predict the locations of others" (Tullis, 1983).

- **Use redundant coding.** That is, have two or more different variables providing the same information. For example, active buttons could always be shown as "shadowed rectangles" and also always be colored turquoise blue.

- **Use symbols or icons for input buttons that are consistent with learners' previous knowledge.** For example, a question mark is relatively well known as a symbol for requesting help or information. **Strive to use words *and* icons.**

NOTE: Too many icons are currently used that are meaningless to the learner. For example, how likely is it that a learner would know the meaning of "\" ? Unfortunately, I have seen button icons that are exactly that symbol. The best icon is one that passes "predictive" user testing. That is, the function of an icon, when it first appears on a screen, should be predictable by a novice user. If you ask them, "what happens when you click on this icon?", they should be able to accurately tell you.

- **Don't use hidden buttons** (where parts of the screen are active hot spots, but you can't tell by visual inspection). Exceptions are when (a) all parts of the screen are hot, as in a map or picture with parts acting as hot buttons, and users know this; or (b) the cursor symbol changes when it is over buttons, giving visual feedback that the area is hot.

- In most circumstances, **start in the upper left-hand corner of the screen,** as this is where people habitually begin search and reading.

- **Use upper and lower case for text.** Use minimal or no text that is all upper case (Tullis, 1988).

- Blinking is a very strong visual cue and can be distracting. **Use only in emergencies or critical situations.**

- **Use reverse video sparingly.** Appropriate uses would include indications that an item has been selected, or to indicate an error field (Mayhew, 1992).

- **Messages should be brief, concise, specific, helpful, and comprehensible.** The following are examples of poor and improved messages from Mayhew (1992):

POOR:	*IMPROVED:*
1. The processing of the text editor yielded 23 pages of output	Output 23 pages
2. Error in DRESS SIZE field	Error: DRESS SIZE range is 4 to 16
3. Cannot exit before saving file	Save file before exiting
4. Bad/illegal/invalid file name	Maximum file name length is eight characters

- **Use instructional prompts** to cue users when they must type or perform other necessary input activity (see Mayhew, 1992).

Use of Color

Based on empirical research, Mayhew (1992) suggests using color sparingly. One should first design in black and white then add color only when there is a reason, that is, when it serves a particular purpose.

- **Don't use bright colors such as red, yellow, pink, orange for large areas.** Use only in small amounts to gain users' attention.

- **Don't use too many colors on a screen,** unless it is a full photo or video image type of situation. Sanders and McCormick (1987) summarize the suggested limit as being nine colors.

- **Use color coding and be consistent in color use.** For example, if light blue is used for video control buttons in one part of a CAI program, they should be used consistently throughout.

- **Limit the color coding to eight distinct colors** (Mayhew, 1992).

- **Use colors to draw attention, communicate, or organize information.** Showing certain types of information in a particular color can help users search for and identify information. Similarly, avoid using color in "non-task-related" ways. This will only make things more difficult and distracting for the user.

- **Do not use color images or text on color background.** Use colors on achromatic backgrounds such as black, gray, white. Or use achromatic images on color background.

- **Avoid using bright blue for text or small, thin-lined graphics.** We are not good at distinguishing blue objects from their background, and our eyesight for blue decreases with age (Murch, 1987).

- **Avoid the combined use of opposing colors, highly saturated colors, or colors far apart on the color spectrum.** This applies within one screen as well as across screens.

- Determine the uses of color for your specific application, write the information as a style guide (job aid), and **follow the style guide rigorously.**

Additional guidelines are provided in Mayhew (1992).

Designing for Interaction

When a user sits down to interact with an instructional computer system, it should exhibit a quality known as **transparency.** That is, at least two things should be relatively apparent to the user: the general nature of the program (what it is used for), and how to interact with the program. In keeping with the second goal, the user should know how to accomplish goals, whether they are vague (I'm totally lost on this topic, can you help me?) or very specific (I want to quit for lunch).

CAI systems lie somewhere along a continuum that has the following end points:

1. The system is a knowledge base and the user is free to browse through it in whatever fashion he or she desires.
2. The system is structured so that the CAI program leads the learner through a carefully designed sequence of instructional modules.

Most CAI systems reside somewhere in between these end points. But whichever approach is taken, the designer must still ensure that the user understands where the system lies along this continuum. If it is a network of knowledge for the user to peruse, this fact must be conveyed, and an interface provided that supports such learner-controlled browsing. If it is a more systematic progression through modules, this must be conveyed also. Even when the system is predominantly computer-directed, the user should still have certain options (such as quiting, stopping for a time, reviewing previous material, etc). Thus, for any interactive system, the designer must explicitly decide what controls will reside in the computer program itself, and what controls the learner will have.

In addition to these general principles of interface design, designers should attempt to follow certain basic guidelines for developing instructional systems. Some are given below.

- At the beginning, **present an overview of how the system works, and how the learner interacts with the system.** This may be explicit information or implicit in the screen design.
- **Provide an advance organizer.** This can be something as simple as a list of topics, or a map of the program.
- **Make use of previously existing knowledge.** This can range from using icons such as a "?" to using an entire metaphor. Some metaphors that have been used in software include the desktop (calendars, appointment books, etc.), a schoolhouse metaphor, or a city. Most users are now familiar with touching a place on the screen or clicking on buttons with a mouse, so these are good input mechanisms. A variety of symbols may also be familiar, depending on the target population. For example, VCR and cassette recorders often use variations of arrows for play, record, reverse, etc. Trainee analysis can be used to determine symbols that will be adequately familiar to learners.
- **In general, a graphical interface will be better than command language input** for training programs. As always, a designer must match the complexity, input mechanisms, and symbology of a system to the trainee population and task characteristics.
- **Use consistent design elements from one screen to the next.** This includes color coding for functions, placement of control buttons, use of symbols for icons, etc. Figure 11.5 shows an example of a CAI system that *does not* follow this principle. In the first screen, the user clicks on diamond-shaped buttons (not apparent buttons), and on subsequent screens the user must click on rectangular topic boxes (also not apparent). Color coding was not used in either case.
- For most hypertext or hypermedia systems, **make the underlying organizational structure hierarchical** rather than an unstructured network of nodes (Gordon & Lewis, 1992).

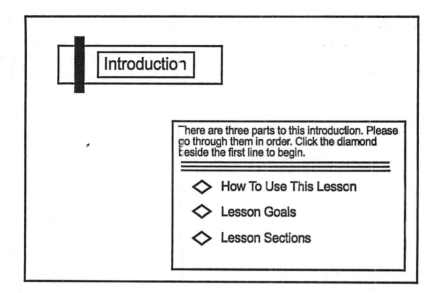

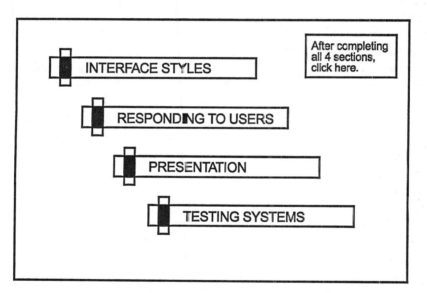

Figure 11.5 Example of failure to use consistency across screens.

In summary, a design team must create a simple and obvious system for the user interface, and stick to that system consistently throughout the instructional program. This doesn't mean that the instructional program has to be simple or plain, but it should have a structure and interface that users can understand easily and use without difficulty

Designing a user-friendly system requires effort and creativity. And, even with those characteristics, at some point it becomes virtually impossible for a designer to see the system from the eyes of a novice learner. This is why the prototyping and user testing phases are especially critical for the development of CAI instructional programs. Nothing else can replace this step, not even asking other designers to judge your program. In Chapter 13 we will discuss how to implement user testing.

REFERENCES

ADAMS, K. A., & HALASZ, I. M. (1983). *25 ways to improve your software user manuals.* Worthington, OH: Technology Training Systems.

BAILEY, R. W. (1989). *Human performance engineering: Using human factors/ergonomics to achieve computer system usability.* Englewood Cliffs, NJ: Prentice Hall.

BETHKE, F. J., DEAN, W. M., KAISER, P. H., ORT, E., & PESSIN, F. H. (1981). Improving the usability of programming publications. *IBM Systems Journal, 20,* 306–320.

BLAIWES, A. S. (1974). Formats for presenting procedural instructions. *Journal of Applied Psychology, 59,* 683–686.

BRITTON, B. K., & GULGOZ, S. (1991). Using Kintsch's computational model to improve instructional text: Effects of repairing inference calls on recall and cognitive structures. *Journal of Educational Psychology, 83,* 329–345.

BROADBENT, D. E. (1977). Language and ergonomics. *Applied Ergonomics, 8,* 15–18.

CAKIR, A., HART, D. J., & STEWART, T. F. M. (1980). *Visual display terminals: A manual covering ergonomics, workplace design, health and safety, task organization.* England: Wiley.

CARROLL, J. M., & CARRITHERS, C. (1984). Blocking learner errors in a training-wheels system. *Human Factors, 26,* 377–390.

CARROLL, J. M., & MACK, R. L. (1984). Learning to use a word-processor: By doing, by thinking, and by knowing. In J. C. Thomas and M. L. Schneider (Eds.), *Human factors in computer systems* (pp. 13–51). Norwood, NJ: Ablex.

CARROLL, J. M., MACK, R. L., LEWIS, C., GRISCHKOWSKI, N., & ROBERTSON, S. (1985). Exploring a word processor. *Human-Computer Interaction, 1,* 283–307.

CARROLL, J. M., SMITH-KERKER, P. L., FORD, J. R., & MAZUR, S. A. (1986). *The minimal manual* (Research Report RC 11637). Yorktown Heights, NY: IBM T. J. Watson Research Center.

CATRAMBONE, R., & CARROLL, J. M. (1987). Learning a word-processing system with training wheels and guided exploration. In *Proceedings of CHI+GI 1987* (pp. 169–174). New York: ACM.

COKE, E. U. (1982). Computer aids for writing text. In D. H. Jonassen (ed.), *The technology of text* (pp. 383–399). Englewood Cliffs, NJ: Educational Technology Publication.

CONRAD, R. (1962). The design of information. *Occupational Psychology, 36,* 159–162.

FOLLEY, JR. J. D. (1961). *A preliminary procedure for systematically designing performance aids* (ASD Technical Report 61–550). Wright-Patterson Air Force Base, OH.

GALITZ. W. O. (1980). *Human factors in office automation.* Atlanta, GA: Life Office Management Assn.

GORDON, S. E., & LEWIS, V. (1992). Enhancing hypertext documents to support learning from text. *Technical Communication,* 305–308.

GOTTFREDSON, C. A., & GUYMON, R. E. (1989). *Style guide for designing and writing computer documentation.* Alpine, UT: The Gottfredson Group.

HANNUM, W. (1988). Designing courseware to fit subject matter structure. In D. H. Jonassen (ed.), *Instructional designs for microcomputer courseware* (pp. 275–296). Hillsdale, NJ: Lawrence Erlbaum Associates.

HARTLEY, J., & BURNHILL, P. (1977). Fifty guide-lines for improving instructional text. *Programmed Learning and Educational Technology, 14,* 65–73.

HEINES, J. M. (1984). *Screen design strategies for computer-assisted instruction.* Bedford, MA: Digital Press.

HELANDER, M. (ed.) (1988). *Handbook of human-computer interaction.* Amsterdam: North-Holland.

HOLSINGER, E. (1991). Ulysses: A new hope for education. *New Media, 1(4),* 24–26.

JONASSEN, D. H., BEISSNER, K., & YACCI, M. (1993). *Structural knowledge: Techniques for representing, conveying, and acquiring structural knowledge.* Hillsdale, NJ: Lawrence Erlbaum Associates.

JONES, P. F. (1978). Four principles of man-computer dialog. *IEEE Transactions on Professional Communication, 21(4),* 154–159.

JOYCE, R. P., CHENZOFF, A. P., MULLIGAN, J. F., & MALLORY, W. J. (1973). *Fully proceduralized job performance aids: Vol II, Handbook for JPA developers* (AFHRL-TR-73-43(II)) (AD775-705). Wright-Patterson Air Force Base, OH: Air Force Human Resources Laboratory.

KAMMAN, R. (1975). The comprehensibility of printed instructions and the flowchart alternative. *Human Factors, 17,* 183–191.

KIERAS, D. E. (1985). Improving the comprehensibility of a simulated technical manual. *Technical Report, 0,* University of Michigan.

MACDONALD-ROSS, M. (1977). Graphics in texts. In L. S. Shulman (ed.), *Review of research in education,* Vol 5, Hasca, IL: Reacock.

MAYHEW, D. J. (1992). *Principles and guidelines in software user interface design.* Englewood Cliffs, NJ: Prentice Hall.

MILLER, L. A. (1981). Natural language programming: Styles, strategies, and contrasts. *IBM Systems Journal, 20(2)*, 184–215.

MOORE, J., & GORDON, S. E. (1988). Conceptual graphs as instructional tools. *Proceedings of the Human Factors Society 32nd Annual Meeting* (pp. 1289–1293). Santa Monica, CA: Human Factors Society.

MORROW, D., LEIRER, V., & ANDRASSY, J. (1993). Designing medication instructions for older adults. *Proceedings of the Human Factors and Ergonomics Society 37th Annual Meeting*, pp. 197–201. Santa Monica, CA: Human Factors and Ergonomics Society.

MURCH, G. M. (1987). Colour graphics—Blessing or ballyhoo? In R. M. Baecker and W. A. S. Buxton (eds.), *Readings in human-computer interaction* (pp. 333–341). Los Altos, CA: Morgan Kaumann Publishers.

PEARLSTEIN, R. B., SCHUMACHER, S. P., SMITH, A. P., & RIFKIN, K. I. (1973). *Information display design*. Valencia, PA: Applied Science Associates.

RIFKIN, K. I., & EVERHART, M. C. (1971). *Position performance aid development*. Valencia, PA: Applied Science Associates.

SANDERS, M. S., & MCCORMICK, E. J. (1987). *Human factors in engineering and design* (6th ed.). New York: McGraw-Hill.

SHEPPARD, S. B. (1987). Documentation for software systems. In G. Salvendy (ed.), *Handbook of human factors* (pp. 1542–1584). New York: John Wiley & Sons.

SLOBIN, D. (1966). Grammatical transformations and sentence comprehension in childhood and adulthood. *Journal of Verbal Learning and Verbal Behavior, 5*, 219–227.

SMILLIE, R. J. (1985). Design strategies for job performance aids. In T. M. Duffy and R. Waller (eds.), *Designing usable text* (pp. 969–977). New York: Academic Press.

STRUNK, W., & WHITE, E. B. (1972). *The elements of style*. New York: Macmillan.

SWEZEY, R. W. (1987). Design of job aids and procedure writing. In G. Salvendy (ed.), *Handbook of human factors* (pp. 1039–1057). New York: John Wiley & Sons.

TINKER, M. A. (1963). *Legibility of print*. Ames, Iowa: Iowa State University Press.

TREISMAN, A. (1982). Perceptual grouping and attention in visual search for features and for objects. *Journal of Experimental Psychology: Human Perception and Performance, 8*, 194–214.

TULLIS, T. S. (1983). The formatting of alphanumeric displays: A review and analysis. *Human Factors, 25(6)*, 657–682.

TULLIS, T. S. (1988). Screen design. In M. Helander (ed.), *Handbook of human-computer interaction* (pp. 377–411). Amsterdam: North-Holland.

WEISS, E. H. (1991). *How to write usable user documentation* (2nd ed.). Phoenix, AZ: Oryx Press.

WRIGHT, P. (1988). Issues of content and presentation in document design. In M.

Helander (ed.), *Handbook of human-computer interaction* (pp. 629–652). Amsterdam: North-Holland.

ADDITIONAL RESOURCES

Writing Documents

DUFFY, T. M., CURRAN, T. E., & SASS, D. (1983). Document design for technical job tasks: An evaluation. *Human Factors, 25,* 143–160.

DUFFY, T. M., & WALLER, R. (1985). *Designing usable text.* New York: Academic Press.

HARTLEY, J. (1985). *Designing instructional text* (2nd ed.). London: Kogan Page.

JEREB, B. (1986). Plain English on the plant floor. *Visible Language, 20,* 219–225.

POLLER, M. F., FRIEND, E., HEGARTY, J. A., RUBIN, J. J., & DEVER, J. J. (1981). *Handbook for writing procedures.* Indianapolis, IN: Western Electric Publication Center.

SCHRIVER, K. A. (1972). *Designing computer documentation: A review of the relevant literature.* Pittsburgh, PA: Carnegie Mellon University, 1986.

Job Aids

DUFFY, T. M., & WALLER, R. (1985). *Designing usable text.* New York: Academic Press.

GEIS, G. L. (1984). Checklisting. *Journal of Instructional Development, 7,* 2–9.

SWEZEY, R. W. (1977). Performance aids as adjuncts to instruction. *Educational Technology, 3,* 27–32.

Human-Computer Interaction

BADRE, A., & SHNEIDERMAN, B. (Eds.) (1982). *Directions in human-computer interaction.* Norwood, NJ: Ablex.

ELKERTON, J. (1988). Online aiding for human-computer interfaces. In M. Helander (ed.), *Handbook of human-computer interaction* (pp. 345–364). Amsterdam: North-Holland.

GALER, M., HARKER, S., & ZIEGLER, J. (1992). *Methods and tools in user-centred design for information technology.* Amsterdam: North Holland.

MCGEE, K., & MATTHEWS, C. (eds.) (1985). *The design of interactive computer displays: A guide to the select literature.* Lawrence, KA: The Report Store.

RUBINSTEIN, R., & HERSH, H. (1984). *The human factor: Designing computer systems for people.* Burlington, MA: Digital Press.

SALVENDY, G. (ed.) (1984). *Human-computer interaction.* Amsterdam: Elsevier.

SALVENDY, G., & SMITH, M. J. (1993). *Human-computer interaction: Software and hardware interfaces.* Amsterdam: Elsevier.

SHNEIDERMAN, B. (1980). *Software psychology: human factors in computer and information systems.* Cambridge, MA: Winthrop.

WICKENS, C. D. (1991). *Engineering psychology and human performance* (2nd ed.). Columbus, OH: Charles E. Merrill.

WILLEGES, R. C., WILLEGES, B. H., & ELKERTON, J. (1987). Software interface design. In G. Salvendy (ed.), *Handbook of human factors* (pp. 1416–1449). New York: John Wiley & Sons.

——————— 12 ———————

Procedure:
Initial Development
and Prototyping

After a preliminary design concept has been identified, several procedures should begin. These include more extensive design, prototyping, and user testing. In this chapter, we will discuss how one goes about developing a system to the point where some formative evaluation and user testing can be conducted. More specifically, I will cover three topics: (a) the origins and rationale for prototyping, (b) methods for prototyping, and (c) procedures for initial development and prototyping for each of the major instructional techniques. In Chapter 13, I will review methods for carrying out the user testing, and in Chapter 14, I cover some of the more extensive methods for full system development. Readers are encouraged to concurrently refer to appropriate sections in Chapters 12, 13, and 14 for information specific to the particular type of instructional program being developed.

ORIGINS OF PROTOTYPING

Traditional Engineering Design

In standard product engineering, a **prototype** is a full-scale, fully operational system or device that is made using a procedure different from the process that will ultimately be used for large-scale manufacturing. Prototype means "first of a type," and this first product is usually made laboriously by hand or some means less efficient than the final manufacturing method.

The prototype is made so that certain data can be collected. Engineering tests are performed, marketing personnel can elicit consumer reaction, user testing is conducted to determine whether the system can be operated by the target population, etc. Once the data confirm that the product fulfills its requirements, plans are made for full-scale production.

In many instances a **mock-up** is built even before the prototyping begins. A mock-up is a preliminary version of the product that has limited functionality, and is probably not made of the same materials. The mock-up is used early in the design stage as a tool for designers to share their ideas among themselves, and with other key people. Generally, designers move in a sequence going from mock-ups that are less similar to the final product, to prototypes that are identical to the final product. This use of multiple mock-ups and prototypes makes the design process proceed more rapidly and efficiently than if such methods were not used.

Human Factors Engineering

In human factors engineering, the distinction between mock-ups and prototypes becomes somewhat blurred. This is because while mock-ups and prototypes are both used, even the prototypes are usually not functionally complete. For example, a prototype of a software application might be complete in its interface characteristics, but not complete in being able to accomplish all tasks. Stated differently, human factors engineers use mock-ups early in the design process (just as in traditional engineering domains), but then designers move to successively more complete versions of a prototype rather than one fully functional prototype. Thus, prototypes have the look and feel of the final product to varying degrees, but almost never have the entire range of functions.

In summary, "sketchy" prototypes are used by all members of a design team to discuss the first cuts at project design. After a design concept is pinned down, it is time to determine whether it will work, and refine the concept as a product. More elaborate and realistic prototypes are built and used for at least three purposes: (1) as a vehicle for the design team to continue product development, (2) as a tool to show other people, such as managers, what the product is shaping up like (thus providing a means for eliciting their feedback), and (3) as a vehicle to perform user testing.

Prototyping and user testing go hand-in-hand. Any time a system is being developed that people must use, it is very important to make sure that they can, in fact, use it. Too often, designers determine this factor on the basis of their own subjective judgment. They don't realize that as design experts, they have lost the ability to see their system from a novice user's point of view. To counteract this problem, human factors engineers ask novices to use the prototype at various stages of development. This allows them to determine what aspects of the system must be modified to attain the goal of system usability.

It should be apparent from the discussion above that prototyping and user testing are very iterative processes. Unless the product is extremely simple, the designer should perform user testing, get feedback, make changes, get more feedback, etc. While it sounds like a lot of work, studies have shown that use of prototyping and user testing can actually reduce product development time and costs. For example, a study by Boehm, Gray, and Seewaldt (1984) found that cost savings were at least 3:1, and the use of prototyping for software design took 45% less development time than standard methods.

PROTOTYPING AND RAPID PROTOTYPING
FOR TRAINING PROGRAMS

In the design of training and performance support systems, the product being developed is the instructional system, a system designed to promote effective and efficient task performance by people. As in other types of product development, our instructional program is frequently not perfect the first time around. At some point, various aspects of the program will need modification and improvement, whether it is the addition of material, modification of how the material is delivered, CAI screen redesign, etc. Designers rarely have a choice about *whether* changes will have to be made, but they do have a choice about *when* changes are made. We will first evaluate when such changes are made in traditional instructional design models.

Traditional Instructional Design Models

In the traditional instructional design model, a design team performs tasks in a relatively sequential fashion. The designer or design team writes instructional objectives, develops scripts or storyboards (specifying the entire instructional sequence), creates the instructional product, and then evaluates the product. As an example of this model, consider Table 12.1. The table shows the major design steps suggested by Imke in the book *Interactive video management and production* (Imke, 1991). The design model formally calls for implementing evaluation as the last step. This is typically known as *summa-*

TABLE 12.1   A Traditional Design Model for Developing Interactive Video

STEP 1: Plan Stage
- Gather information about the project
- Develop a project plan (contract with client)

STEP 2: Design Stage
- Target audience analysis
- Define learning objectives
- Content analysis

STEP 3: Specification Stage
- Define treatment and instructional strategy
- Write functional specifications

STEP 4: Script/Storyboard Stage
- Write script
- Create storyboards
- Review (by team, technicians, client)
- Record narration

STEP 5: Video Production Stage
- Shoot video
- Create disc geometry
- Edit premaster
- Review by technical expert

STEP 6: Program Production Stage (concurrent with step 5)
- Write course code
- Develop graphics and animation sequences

STEP 7: Merge/Debug Stage
- Merge computer based graphics, course code, and video disc
- Debug course

STEP 8: Evaluation Stage
- Design team members take course to evaluate it
- Audience members (learners) take course

---

tive evaluation. In some cases, *formative evaluation* also takes place during product development, but is usually minimal and performed in an informal manner (consisting of activities such as showing the design to other designers or clients, and asking for an opinion on the sufficiency of the design).

As researchers have previously noted (e.g., Merrill, Li, & Jones, 1990), traditional instructional design models are characterized by several traits: (a) they promote separation between program design and program development, the phases are chronologically separate, and rely on different tools, knowledge, and people; (b) each of the phases are extremely work intensive; and (c) the major evaluation of the program occurs after a large portion of the product development has been accomplished. These characteristics result in several problems. First, for training projects in industry, the client or manager often doesn't see anything resembling the actual training program for many months. This frequently results in an anxious client looking over the shoulder of the design team. And when they finally see the program, clients or instructors may tell the design team that it is not really what they wanted or needed.

Because feedback from clients, instructors, and learners occurs mostly after product development, which is the last procedure in a long sequence of steps, the need for changes to the *design* becomes apparent relatively late in the product development life cycle. Making the changes therefore requires large amounts of redevelopment time and money. It is not surprising that system designers, and even managers, are resistant to change at this point. From all points of view, it is not desirable to go back and make changes in the program.

Because of this characteristic of instructional system design, a frequent result is that (a) the instructional product undergoes changes at a great expense and delay, or (b) the instructional system does not undergo changes and is therefore unsatisfactory from the instructor, management, or learner viewpoints.

In summary, use of the traditional instructional design model often results in products that are unacceptable to trainers and learners, and/or results in lengthy and expensive development life cycles. At the same time, businesses are currently attempting to speed up product development times (termed "fast cycle time" methods). Training program designers are similarly being pressured to reduce development times. What is needed are design methods that support simultaneous design and development with rapid feedback for easy and early design modifications. This need is filled by incorporating prototyping and user testing.

Incorporating Prototyping into the Instructional Design Model

The instructional design model presented in this text incorporates prototyping and user testing in the middle design and development phase. The purpose is to acquire extensive feedback about the design very early in the design process. By acquiring feedback and making program modifications early, the design process becomes more streamlined, more efficient, and more effective. The goal is to perform formative evaluation as *early* and *extensively* as possible. In this chapter, we will discuss the nature of prototyping and user testing, and suggest how these procedures should be carried out for various types of instructional systems. In this and the next chapter, we will show how prototyping and user testing can often be performed using the same materials that are being developed for the actual instructional system.

We have been talking about prototyping and the need to perform it early in product design stages. What exactly is a prototype in instructional design? A prototype is a facsimile of the ultimate program; a "system" with the look and feel of the ultimate instructional program or performance support system. For example, if the instructional strategy is to be a videotape, it could be a relatively rough, unedited version of the videotape. A prototype of a CAI program might be the initial interface with only one topic partially completed. Prototypes for the various instructional strategies will be described shortly.

Depending on the type of instructional delivery system, prototypes will vary in their complexity. Documents, videotapes, and CAI systems are actual deliverable products, and therefore will generally need more user testing during development than person-based methods such as lecture, tutoring, or on-the-job training. CAI usually needs the most extensive user testing and modification because (a) it is generally the most difficult and/or complex delivery system, and (b) learners know less about using that technology than using documents or videotapes. In addition, we have a greater store f knowledge about the design of text and video than we do about the design of CAI systems.

Rapid Prototyping

For CAI instructional systems, usability is a central issue. We don't question whether a learner knows how to use a videotape, but we do question whether he or she can use a CAI instructional system. Because CAI systems are usually complex in terms of controls and display elements, prototyping and user testing often require many iterations and changes. This is extremely difficult to do if the prototype has to be coded from scratch. This problem has led to a technique called **rapid prototyping** (Wilson & Rosenberg, 1988). It is a method of making a computer-based mock-up of the user interface where much or all of the instructional system isn't yet functional. The goal is to have a prototype that is extremely easy to modify. This allows many iterations of design in a short period of time, often within a single session with a user. As Wilson and Rosenberg (1988) state:

> Since it is impossible to solve any significant design problem in a single iteration, designers require flexible tools which facilitate multiple iterations. . . . In the design of a user interface the rapid prototyping system is the equivalent to the sculptor's clay. It fosters speedy implementation of alternative user interface concepts at both a sketching and a simulated implementation level.

Thus, rapid prototyping is prototyping that requires very short periods of time for development and modification of the prototype. We will be discussing prototyping as a general procedure, with the assumption that rapid prototyping is functionally the same, but just requires a relatively short amount of time.

There have been several software tools designed over the last ten years to support rapid prototyping. Some of these interface design tools are listed in Table 12.2. Most of the tools create empty interface shells that do not have any true functionality. However, this separation between prototyping and actual software development is changing. This is because there are now a number of software development tools that are based on a graphical user interface (GUI) requiring little or no programming. These tools make building actual working application prototypes as simple and fast as the older rapid prototyping tools that built only nonfunctional interface shells. Examples of GUI development tools include *HyperCard, SuperCard, Quicktime, ToolBook, Compel, IconAuthor,* and *Authorware.* As these tools proliferate, we may see the term rapid prototype drop out of use, because normal prototyping will be able to proceed just as quickly as the "rapid prototyping" using special tools.

Prototyping for Instructional Design

Purposes of prototyping and user testing.   There are many reasons for using prototypes during design and development; all have to do with formative evaluation. That is, the design team evaluates the product as it is

TABLE 12.2   Some of the More Well-known Rapid Prototyping Tools

Software Name	Reference
ACT/1	Mason & Carey, 1983
Author's Interactive Dialogue Design Environment (AIDE)	Hartson & Johnson, 1986
Dialog Management System (DMS)	Hartson, Johnson, & Ehrich, 1984
Forms Management System (FMS)	Digital Equipment Corp., 1984
Functional Language Articulated Interactive Resources (FLAIR)	Wong & Reid, 1982
Interactive Dialogue Synthesizer (IDS)	Hanau & Lenorovitz, 1980
Protoscreens	Bailey & Bailey, 1991
Rapid Prototypes of Interactive Dialogues (RAPID)	Wasserman & Shewmake, 1982

developed, not just at the end. Three of the most important reasons for using prototypes are:

1. **Design team communication.** Prototypes provide a tool for the members of the design team to communicate elements of their design concepts with one another. Having a prototype "model" of the system makes working on the design easier than trying to do it in abstract fashion.

2. **Project approval.** Prototypes are used to communicate a design concept or its elements to a client, manager, instructor, or other personnel who are responsible for approval of the project. This allows the design team to acquire approval or comments before getting too far into the development process.

3. **User testing.** Prototypes are used as a product and context for user testing. User testing is conducted to determine the effectiveness of the program in meeting the basic goals, objectives, and requirements in the system specifications document.

The first two purposes are relatively self-explanatory. The last purpose, prototyping to support user testing, needs some elaboration. There are two types of information needed for formative evaluation that can best be obtained from learners themselves. One is the usability of the system, the other is the instructional effectiveness of the system.

Many designers feel that they are capable of determining the usability and instructional effectiveness of a system, but that is generally not the case. The reason is that the designer has become an expert in several areas. He or she has become an expert in the specific content of the instructional program (although maybe not in the *domain* per se), and has become an expert in terms of knowing the instructional program itself. By definition, he/she is not a novice. Even if they might try, it is almost impossible for designers to take the point of view of a novice.

It is for this reason that the field of human factors has embraced the process of user testing. Although human factors engineers know the basic interface design errors to avoid, they still rely strongly on user testing, because they realize that no one else can approach the system with the same cognitive set and knowledge as a learner interacting with the system for the first time.

One rarely hears about prototyping for instructional design except in the case of CAI systems. This is unfortunate, because almost all of the instructional strategies can benefit from some type of user testing before the entire system is developed and submitted for final evaluation. However, it is certainly the case that some types of systems require more *extensive* user testing than others. Methods for proto yping all types of instructional strategies will be reviewed shortly. First I will briefly describe the general types of prototyping methodologies.

Prototyping methods. There are numerous ways to develop prototypes, and the nature of a prototype will depend largely on the specific design concept that is being evaluated However, in spite of these differences there are certain traditional methods of doing prototyping. Several common methods for prototyping are:

- Document headings plus cne full section
- Storyboard/sketch
- Slide show techniques
- Linked computer screens
- Fully animated prototypes
- Wizard of Oz

The first method is commonly used to prototype standard documents. A list of headings and subheadings is developed. For one topic or section, the entire text with graphics is developed in rough draft form.

A **storyboard/sketch** is a series of static pictures or graphics that sketch out the nature of the program. Sometimes designers will cover an entire wall with such sketches. This gives a general overview of what the program will look like when it is finished.

The use of **slide shows** for prototyping is relatively self-explanatory; they can be useful for conveying complex or realistic visual information. **Linked computer screens** can be used to prototype lectures, workshops, simulations, CAI, videotape, in short, almost any type of training program. They can also be useful for prototyping job aids, which are often on-line information bases. For more dynamic programs such as CAI simulations or games, simple **animated prototypes** are more appropriate, and can be developed relatively quickly and easily with the new animation software on the market (e.g., Autodesk *Animator^{TM}*).

Finally, for ICAI applications that have complex interactive capabilities, prototyping can be done using the *Wizard of Oz technique* (Wixon, Whiteside, Good, & Jones, 1983). In this method, the learner interacts with a computer that appears to be a fully functional system. However, a human is actually sitting behind a screen or wall, and providing the reactions that are supposedly coming from the computer itself. This method allows a variety of instructional strategies to be quickly tried out with the learner, without requiring any programming.

Advantages of prototyping. We have briefly touched on some of the reasoning behind prototyping, and therefore indirectly on its advantages. Wilson and Rosenberg (1988, p. 861) give seven benefits of prototyping in software development, and these are equally applicable to instructional or performance support programs:

1. It provides a means for testing product-specific questions that cannot be answered by generic research and guidelines.
2. It provides a tangible means of evaluating an interface concept.
3. It provides a common reference point for all members of the design team, management, instructors, etc.
4. It allows the solicitation of meaningful feedback from users.
5. It improves the quality and completeness of a product's final functional specification.
6. It increases the probability that the product will perform as expected.
7. It substantially reduces the total development cost.

With recent technological advances related to the computer industry, training analysts now have a number of tools to support prototyping activities. There is good reason to embed prototyping into the early design phase, and few excuses not to do so.

DESIGNING INSTRUCTIONAL MATERIALS

As a result of identifying a design concept, the analyst will have written a summary of both the content and general presentation strategy for the instructional program. This concept is now transformed into a product. In this section, I will briefly discuss three topics that pertain to a majority of the instructional methods:

- Determining the overall instructional content
- Determining the sequence of instructional activities or units
- Determining the instructional content at the unit level

After those topics, I will review specific methods for each type of instructional design. Keep in mind that this chapter covers only the initial development and prototyping steps. Chapter 14 reviews procedures for full-scale development.

Determining Overall Instructional Content

The first task in designing instructional materials is to determine the overall content for the program. This means identifying the tasks to be taught and conceptual knowledge to be acquired. If the task analysis has been done adequately, the information will be contained in that database. It is now a matter of going through the information to find the major topics. If a conceptual graph or other network has been used, the central topics will be the ones that are high in the hierararchy, and the ones with many associated links. Gordon and Lewis (1990) gave an algorithm for determining the central nodes in a conceptual graph structure.

Besides determining which tasks or skills are to be learned, the analyst must also identify the breadth to be covered for each task. That is, if it is important that the trainee be able to transfer the skills to a wide range of problems, then a representative wide range should be covered through training.

Other topics to be identified at this time include necessary supporting materials. These include examples, case studies or expert stories, illustrations, problems, etc. In determining the basic structure and content of the instruction, an analyst can use many of the tools we discussed in task analysis. In particular, concept networks, goal structures, flow charts, and hierarchical lists are appropriate representational formats (Gill & Gordon, in press; Cook, 1991).

Determining Instructional Sequence

Most types of instruction require the analyst to identify a sequence of activities for the training program. Whether we are presenting conceptual information, or providing a context for skill practice, we must identify a particular order in which activities will take place. The designer usually performs this task by choosing one characteristic or aspect of the information to use for sequencing the instructional modules.

There is no one way that has been shown to be most effective for sequencing instructional programs. Each instructor will have favorite methods for organizing material. However, most instructors first divide the material by general topics. Once topics have been identified, there are some general rules that can be followed. As a starting point, the reader should go back and review the instructional strategies listed at the end of Chapter 6 on Learning and Motivation. Many of the guidelines written for instructional sequencing will reflect the strategies presented there. For example, a guideline written by General Physics Corporation (1983) suggests the following five principles:

- Start with what is known or familiar
- Work from simple concepts to more complex understandings
- Use concrete instances to lead to abstract principles
- Present the "big picture" first; then focus on the details
- Structure the flow of ideas

Notice that all of these suggestions are consistent with learning principles discussed earlier in the text. The fifth principle in the list refers to principles based on information-processing theories, such as grouping similar or associated ideas together, giving labels to each group of ideas, and ordering groups of ideas so as to reinforce their associations. There are several authors who have presented guidelines for sequencing instructional elements, including Dick and Carey (1985), Gagne, Briggs, and Wager (1988), Reigeluth (1987), and Scandura (1971, 1977a, 1977b, 1980).

The instructional designer should work with information obtained from the task analysis and also with the subject matter expert(s) to determine the most effective order of information for each topic. This order will be based on principles of design but also on goals of the program. A good example is provided by General Physics Corporation's *Principles of Instructional Design* (1983). The example concerns ordering of information about employees and radiation exposure. Imagine that the designers had developed a conceptual graph with the primary concepts or nodes being those shown in Figure 12.1. Based on that information, a typical instructional sequencing might be something like:

- What is radiation?
- How is radiation measured?
- What are the biological effects of radiation?
- How are employees exposed to radiation?
- How is exposure controlled?

These topics would organize the information from Figure 12.1. Notice that this plan is conceptually well-organized and comprehensive. However, it starts with abstract and unknown ideas rather than concrete, familiar ideas. As the authors state, "For the general employee, the main purpose of learning about radiation is so that he or she will recognize the radiation risks . . . and observe the mandatory precautions" (General Physics, 1983). A better organization puts concrete information most relevant to the trainee *first*:

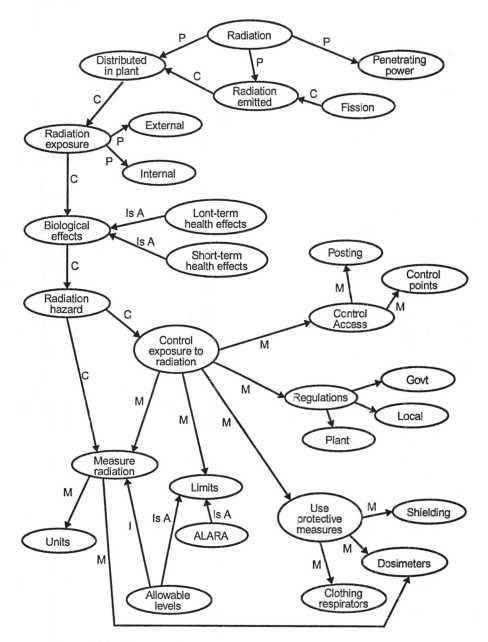

Figure 12.1 Primary nodes in conceptual graph for training about employee exposure to radiation.

- The problem of radiation
- The causes of the problem
- The employee's response

Notice that both plans would cover the material contained in the conceptual graph, but in a different order. The second order is more in line with the instructional guidelines presented in Chapter 6.

After you sequence the information to be presented, add information concerning how and where previous knowledge will be brought in, where examples will be placed, where outside materials will be used, where trainees will perform tasks, etc. At this point, it is a good idea to have other people check the topics you are covering and your lesson sequencing. These people can be peers or other members of the design team, the person who will do the training, and learners. For the learners, ask them whether the material sounds interesting and relevant, whether the topics include information that they feel they need, and whether the order makes sense to them. Obviously, if it is a completely new area for learners, you may not be able to get all of these types of feedback from them.

Determining Unit Content

After you have determined the basic sequencing, you will need to determine the content for each instructional unit. This can be done in a number of ways. The first is to analyze the material and organize the content using the guidelines presented in Chapter 6. For declarative conceptual knowledge, provide definitions with an adequate number of examples, nonexamples, and demonstrations. For causal information, try to allow learners to experiment with the system by asking "what if?" questions. While it is easiest to do this with simulations, it can be done with lecture or other methods as well.

In particular, remember to generate activities that will involve learners if there is sufficient time. Some researchers suggest asking learners to try things *before* you give them specific information (e.g., Esler & Sciortino, 1991). By embedding the subsequent information into a task you have given the learners, this method causes them to process the information more deeply. As before, follow the guidelines in Chapter 6. As you develop the materials, don't hesitate to use prospective learners to "user test" your ideas using some of the methods discussed in the next chapter.

Content guidelines.   There are several prescriptive instructional guidelines that have been developed over the years to help analysts identify the types of instructional materials that should be included. One such guideline is component display theory (Merrill, 1983). Component display theory is based on a building blocks approach to instructional design. According to this model, there are certain types of information that may be included in an instructional program:

- **Facts:** A fact is an arbitrary or rote or historically determined relationship between objects, symbols, and events.
- **Concepts:** A concept is a set of objects or events which share common characteristics and a common name.
- **Processes:** A process is a description of how a system "works" that emphasizes the timing and sequence of events.
- **Procedures:** A procedure is a set of sequential steps designed to solve a problem or achieve a goal.
- **Principles:** Principles are guidelines or factors which often incorporate descriptions of cause and effect relations and which can be used to solve problems that are novel or not previously encountered.

Merrill (1983) suggests that in addition to presenting these types of information, learners need examples, nonexamples, and case studies. In addition, for each type of information, learners can be given information (exposition), or can be asked questions regarding the information. Table 12.3 shows the types of information down the left column and the method of dealing with

TABLE 12.3 Primary Types of Information and Presentation Formats

Type/Level of Information	Present Information	Ask Information
GENERALITY Concept Process Principle Procedure	Present the main idea (concept, process, etc.)	Ask a question about the main idea
SPECIFIC Examples Nonexamples Case studies	Present example or non-example of the main idea	Ask a question about an example or nonexample
FACTS	Present a fact	Ask question about a fact

the information across the top. The content of the cells thus becomes the specific molecular units of which the instruction is made up. The information shown in the cells of Table 12.3 is known as the **primary presentation forms** for instructional design. According to the theory, the instructional content of a program should be a well-balanced mix of the six types of instructional units shown in the table cells.

According to Merrill (1983) there are also certain types of additional information that can be used to support and enhance the primary forms. These include:

- Review or explain terms or components in the generality.
- Relate the generality to a specific example or nonexample.
- Present an alternative representation of the generality (such as paraphrase or diagram).
- Present an analogy that likens the relationship in the generality to a similar situation already understood by the learner.
- Provide a context for a specific example or set of examples.
- Provide the correct answer to a problem or question.

Comments.   These "building blocks" type of guidelines provide a good reference for instructional content ideas, and a useful checklist as the system is being designed. For example, the analyst can make sure that examples and nonexamples have been given for concepts or principles being presented.

One disadvantage in using such guidelines is that the analyst can be lulled into thinking that the program will be effective simply because it contains each type of instructional unit. This is not at all necessarily the case. We saw earlier that acquiring pieces of knowledge is not enough, learners need "a deeper understanding of why, how, and when various concepts and procedures are useful, and they need the kinds of experiences that will allow them to develop organized knowledge structures that are richly integrated" (Cognition and Technology Group at Vanderbilt, 1993).

The best way to illustrate the difference is through an example of a program that has the basic instructional units but not the overall desired effect. Suppose that we create a computer-based instructional program for learning about basic mechanics. We provide a set of wheels, cogs, chains, ropes, and pulleys in a toolbox (on the computer). The learner is asked to assemble no more than five pieces so that a 500-pound weight is lifted to a point halfway between the floor and ceiling of a room. In order to do the task, the learner can use a mouse to assemble the pieces on the computer monitor and see how far the weight moves up from the floor as a result of the assembly. In addition, to help in the task, the learner can "ask" the computer to provide some basic concepts and principles in physics and engineering that are relevant to the problem. It is possible to design such an instructional system in accordance with component display theory, with a correct mix of instructional components and forms. In this particular case, the concepts and principles would be complete and correctly presented.

However, the system may still fail. For example, if the learner asks about concepts and principles, he or she may not be able to see how the information is useful for the particular problem at hand. Finding that making all of the necessary inferences is too difficult, the learner simply assembles and modifies the unit in trial and error fashion until the problem has been solved. The learner saw and read the material, and "solved" the problem. The trainee did not really learn and *comprehend* the information, and certainly did not learn to *apply* it, either in the problem given or to others. While the molecular instructional units were correct, the overall instructional system failed.

In summary, the designer must strive to keep in mind the *overall goals* of the instructional system and design a program to achieve those goals. If the analyst does happen to develop a program that fails to meet the goals, this fact should become apparent during user testing. This is why correct user testing is so critical.

DEVELOPMENT METHODS FOR MAJOR INSTRUCTIONAL TECHNIQUES

Depending on the type of strategy you have chosen, some of the *possible* steps for developing the system are:

1. Determine general content
2. Sequence and structure
3. Write a flow chart if the program is interactive
4. Develop interface prototype(s)
5. Write script
6. Develop storyboards
7. Develop materials
8. Write computer code
9. Compose final product
10. Debug

Table 12.4 shows the steps that are most likely to be included for each of the instructional design strategies. While the general procedure is the same for all strategies, it can be seen that some strategies require much more extensive development processes than others.

For some methods, such as videotape, a prototype cannot really be started without actually beginning on the project itself. For example, one of the first steps in developing a videotape is to write the storyboard. This process is required for prototyping as well. Therefore, this chapter provides most of the details for system development as well as prototyping. Processes that are specifically involved at later points after user testing will be covered in Chapter 14. In the following sections, we will review the initial development steps required for each instructional strategy in the order shown in Table 12.4. In each section I successively focus on the new additional procedures.

Tutoring, OTJ Training

Development of tutoring and OTJ training systems requires less in the way of prototyping and user testing than other instructional methods. This is because (a) there is usually not a great deal of materials development, and (b) a human performs the instruction rather than a product such as a videotape or CAI program. This means that ideally, the human can adapt to learner needs and misunderstandings, a characteristic uncommon for physical training "products." However, like lectures, these systems should be subjected to at least minimal

TABLE 12.4   Procedures Included in Different Instructional Strategies

	Instructional Design Steps									
	1.	2.	3.	4.	5.	6.	7.	8.	9.	10.
Tutoring	X						X			
OTJ Training	X						X			
Lectures and Workshops	X	X					X			
Linear Documents and Job Aids	X	X					X		X	
Videotape	X	X			X	X	X		X	
CAI	X	X	X	X	X	X	X	?	X	X
Nonlinear Documents and Job Aids	X	X	X	X	X	X	X	?	X	X
Simulations	X	?	X	X	?	?	X	?	X	?

formative evaluation and user testing. The analyst might proceed as follows in developing the system:

1. Determine Instructional Content and (perhaps) Sequence. Traditionally, both tutoring and on-the-job training have relied on the instructor's implicit knowledge of what to cover and how to cover it. This means that often there is no preparation of an instructional sequence. However, if you have decided that tutoring or OTJ training is the best strategy, you will make the program more successful if you provide content guidance for the trainer or instructor. This guidance should come in the form of topics to be covered and the types of tasks that are to be mastered.

2. Develop Materials. For both tutoring and on-the-job training, checklists should normally be provided, and depending on the particular situation, exit tests as well. Checklists should include important information for each subtask to act as reminders for the trainer. Examples include conditions under which the subtask is performed, special exceptions, expected and unexpected outcomes, etc. These materials will come from your task analysis and instructional objectives. When there is any doubt, insure that the person to be doing the training is aware of proper procedures for carrying out all of the tasks to be trained.

If you are not the person to be doing the training, a short introduction to the checklist should be included for the person who will be doing it. This introduction explains the importance of the checklist and how to use it most effectively. Items on a checklist can be marked if they have been covered during instruction, or alternatively, they can be marked when the trainee successfully performs the task. Make sure your instructions are clear on which of these alternatives is to be used.

3. Develop prototype materials. The materials to be used for formative evaluation and user testing include the list of topics and tasks to be covered during the instruction, a description of how the material will be covered (whether by demonstration or having individual trainees perform the task, etc.), and example(s) of exit test items.

4. Use the materials for formative evaluation and user testing as described in the next chapter.

Lectures and Workshops

For lectures and workshops, system development generally consists of preparing outlines, handouts, visual aids, any job aids to be provided, and guided learning or practice activities. The prototype should be an outline of the topics to be covered plus a complete set of instructional materials for one module or component of the program. Probably the most efficient method for initial development and prototyping is as follows:

1. Divide the material for the training program into units along naturally occurring breaks in your task analysis data. A unit might represent 10 minutes of lecture or it could represent a day. Most people prefer to break the material down into relatively small sizes, for example, a module might take one-half to one hour to present.

2. Determine the types of instructional strategy for each of these units. If all of the units are the same type of information, such as system components and causal relationships, you can identify one strategy for the entire program. It is more likely that there will be more than one type of information. As described above, identify the instructional strategy to be used for each unit or module.

3. Decide on the sequence for presenting the units. See guidelines presented earlier in the book.

4. Choose one to three units to develop as prototype(s) for formative evaluation and user testing. One unit will suffice if the material is relatively homogenous in type and the same instructional strategies are being applied throughout the program. If there is wide variety in type of material, difficulty of material, or type of instructional approach, choose two or three units to cover the differences (e.g., one conceptual unit, one causal system, and one goal/procedural unit).

5. Develop the lecture outline and associated materials using the guidelines in Chapter 6 or a methodology such as *component display theory*.

6. Use the fully developed unit for formative evaluation and user testing as described in the next chapter. If there are major modifications to be made after user testing, modify the unit and develop a second one. Develop additional units and repeat user testing if possible.

Linear Documents

While prototypes of documents are almost always developed (e.g., a working draft), it is probably safe to say that most documents published in both academia and industry do not undergo user testing during product development. This is extremely unfortunate because, as Wright (1988) points out, current documentation is seriously deficient in serving the needs of its readership. In discussing the seriousness of the problem, Wright cites a quote from Houston (1984), who stated that "Data such as the number of times per week the document is thrown against the wall might be useful in evaluating a product."

There is certainly a need for improved design and development of documents of all types. And while there are many guidelines for this process, prototyping and user testing is still necessary. One procedure for prototyping documents, whether hardcopy or electronic, is as follows:

1. Identify the major topics to be covered in the document. The topics are usually complex enough that you will need to develop a hierarchical list.

2. Determine the sequence of materials for text. Instructional sequencing for documents will proceed in essentially the same manner as for lectures. Determine the order of topics and subtopics, then outline/describe all supporting materials. Use the document design guidelines given in Chapter 11 as well as other guidelines such as Clark's (1989) book on developing technical training materials. Make notes regarding where to add graphics when they would be useful for expanding or clarifying information in the text. Once a detailed outline has been made, this is a good time to obtain approval from clients and/or instructors. Users may be helpful at this point also, depending on the application.

3. Write a style guide if you have not already done so (specifying the layout, format, etc.).

4. Develop one of *each type* of section or unit. The text should be complete in first draft form. For figures or graphics, sketches will do as long as the readers can comprehend them.

5. Perform formative evaluation and user testing using the instructional sequence, such as a table of contents, plus the units that have been developed. Based on user testing, modify the prototypes and perform user testing again.

Hardcopy Job Aids

Job aids are a rather unique class of system. Because of the wide variety of job aids and performance support systems, it is difficult to give specific procedures for prototyping and user testing. For computer-based job aids and performance support systems, they should be treated like the CAI systems described below. When job aids are relatively small, such as the ones shown in Figures 11.1 through 11.3, prototyping and user testing can proceed as follows:

1. **Determine the basic topics and content for the job aid.**
2. **Determine the organization and sequencing of the job aid.**
3. **Develop a prototype.** Write a rough draft of the entire contents in a form that users can read and comprehend.
4. **Submit the prototype to formative evaluation and user testing.**
5. **Modify the prototype and continue user testing.**

When the job aid is long, perform the prototyping and user testing as described for documents.

Videotape

Making a high-quality videotape takes time, resources, and knowledgeable people. If your videotape will have many types of information, such as text, animation, and full-motion video, you will need people on your design team to perform the following roles:

- Project Leader
- Subject Matter Expert
- Instructional Designer
- Scriptwriter
- Video Producer
- Graphic Designer
- Programmer
- Learner

Sometimes one person has the capabilities to assume more than one of these roles. As we discuss the various processes needed to develop videotapes, it should be apparent who will be doing the majority of the work in that phase.

1. Instructional sequencing. Determining the sequencing for a videotape is relatively similar to the previous instructional strategies, in that the designer must determine a linear sequencing of the information that is well-organized, comprehensive, coherent, and that will draw the learner into the material. For that reason, the guidelines given above for lecture and other methods are applicable. In determining the sequencing of a videotape, it is somewhat easier to draw the learner into the material because film and animation can be used at the beginning. Remember not to use these media for their own sake, but as a tool to bring the learner "into" the subject, for example showing interesting scenes that are relevant to their own personal lives.

Also, remember that videotape is sometimes not a medium through which the learner can progress at an individual pace. Intersperse primary critical information with elaborative, interesting film clips and animation sequences. This will give the learner time to encode the primary information. *Many* instructional videotapes present too many independent pieces of information at a rapid pace, making it difficult to remember the material. Finally, remember that videotape now frequently includes computer-generated text, graphics, and animations. Use these media where appropriate for the type of material being conveyed (e.g., a list of correct procedures), and intersperse them with video clips.

2. Write production script. After determining the sequence of topics, a production script is written. In larger projects, this task is typically performed by a professional scriptwriter. Scripts are predominantly used to develop audiovisual delivery systems such as videotapes and CAI programs. A script specifies narrative or audio content, and the types of visuals to be used, whether computer-based graphics or dynamic video. It gives direction to production of materials for the final product.

Scripts are written iteratively, with the first pass being relatively rough. They are written with what one might think of as a storyline; a beginning, a middle, and an end. In addition, Swain and Swain (1991) suggest that the video part of the script carry the major burden, with audio filling in only where necessary. Other researchers warn that while use of dynamic graphics and video can heighten the persuasiveness of a message, designers should guard against using video that is irrelevant to the message simply for the sake of using catchy visuals (e.g., King, Dent, & Miles, 1991). Anderson and Veljkov (1990) provide a good overview for scripting and storyboarding for multimedia systems, which is equally applicable to videotape.

The visual elements in scripts are simple sketches or just verbal descriptions of scenes. Each can be labeled as computer graphics, animations, or video clips. Specify text where needed. The audio side is also first written in a sketchy fashion. Designate sounds such as music or sound effects, narration, or other types of audio. In the case of narration, put down the gist of what should be stated. Try to avoid redundancy between visual and audio information.

There will invariably be times when you will be writing your own script. Sometimes the script will be a straightforward translation from information obtained through task analysis. Other times, the application will allow some creativity. When in doubt about your writing abilities for a project, seek out appropriate sources of help. For example, Swain and Swain (1991) have a very readable book on scriptwriting with numerous helpful examples.

3. Do early formative evaluation and user testing (optional). After a good part of the script has been written, the analyst may want to do some very simple and preliminary formative evaluation and user testing (see Chapter 13). Expert opinion can be sought, supervisor approval is obtained, and users can be interviewed to get their general reaction to the program. This avoids finding the deficiencies after the costly process of storyboarding.

4. Write the storyboard. After the script has been revised based on user testing, the analyst goes on to develop the project further through the use of storyboarding. User testing can be done relatively conveniently using storyboards.

Storyboards are individual still frames that show the specific content of the scripts. They may contain only the audio and visuals that the learner will see (and hear), or they may also give specific instructions to producers and programmers. Storyboards are either hardcopy "paste-ups" or on-line storyboards. Storyboards usually have visual images in a form that is still somewhat sketchy and incomplete. However, with more powerful computer graphics and animation packages, this does not always have to be the case. Figure 12.2 shows an example of the first two storyboards for a home safety videotape. We should note here that some authors consider storyboarding to be just one way of performing scripting (e.g., Anderson & Veljkov, 1990).

Advantages of hardcopy versions such as those on cards or sheets of paper is that they can be placed on a large board or table and rearranged or replaced. It is often helpful to evaluate individual frames in the context of the big picture. However, an advantage of having the storyboard on a computer (and integrated with a prototype), is that it makes a better testbed for user testing and showing the product to a client.

5. Develop prototype for formative evaluation and user testing. Most videotapes can be and should be submitted to prototyping and user testing before final production. There are basically two ways of doing this, with storyboarding and with an actual video prototype. The easiest way to do prototyping is with storyboarding—either on paper or on the computer. Using a computer, roughed out segments can be put together using development software such as *ToolBook, SuperCard,* or *Authorware.* The segments that are real-world video clips can be described, animated, or actually shot as video clips and digitized.

A second way to do video prototyping is to make a videotape using the storyboard but make it a rough version. That is, have a person read the script

Christmas Tree Hazards
Draft 1, 10/12/91
Board 1: Context

VIDEO   AUDIO

Family in home...
mom, dad, two girls.
Parents are sitting.
Girls walk to parents
in front of tree,   No dialog or narration
talking and looking
at it.

                                                         SFX: Soft Christmas music in
                                                                background

Christmas Tree Hazards
Draft 1, 10/12/91
Board 2: Introduction to subject

VIDEO   AUDIO

Continuation of family   "Christmas trees bring light
by tree   and joy into our homes at the
                                                         darkest time of the year. But
                                                         they can also bring tragedy..."

Figure 12.2 Storyboards for first part of videotape on home fire hazards associated with Christmas trees.

or storyboard and use rough sketches for graphics. The content should be similar to the final version, but the production glitz will still be missing. Certain
types of instructional material merit this level of prototyping. They include
difficult or complex material, and material that is dynamic in nature (and can't
be portrayed well in static storyboard form).

CAI and Nonlinear Electronic Documents

In this section, I will review the basic methods used to develop computer-based systems such as CAI and nonlinear documents (which will include on-line job aids).

1. Instructional content and sequencing. Before beginning system development, the design team specifies the topics to be covered, and whether there is any information that should only be covered in a specific order. This is really a question of the degree to which the learner should control access of information. While some have argued for increased learner-control and self-directed learning (e.g., Rasch, 1988), recent research has shown that novices within a given domain acquire knowledge much more poorly with a learner-controlled system than a program-controlled system (e.g., Lee & Lee, 1991). Presumably this is because they don't have enough domain-specific information to guide their search activities. In support of this hypothesis, Lee and Lee (1991) found that use of learner-control systems for *review* was not problematic. This means that a training program designed to deliver information on a new subject should have relatively less built-in learner control. Information systems for people who have a reasonable amount of domain knowledge can provide more learner control. For example, a simulation that covers material that has *already* been taught can allow a large degree of learner control (Lee & Lee, 1991). Or as another example, most job aids assume flexible use by a relatively knowledgeable person.

2. Determine degree and type of learner control. Finally, if there is information to which all trainees should be exposed, such as safe equipment operating procedures, a system allowing learners to choose which topics they wish to study would not be a good choice. There is a reasonably large amount of instructional applications that are not particularly amenable to learner-controlled access (a characteristic which is typical of most hypertext applications). That is because much information has a "building blocks" structure; certain basic information must be learned before more complex concepts. Also, it is often the case that the design team has determined the information necessary for training, and allowing learner free access means that certain parts of that information will not be seen. Research has shown that this can lead to poorer learning of the material itself, and also poorer performance on realistic tasks requiring use of that knowledge (Gordon & Lewis, 1992). One solution that has been suggested is to divide the information into an organized hierarchy, but to provide the novice with a "guided tour" of the information (Shneiderman, Kreitzberg, & Berk, 1991).

3. Determine types of interaction. In addition to determining the degree to which the learner should control the order of information, the designer must also decide what types of interaction will be supported within each topic. Will the learner be interactive at just the information access level (as in hypertext and hypermedia) or will the program be interactive in the sense of simulations, instructional games, and learner activities such as summarizing, making concept maps, etc.? (These questions may have already been resolved back in the design concept stage.) The first type of interactivity, which is interactive access of information, is discussed in the next section.

High-level Structure of Nonlinear Programs. When developing a nonlinear document, job aid, information base, or any type of hypertext or hypermedia environment, the designer must be extremely careful how he or she structures the information. The underlying assumption that linking information together will allow easy traversal by learners has been shown to be false (e.g., Bernstein, 1991; Gay & Mazur, 1991). The structure of the information must make it easy for the user to navigate through the information. Many researchers strongly encourage the use of only strictly hierarchical hypertext structures with *no links* between different branch structures of the hierarchy (e.g., Akscyn, McCracken, & Yoder, 1987; Gordon & Lewis, 1992). In addition, the top-level hierarchy should be an access menu that is easy to understand, and that also gives the learner an idea of the structure of the underlying database (Kearsley & Shneiderman, 1988).

Information bases can also be designed to have multiple access hierarchies (Garzotto, Paolini, Schwabe, & Bernstein, 1991). An example is the on-line help system for the Microsoft Windows Paintbrush program. Users can access essentially the same information via a list of commands, procedures, the specific tools used to create graphic designs, or an index (which simply links to the screens in the other four hierarchies).

In summary, keep in mind that access to nonlinear documents can be very complex. The danger is that to the designer, who has a very clear *designer's mental model* of the system, the access hierarchy or network may seem very straightforward. However, a user may have trouble developing *any* mental model of the system organization. For these types of systems, user testing is absolutely critical.

System Flow Charts. After determining the general sequencing or structure of the program, a more detailed flow chart is developed. Sometimes these are constructed entirely before scripting and storyboarding. Other times, they may be designed as the frames are scripted (to be discussed shortly).

Flow charts describe the organization of the instructional content; what choices are available to the learner, and what the system will do in response to those choices. The flow chart graphically depicts all of the different paths which the user could follow through interaction with the instructional system. With flow charts, we specify "certain portions of a lesson to display information, others to present questions and solicit responses, and still others to skip, repeat, or continue lesson execution based upon the accuracy of a student response" (Hannafin & Peck, 1988). Each module in an interactive program may contain a variety of elements including instructional objectives, presentation of information or procedures, games or simulation elements, questions or tasks for the learner, demonstration and practice, and tests. These are identified in the flow charting process.

Figure 12.3 shows a flow chart syntax used by Hannafin and Peck (1988), with terminal nodes and input nodes. Although there are many types, these flow charts illustrate forward and backward branching (see Hannafin & Peck, 1988, for an extensive review). It can be seen that these flow charts specify the functional interaction between learner and computer, but don't specify screen content or design.

4. Develop interface prototypes. At the same time that you are developing the flowcharts, you can begin to develop the system interface. The interface should have the "look and feel" of the final system. This includes screen layout, icons, buttons, etc. Remember to use the interface design guidelines given earlier in Chapter 11. We mentioned several rapid prototyping tools available for this task. The hypertext authoring tools make it easy to sit down with the user and actually make changes to the interface during user testing.

As noted in Chapter 11, one should develop a CAI system that is easily and immediately usable by the learner (assuming a stand-alone system). That means that the analyst should design the system such that when the learner uses it, *at any given point in time,* the learner knows the following information:

- What options are currently available to the user.
- How to achieve their current goal(s).
- The meaning of all information displayed on the screen.
- Whether or not the system is ready for user input.
- What the system is doing (such as waiting, loading files, etc.).
- How much of the system operation has been completed (if it will take a long time).

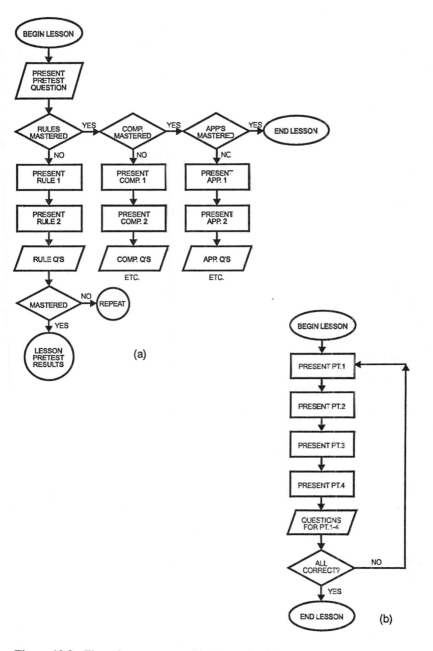

Figure 12.3 Flow chart syntax used by Hannafin & Peck; (a) Forward-branching flow chart based on pretest scores, (b) backward-branching flow chart. (Reprinted with the permission of Macmillan College Publishing Company from THE DESIGN, DEVELOPMENT, AND EVALUATION OF INSTRUCTIONAL SOFTWARE by Michael J. Hannafin and Kyle L. Peck. Copyright 1988 by Macmillan College Publishing Company, Inc.)

Keep these goals in mind as the program is designed. They can also be tested directly during user testing.

As we have said earlier, some systems such as *Authorware* will allow you to do these things simultaneously. Develop the flow charts and interface prototypes *very early* before any extensive scripting, graphic design, and so forth. You may need to go back and perform major changes on your original design concept, interactive mechanisms, screen designs, and so forth. It is easiest to do this before much work goes into the instructional materials themselves.

5. Do first round of user testing. Perform the first round of formative evaluation and user testing before scripting, storyboarding, and full-scale development.

6. Write the necessary scripts. After the outline or flow chart has been developed and the interface has passed user testing, you will begin fleshing out the instructional content. This is done by scripting, a process that was described above for developing videotapes. Scripting for CAI or computer-based systems is similar in the types of materials that are written. The scripts are different in the sense that for linear programs, the script can be designed as a "whole," with continuity between elements. For nonlinear, computer-based systems, each component must be designed and written as a complete and autonomous segment. Thus, each component is a small vignette or idea, able to stand on its own. However, other than this one difference the content and manner of scripting is much the same as for linear systems.

7. Develop storyboards. Development of storyboards proceeds in the same fashion as described for videotapes. Each storyboard specifies text, sound or narration, graphics, photos, animation, and video. These are roughed in at first to obtain approval and conduct initial user testing. Usually the storyboards can be developed directly on the computer, although some people prefer to do it by hand for the first draft. As noted before, this allows more of a global view of many storyboards, laid out in the organization of the application being developed. Storyboards for interactive systems may be more complex; they often must specify information for programmers and other specialists.

8. Develop one instructional unit in entirety to use for formative evaluation and user testing.

9. Perform further formative evaluation and user testing, as specified in Chapter 14.

Simulations

There is no hard-and-fast sequence of steps for developing a simulation, because each will be unique. In general, one determines the tasks which will be simulated, and in what level of detail. Then the analyst determines whether there are tutorial or other instructional elements to be delivered either by a trainer or an embedded CAI system. In most cases, the design activity can proceed as follows:

1. **Determine general nature of simulation.** This step involves identifying the range of tasks that will be performed on the simulation. Good simulations have an opportunity for the trainee to perform behavioral and cognitive subtasks, and receive feedback on those subtasks. Often, a simulation can provide an opportunity for the trainee to externalize his/her cognitive processes providing better training than real-world situations (see previous discussions of this, for example, the TMT).
2. **Develop a prototype or mock-up of simulation,** including basic interface characteristics.
3. **Use the prototype or mock-up for formative evaluation and user testing** as described in the next chapter.
4. **Develop the simulation in greater detail,** and add a representative section of the declarative knowledge to be contained in any information databases for learner access
5. **Repeat user testing.**

COMBINING PROCEDURES

The various procedures listed above have historically come from days when computers were less powerful and many of the processes were done by hand with pencil and paper or typewriter. The processes were conducted sequentially in the order I have shown. That is, the script was written by a scriptwriter and then handed over to the instructional designer, project lead designer, or someone else for storyboarding. Once storyboarding was finished, the materials were handed over to a producer (for videotape) or other appropriate technicians (such as graphic designers, programmers, etc.).

Using such a stepped sequence for the development process has several disadvantages. The first is that the process makes it too hard to get feedback from learners early enough in the development cycle. That is, it is often difficult to perform user testing with scripts or even storyboards. Second, clients and instructors may become anxious waiting for months while the scripts and storyboards are developed (Gayeski, 1991). In addition, when they finally see the product, any changes they request (which are virtually certain) may have serious financial and time consequences because so much work has already gone into the product. Third, passing the project from one specialist to another reduces the team effort. Each person works in an isolated fashion, and not only loses out on other input, but may become attached to "their" work and become resentful as it is changed down the line. And fourth, there are methods of combining some of the procedures with computer technology that simply saves time and effort, and make development a more natural and integrated effort.

As an example of how several procedures can be combined and performed concurrently, consider developing a nonlinear computer-based document or CAI program. Regardless of how the program will finally be implemented (type of language, etc.), the flow charting, scripting, and storyboarding can be performed concurrently. This can be done on paper, or even more easily, by using one of the software tools now available. For example, the design team could work together in the *Authorware Professional* programming environment. The flow chart function could be used to begin the flow charting process. At the same time, individual screens can be started. Some of these screens might be storyboards representing ultimate sets of many screens. Each screen can act as a storyboard with sketches, designer notes, and text. Or alternatively, designer notes can be appended separately and accessed by special buttons.

Notice that this provides one integrated tool for flow charting, scripting, storyboarding, and *prototyping*; the working product can be used for user testing as it develops. This will result in rapid prototyping with both client and user input as the project proceeds. In addition, graphics and animation developed by technicians can be easily brought into the design and moved around as the team sees fit.

If the instructional program is ultimately to be some other format such as a videotape or programmed CAI, much of the materials such as animation or graphics can still be used. Even for videotapes, the same computer-based methods can be used without any branching. Digitized photos and videotape segments can be pulled in for prototyping and early design work.

One of the methods we have found effective is for the instructional designer to just sketch out brief and incomplete flow charts of boxes referenced to roughed out storyboards (see Figure 12.4). This is often faster than developing them on the computer and gives a broad overview. Then a technician or assistant can start screen design on the computer using tools such as *Hyper-*

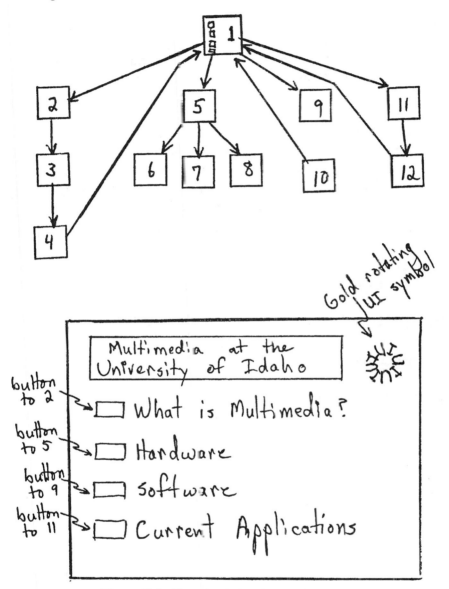

Figure 12.4   Flow chart indexed to storyboards.

Card, ToolBook, IconAuthor, or *Authorware Professional.* Alternatively, if screen design is really uncertain, we design directly on the computer to try out different options in a more "high fidelity" type of environment. The point is to get the concept down in a form that can be expanded and also used for quick turnaround in user testing.

SUMMARY

In summary, for all programs other than tutorials and OTJ training, there are certain development activities that are combined with prototyping. These include:

1. Determine instructional content
2. Determine instructional strategies for each topic or unit
3. Determine sequencing of material
4. Develop materials for one or more instructional units for prototyping
5. Develop computer interface if CAI or other computer-based system
6. Begin formative evaluation and user testing as early in the development cycle as feasible

The next chapter on user testing should be read before development and prototyping begins, as a full understanding of the purpose and methods of user testing will shed light on the design of a prototype.

REFERENCES

AKSCYN, R., MCCRACKEN, D. YODER, E. (1987). KMS: A distributed hypermedia system for managing knowledge in organizations. *Proceedings, Hypertext '87* (pp. 1–20). Baltimore, MD: ACM.

ANDERSON, C. J., & VELJKOV, M. D. (1990). *Creating interactive multimedia: A practical guide.* Glenview, IL: Scott, Foresman.

BAILEY & BAILEY SOFTWARE (1991). *User's manual for Protoscreens 4.0.* Ogden, Utah: Bailey & Bailey Software Corporation.

BERNSTEIN, M. (1991). The navigation problem reconsidered. In E. Berk and J. Devlin (eds.), *Hypertext/hypermedia handbook* (pp. 285–297). New York: McGraw-Hill.

BOEHM, B. W., GRAY, T. E., & SEEWALDT, T. (1984). Prototyping versus specifying: A multiproject experiment. *IEEE Transactions on Software Engineering, 10(3),* 290–302.

CLARK, R. C. (1989). *Developing technical training: A structured approach for the development of classroom and computer-based instructional materials.* Reading, MA: Addison-Wesley.

COGNITION and TECHNOLOGY GROUP AT VANDERBILT (1993). Toward integrated curricula: Possibilities from anchored instruction. In M. Rabinowitz (ed.), *Cognitive science foundations of instruction* (pp. 33–55). Hillsdale, NJ: Lawrence Erlbaum Associates.

Cook, J. M. (1991). *Using mapping for course development.* Alexandria, VA: American Society for Training and Development.

Dick, W., & Carey, L. (1985). *The systematic design of instruction* (2nd ed.). Glenview, IL: Scott, Foresman.

Digital Equipment Corporation (1984). DEC's VAX11 form management system. (Document SPD AE-R440C-TE). Burlington, MA: Digital Equipment Corporation.

Esler, W. K., & Sciortino, P. (1991). *Methods for teaching: An overview of current practices.* Raleigh, NC: Contemporary Publishing.

Gagne, R. M., Briggs, L. J., & Wager, W. W. (1988). *Principles of instructional design* (3rd ed.). New York: Holt, Rinehart, and Winston.

Garzotto, F., Paolini, P., Schwabe, D. & Bernstein, M. (1991). Tools for designing hyperdocuments. In E. Berk and J. Devlin (eds.), *Hypertext/hypermedia handbook* (pp. 179–207). New York: McGraw-Hill.

Gay, G., & Mazur, J. (1991). Navigating in hypermedia. In E. Berk and J. Devlin (eds.), *Hypertext/hypermedia handbook* (pp. 271–283). New York: McGraw-Hill.

Gayeski, D. (1991). *Rapid prototyping of interactive media.* Workshop given at 33rd International ADCIS Conference.

General Physics Corporation (1983). *Principles of instructional design.* Columbia, MD: GP Courseware.

Gill, R. T., & Gordon, S. E. (in press). Conceptual graph analysis: A tool for curriculum development, instructional design, and trainee evaluation. *Proceedings of the 1993 Interservice/Industry Training Systems and Education Conference.*

Gordon, S. E., & Lewis, V. (1990). Knowledge engineering for hypertext instructional systems. *Proceedings of the 34th Annual Meeting of the Human Factors Society* (pp. 1412–1416). Santa Monica, CA: Human Factors Society.

Gordon, S. E., & Lewis, V. (1992). Enhancing hypertext documents to support learning from text. *Technical Communication,* 305–308.

Hanau, P. R., & Lenorovitz, D. R. (1980). Prototyping and simulation tools for user/computer dialogue design. *Proceedings of computer graphics SIGGRAPH 1980 conference.* New York: Association for Computing Machinery, Vol. 14.

Hannafin, M. J., & Peck, K. L. (1988). *The design, development, and evaluation of instructional software.* New York: Macmillan.

Hartson, H. R., & Johnson, D. H. (1986). Human-computer interface development: Concepts and systems for its management. (Technical Report TR-86-07). Blacksburg, VA: Virginia Polytechnic Institute and State University.

Hartson, H. R., Johnson, D. H., & Ehrich, R. W. (1984). A computer dialogue management system. *Proceedings of INTERACT'84 first IFIP conference on human-computer interaction* (pp. 57–61). London.

Houston, J. (1984). Don't bench me in. *BYTE, 9,* 161–164.

Imke, S. (1991). *Interactive video management and production.* Englewood Cliffs, NJ: Educational Technology Publications.

KEARSLEY, G., & SHNEIDERMAN, B. (1988). *Hypertext hands-on!* Reading, MA: Addison-Wesley.

KING, W. C., DENT, M. M., & MILES, E. W. (1991). The persuasive effective of graphics in computer-mediated communication. *Computers in Human Behavior, 7,* 269–279.

LEE, S., and LEE, Y. H. K. (1991). Effects of learner-control versus program-control strategies on computer-aided learning of chemistry problems: For acquisition or review? *Journal of Educational Psychology, 4,* 491–498.

MASON, R. E. A., & CAREY, T. T. (1983). Prototyping interactive information systems. *Communications of the ACM, 26,* 347–352.

MERRILL, M. D. (1983). Component display theory. In C. M. Reigeluth (ed.), *Instructional-design theories and models: An overview of their current status* (pp. 279–334). Hillsdale, NJ: Lawrence Erlbaum Associates.

MERRILL, M. D., LI, Z., & JONES, M. K. (1990). The second generation instructional design research program. *Educational Technology,* March, 26–31.

RASCH, M. (1988). Computer-based instructional strategies to improve creativity. *Computers in Human Behavior, 4,* 23–28.

REIGELUTH, C. M. (1987). Lesson blueprints based on the elaboration theory of instruction. In C. M. Reigeluth (ed.), *Instructional theories in action: Lessons illustrating selected theories and models* (pp. 245–288). Hillsdale, NJ: Lawrence Erlbaum Associates.

SCANDURA, J. M. (1971). Deterministic theorizing in structural learning. *Journal of Structural Learning, 3,* 21–53.

SCANDURA, J. M. (1977a). *Problem solving: A structural/process approach with instructional implications.* New York: Academic Press.

SCANDURA, J. M. (1977b). Structural approach to instructional problems. *American Psychologist, 32,* 33–53.

SCANDURA, J. M. (1980). Theoretical foundations of instruction: A systems alternative to cognitive psychology. *Journal of Structural Learning, 6,* 347–394.

SHNEIDERMAN, B., KREITZBERG, C. & BERK, E. (1991). Editing to structure a reader's experience. In E. Berk & J. Devlin (eds.), *Hypertext/hypermedia handbook* (pp. 143–164). New York: McGraw-Hill.

SWAIN, D. V., & SWAIN, J. R. (1991). *Scripting for the new av technologies* (2nd ed.). Stoneham, MA: Butterworth-Heinemann.

WASSERMAN, A. I., and SHEWMAKE, D. T. (1982). Rapid prototyping of interactive systems. *ACM SIGSOFT Software Engineering Notes,* 1–18.

WILSON, J., & ROSENBERG, D. (1988). Rapid prototyping for user interface design. In M. Helander (ed.), *Handbook of human-computer interaction* (pp. 859–875). Amsterdam: North-Holland.

WIXON, D., WHITESIDE, J., GOOD, M., & JONES, S. (1983). Building a user-defined interface. *Proceedings of CHI'83 Conference on Human Factors in Computing Systems,* 24–27. New York: Association for Computing Machinery.

WONG, P. C. S., & REID, E. R. (1982). FLAIR—user interface dialog design tool. *Computer Graphics, 16*, 87–98.

WRIGHT, P. (1988). Issues of content and presentation in document design. In M. Helander (ed.), *Handbook of human-computer interaction* (pp. 629–652). Amsterdam: North-Holland.

ADDITIONAL RESOURCES

ARWADY, J. W., & GAYESKI, D. M. (1989). *Using Video: Interactive and linear designs.* Englewood Cliffs, NJ: Educational Technology Publications.

MICROSOFT CORPORATION (1991). *Microsoft Windows Multimedia: Authoring and tools guide.* Redmond, WA: Microsoft Press. (A good basic description of concepts, tools, and techniques for creating multimedia, many of which are not confined to Windows alone.)

MUNSON, L. S. (1992). *How to conduct training seminars* (2nd ed.). New York: McGraw-Hill.

NILSON, C. (1989). *Training program workbook and kit.* Englewood Cliffs, NJ: Prentice Hall. (Lots of guidelines and worksheets; includes setting up conferences and workshops.)

ROMISZOWSKI, A. J. (1984). *Producing instructional systems: Lesson planning for individualized and group learning activities.* New York: Nichols Publishing. (Provides a detailed account of how to develop instructional lessons.)

13

Procedure:
Formative Evaluation
and User Testing

By now the instructional system has received some initial design and development. As early as possible, the analyst needs to get feedback on the project. Traditionally, in instructional design this process has been termed **formative evaluation** (Goldstein, 1986). In formative evaluation one seeks the opinion and approval of supervisors, project leads, team members who may not have been directly involved to that point (if there are any), subject matter experts, and peer instructional designers. While this process is important, there is another process which has historically not been given enough emphasis, **user testing.** User testing is the process of asking people who are representative of the ultimate trainee population to view and "use" a prototype of the instructional or performance support system and provide feedback of various kinds. Because of the importance of user testing, I treat it as a category separate from formative evaluation.

FORMATIVE EVALUATION

Once a prototype has been developed, it should be evaluated in a number of ways. First, the company and specific people who have requested or consigned the work should receive a briefing using the prototype. Their approval and recommendations for the program should be actively sought at this time.

It is better to have to make modifications at this point rather than later downstream.

Second, the designer may have faithfully implemented and/or reproduced the content from the task analysis data, but still end up with content flaws in the final instructional program. If the analyst has not been working with a subject matter expert on the materials development all along the way, they should ask at least one expert to review the prototype. Another review should be performed once the detailed content for the entire program has been determined (e.g., the final storyboard).

Finally, it is sometimes useful to obtain opinions from various peer professionals, such as other people who are familiar with the project, instructional designers, ergonomists, etc. When you have been intimately working on a project for some time, it becomes difficult to keep an outside perspective. While user testing has its functions, it can also be helpful to get other opinions regarding how you might improve the instructional strategies and content of your program.

USER TESTING

Prototypes are important for communication among design team members and with clients or instructors, but where they really shine is in providing a testbed for user testing. This is where data is directly obtained regarding whether the product or program will work for learners. User testing consists of asking representative learners or trainees to interact with the prototype; where the term "interact" is used in the broadest sense. Learners may be asked for a general opinion about whether they like the program, they may be tested for comprehension of the text, they may be asked to use the program to determine its usability, and so forth. There are a variety of reasons why we perform user testing, and all of them revolve around the different types of questions we are asking. These questions include:

- What is the learner's attitude toward the training program?
- What is the usability level of the program?
- Does the person learn from the program?
- Does the program affect the learner's attitude about the subject?

The particular questions of concern to the designer will determine how user testing is conducted and what variables will be measured.

Variables Measured in User Testing

Based on the questions just outlined, we can categorize the variables to be measured in user testing as follows:

1. Attitude toward the training program. The first variable is the reaction of learners to the specific instructional design we have selected. A person's attitude toward the training program will operate on different levels. That is, they will react anywhere from very negatively to very positively toward the general approach. For example, do they like the idea of a workshop, or a self-paced CAI program? The attitude will also operate at a more specific level. They may like the program in general, but have mixed reactions to specific components of the program. Both of these levels should be measured. The most common method for measuring general attitude toward the program is to simply ask the learners. Questionnaire items can be open-ended, or more likely, a rating scale such as a seven-point Likert type of scale. Trainees can also be asked to rate each component of the program.

2. Usability of the instructional program or performance support system. For products such as documents, on-line help systems, and especially CAI, usability testing will be important. There are many ways to measure the usability of a system. The most common are to ask for a subjective opinion, and to measure performance of a person using the system. Table 13.1 gives some examples of variables that can be measured to assess system usability (many of which were previously listed by Whiteside, Bennett, and Holtzblatt, 1988). All of the method categories in Table 13.1 have been used by numerous researchers. As far as which methods provide the most information for analysts, recent empirical evaluation showed that think-aloud protocols and performance measures were both effective, however, think-aloud protocols were more sensitive, providing a broader array of underlying problems (Virzi, Sorce, & Herbert, 1993).

The list of variables in Table 13.1 should give the designer a good start on assessing system usability. Readers are also referred to outside resources such as Whiteside et al. (1988), Coleman, Williges, and Wixon (1985), and Meister (1987). Computer-based instructional systems and job aids generally require much more extensive user testing *for usability* than do other types of programs. In a later section, we will discuss some simple prototyping and user testing methods for each type of instructional system.

3. Comprehension, retention, and transfer. It is generally useful to determine early in the design and development phase whether the training program is going to be effective in terms of instructional goals. That is, does

TABLE 13.1 Methods for Assessing Instructional System Usability

Ask Learner After Interaction How Easy the System Was to Use
- Open-ended questions
- Rating scales (see Colemar et al., 1985)
- Focus groups
- List good and bad features (measure number of positive and negative features recalled)

Measure Errors Committed in Using the System (predominantly computer systems)
- Frequency of errors
- Type of errors
- Percentage of errors
- Severity of errors
- Relative time spent in errors

Note Verbal Expressions During Performance
- Number of times learner expresses difficulty or dissatisfaction
- Think-aloud verbal protocols, where learners work alone or in pairs
- Answers to question probes administered during system use

Measure Performance on Tasks (especially using Performance Support Systems and Documents)
- Number of tasks accomplished
- Time to complete task(s)
- Percentage of task(s) completed
- Percentage of task(s) completed per unit of time
- Ratio of successes to failures
- Number of repetitions of failed commands
- Number of times needed help from outside source

---

it promote comprehension, retention, and use/transfer of the information? Assessment of these factors can be done both early and late in the design and development phase. These factors can be tested early by developing and testing a *subset* of the learning modules. For example, if an instructor were writing a text, he or she could test comprehension and retention for representative sections. Measures of retention and use/transfer can be developed on the basis of instructional goals and objectives.

Often, questions of learning and retention are left until after the system has been completely developed. However, this is less efficient that performing early prototyping and user testing. Comprehension can be measured by asking learners to read, watch, or listen to instructional materials. They can be asked to verbalize, during or after these sessions, areas where they did not understand, or had questions. The most extensive testing for retention and transfer should be performed after the system has mostly been completed. This is because testing the materials within the context of a fully developed system will provide the most valid data regarding comprehension, retention, and transfer. This final phase of prototyping and user testing should be performed before the system is fielded. Once fielded, additional evaluation will be performed.

4. Attitude toward the subject matter. Some instructional programs will explicitly have as a goal to increase the motivation of the learner with respect to some behavior. For example, a videotape on safe logging procedures might aim to increase the "safety awareness and concern" of loggers. It might also attempt to increase the performance of certain behaviors that are safe methods of accomplishing tasks. It is not enough that the trainees learn the methods shown in the videotape; it is also important that the trainees leave with a positive attitude toward the material taught in the program. This type of attitude can be measured with a questionnaire assessing the perceived *intentions* of the trainees.

Collecting Data for User Testing

The most common methods for user testing are interviews, questionnaires, asking the person to use the system by performing benchmark tasks, and monitoring free use of the system. The designer must determine which of these is most appropriate depending on the type of data being collected. For example, to measure learner attitudes (#1 and #4), the most likely data collection method would be a questionnaire or interview. To measure error rates, learners would be asked to perform a variety of benchmark tasks using the system, etc.

Keep in mind that the people who act as trainees during the user testing *should not* be people who are familiar with the instructional system. For example, employees working with you on development should not be the same people you use to perform the user testing described in this chapter. The goal is to collect data about the effect of your instructional system on its final trainee population. Your trainees for user testing should have the *same characteristics* as that population, including a lack of knowledge about the training program.

USER TESTING
FOR INSTRUCTIONAL STRATEGIES

In the following sections, we will briefly review some of the ways that *formative evaluation* and *user testing* can be conducted for alternative instructional strategies.

Tutoring and OTJ Training

Final approval should be obtained from supervisors or anyone else directly involved in the project. If subject matter experts are being used to develop the training program, they should be asked to review the task list and examples of materials or exit tests. They should also be asked for any other types of information that might be relevant, if such information was not collected during the task analysis. These include:

- The length of time required to train the task and its subtask
- Any hazards in teaching each task
- Whether the task should be automated
- How often the trainee will have to perform the task during the job
- Difficulties trainees might have in learning the task

User testing can consist of at least two activities. First, ask trainees to look over the list of activities to see if there are any *other tasks* for which they think they will need instruction. Ask if they foresee any difficulties with training any of the tasks on the list. Second, if feasible, have the person who will be doing the training perform a subset of the instructional program, even if it is just one or two tasks. Have the instructor use the method (e.g., demonstration) that will be the predominant method in the final program. Collect the relevant types of data listed earlier, such as attitude toward the training program, comprehension, retention, and transfer.

Lectures and Workshops

In performing user testing for a lecture or workshop, the product can be fully developed or close to it. If the instructional program is a long one (for example, ten weeks as opposed to two hours), it is most efficient to initially prototype and perform user testing for only one or two modules. Try the module on a few representative learners to evaluate their general reactions, comprehension, etc. This will give you a good idea whether the general approach is a sound one. Depending on how much time is available, the entire completed instructional system should be tried on representative trainees. When skill training or other activities are to be used, these should all be user tested. This will show where instructions are unclear, materials need improvement, or trainees have some type of difficulty in successfully performing the tasks.

Linear Documents

While there are many guidelines for document design, user testing is still essential. Wright (1988) describes one type of user testing conducted by manuscript writers as follows:

> [a] user was invited to carry out some task relying on the information provided in the document. The user's difficulties were noted and the document was revised. Using this technique, Sullivan and Chapanis (1983) found that initially users encountered problems so frequently that the writers needed to remain alongside in order to provide instant fixes when confusions arose. But gradually, as the documentation was revised, the problems became fewer, and it was possible for the writers to monitor users at a distance (p. 641).

Having document writers themselves perform user testing has two benefits. First, the product itself becomes better over the course of the user testing iterations. Second, there is evidence that user testing provides feedback to make writers ultimately better at document design. For example, Schriver (1986) found that writers who had to repeatedly predict likely trouble spots and then watch videotapes of users talking aloud during document use eventually became better at spotting potential problems in their documents.

The types of user testing that should be implemented will partially depend on the system goals. In general, one would assess the following:

1. General reaction to the document. Users can be asked via interviews or questionnaire techniques for their general reaction to the document. This could be done during reading but is usually done immediately afterward. A frequently used measure is the five- or seven-point bipolar scale. End points for a general assessment can be concepts such as LIKED IT VERY MUCH/DID NOT LIKE AT ALL, or VERY GOOD/VERY BAD, SATISFIED/DISSATISFIED, etc. In addition, users can be asked to evaluate the document using adjectives. For example, end points on seven-point scales for rating the document could include items such as USEFUL/USELESS, CLUTTERED/UNCLUTTERED, HELPFUL/UNHELPFUL, CLEAR/UNCLEAR, etc.

2. Usability. During use, designers can have learners think aloud where they are asked to read and/or use the document for a task and verbalize things such as:

- What parts were unclear?
- Were there items they couldn't find?
- What questions did they have?
- What places did they need more information?

After reading the document, users can be asked where they had difficulties. In addition, seven-point rating scales can be used in several ways. First, a general question can ask users to rate the document, where one is EASY TO USE and seven is DIFFICULT TO USE. Second, a set of all relevant tasks and/or document characteristics can be identified and paired with seven-point scales with end points such as BAD/GOOD or UNIMPORTANT/IMPORTANT (Coleman et al., 1985). Designers should try to include these types of scales because some research has shown that they provide distinctly different types of information than the more general reaction questions (Coleman et al., 1985).

3. Test of knowledge/skills. As discussed earlier, effectiveness of the document can be tested by evaluating *understanding, recall,* and *use* of the information. For good diagnostic data, question probes can be used to pinpoint where the document is not being effective (see Chapter 4). In addition, users should be asked to perform relevant tasks after waiting an appropriate amount of time. This time will depend on the particular domain, and should reflect the mean expected period of time between training and actual task performance. In the case where this period of time on the job is many months, the test delay should be at least 2–3 weeks.

4. Attitude toward the subject matter. If the goal is to change trainees' attitude regarding some topic or behavior, their attitude should be evaluated. A questionnaire is usually most appropriate. It is best to compare readers' scores with the scores of others who did not read the document (we will discuss these types of evaluation further in Chapter 15). As a final note, it may not be possible to evaluate attitude change until the document is in its final form.

Videotape

In showing learners a videotape prototype, two options are possible. First, the designer can go through the material with the learner slowly, discussing each segment one at a time. Often, the designer can actually change the order of materials during this process. The variables most likely to be evaluated at this time will be comprehension and degree of interest. Various options can be shown to the learner and their reaction assessed.

When the designer has finally identified a set of materials and sequences using the methods described above, the final sequence should be shown to a new group of learners. This should be done without conversation during the presentation. After prototype presentation, the designer can use interview or questionnaire methods to assess the types of information described above (e.g., attitude toward the instructional program, comprehension and retention, task performance, attitude toward the subject matter, etc.).

CAI and Nonlinear Documents

In performing user testing for CAI systems, the designer will probably want to evaluate all of the variables described earlier. It is critical that CAI systems have extensive testing in terms of system usability. Recall that one goal of development was that at any given time, learners would know:

- What options are currently available to the user.
- How to achieve their current goal(s).
- The meaning of all information displayed on the screen.
- Whether or not the system is ready for user input.
- What the system is doing (such as waiting, loading files, etc.).
- How much of the system operation has been completed (if it will take a long time).

If the user should know all of these things at any given point in time, that means that he/she should be able to verbalize them. Assessing the extent to which you have accomplished this goal forms the heart of user testing for computer-based systems.

Based on this foundation, we can now specify how to conduct user testing. First, for CAI systems you should normally collect data both *during* and *after* use of the system, and the data collection should start early in the design process. When the system is refined and near completion, user testing can be done exclusively *after* the learner has used the system. Two methods described below are suitable for user testing, think-aloud verbal protocols and question probes. Each method gets at different information and both should be used during early testing.

To obtain the maximum information from think-aloud protocols, some preliminary training should be provided for users. Otherwise, they may not verbalize enough information. Users can be told that you will give them some tasks to perform, and that you would like them to vocalize all of their thoughts as they use your system. These tasks may be as simple as "looking at the different parts of the instructional system." The information you would like them to verbalize includes:

- Their **goals** (what they are trying to do at the moment)
- What **actions** they think they should do to accomplish those goals
- **Why** they are trying those actions
- The **meaning** of different parts of the screen

One way to help users learn to perform think-aloud verbalizations is to model the behavior for them using software that is completely different from that being tested (it doesn't even have to be an instructional system).

When users perform think-aloud verbalization during task performance, they may only state bits and pieces of the types of information listed in the previous paragraph. To alleviate this problem, you can perform user testing with question probes. Have users begin the same tasks, only this time stop them at various points along the way. When you stop them you can ask any or all of the following questions:

- What is your current goal (what are you trying to accomplish)?
- What action(s) do you think you would have to perform to accomplish the goal?
- What actions or options do you think are available to you right now?
- What do you think would happen if you _____ (e.g., click on the right arrow)?
- If I asked you to _____ (goal), how do you think you would do that?

If the system is well-designed, users will be able to accurately state the meaning of all of the icons, buttons, and menus the first time they see them, and predict what will happen when they are activated. They will also be able to determine how to accomplish their goals. For example, at any point during the system use, the user should be able to tell you how to quit the system.

After the early testing for usability, we can begin to focus on other issues such as general attitude toward the system, comprehension, retention, and so forth. It doesn't make any sense to begin measuring these variables before you are sure that the system is easy to use. One reason is that learners won't like systems as well if they aren't easy to use, and they will also show poorer learning to the extent that cognitive resources must be devoted to using the interface (Gordon, Gustavel, Moore, & Hankey, 1988).

Job Aids and Performance Support Systems

If the job aid or performance support system is a hardcopy document, use the methods described above under documents. If the job aid is computer based, use the methods described for CAI. For either case, users should be asked to perform critical **benchmark tasks** using the job aid. Early in the design phase, these benchmark tasks should be the most frequently performed or "prototypical" tasks for which the job aid was designed. Later, when the system is essentially complete, the set of benchmark tasks should be expanded so that they cover basically all categories of tasks to be performed by final system users. The best methods for user testing on benchmark tasks are: have users give think-aloud protocols, use question probes, measure performance and errors, and use questionnaires afterward.

One might ask why think-aloud protocols and question probes are necessary. Why can't a designer just measure whether the person can do the task without error? The reason is that people are pretty good at guesswork. If they read an instruction and aren't really sure what it means, they can guess. Often they are right even if they don't really understand what they are doing. And when they are wrong, they can often keep trying until they hit on the correct action. In such a case, measuring performance would reveal no deficiencies in the job aid. However, a think aloud protocol might reveal something like:

> OK, it says to click on the menu item that says *line printer*. I don't see a menu item that says *line printer* but I guess I'll just look through all these pull-down menus up here until I see it. Oh yea, I guess they put it up here.

Question probes also reveal circumstances where something is not clear. By asking, "How would you X?" you might get an answer that the person doesn't really know.

The nature of user testing is likely to change as the job aid or performance support system undergoes modification and refinement. Early in the development phase, the designer may have to give helpful hints to the user just to get them through the entire task. However, during later iterations of user testing, the analyst should absolutely refrain from any discourse or helpful hints.

FINAL COMMENTS

Before closing this chapter, we will discuss two final topics, use of human subjects and documenting the user testing process. User testing is a form of experimentation using human subjects. As such, certain protocols are appropriate. Readers are encouraged to request a document published by the American Psychological Association detailing the correct procedures for using human subjects in studies. In brief, designers should:

1. Explain the nature of the study at the beginning and assure the person that they do not have to participate unless they so desire. Ideally, you should ask the person to read a description of the study and obtain their signature on this "informed consent" form. Do not use deception without approval of an institutional review board (if you have one).

2. Treat all subjects with dignity and respect. An example of what *not* to do would be to tell a user, "Gee, I thought that was an easy one . . . all fifteen of my user group before you figured it out right away."

3. Explain that all data will be treated with confidentiality.

4. Do not coerce the user into continuing if they become dissatisfied and wish to quit.

As we perform prototyping and user testing, we usually collect data in both formal and informal fashions. When collecting quantitative data, you should use appropriate statistical analyses (e.g., see Maxwell & Delaney, 1990). All of the data you collect during user testing, as well as the decisions you make as a result of that data collection, should be carefully documented. This can be done in a separate information database, such as a "user testing document." An alternative is to do the design on a computer and tag the user testing data/decisions directly to the relevant parts of the document. This can be done relatively easily with hypertext programs. This documentation is important for later times when you need to go back and determine why you did certain things the way you did. However, it is even more critical in any instances where there may be liability issues.

SUMMARY

To summarize, use the initial prototype to obtain feedback from clients, peers, and other relevant people. Use the prototype to perform user testing via the methods described. User testing will generally involve the following activities:

1. **Have a sample of representative users try out the training system.** These users should represent the future trainee population and be "naive" with respect to the training system.

2. **During use, assess factors such as ease of use and comprehension of the material.** This can be done with methods such as think-aloud protocols and question probes.

3. **After users have completed interacting with the prototype, assess retention, transfer of knowledge and skills, general liking and ease of use.**

4. **When attitude change is a goal, measure users' attitude toward the subject matter and future intended behavior.**

REFERENCES

COLEMAN, W. D., WILLIGES, R. C., & WIXON, D. R. (1985). Collecting detailed user evaluations of software interfaces. *Proceedings of the Human Factors Society 29th Annual Meeting* (pp. 240–244). Santa Monica, CA: Human Factors Society.

GOLDSTEIN, I. L. (1986). *Training in organizations: Needs assessment, development, and evaluation* (2nd ed.). Monterey, CA: Brooks/Cole.

GORDON, S. E., GUSTAVEL, J., MOORE, J., and HANKEY, J. (1988). The effects of hypertext on reader knowledge representation. *Proceedings of the Human Factors Society 32nd Annual Meeting* (pp. 296–300). Santa Monica, CA: Human Factors Society.

MAXWELL, S. E., & DELANEY, H. D. (1990). *Designing experiments and analyzing data: A model comparison perspective.* Belmont, CA: Wadsworth.

MEISTER, D. (1987). System effectiveness testing. In G. Salvendy (ed.), *Handbook of human factors* (pp. 1271–1297). New York: John Wiley & Sons.

SCHRIVER, K. A. (1986). *Teaching writers to predict reader's comprehension problems with text.* Paper presented at the American Educational Research Association (AERA) Convention, San Francisco, CA.

SULLIVAN, M. A., & CHAPANIS, A. (1983). Human factoring a text editor manual. *Behavior and Information Technology, 2,* 113–125.

VIRZI, R. A., SORCE, J. F., & HERBERT, L. B. (1993). A comparison of three usability evaluation methods: Heuristic, think-aloud, and performance testing. *Proceedings of the Human Factors and Ergonomics Society 37th Annual Meeting* (pp. 309–313). Santa Monica, CA: Human Factors and Ergonomics Society.

WHITESIDE, J., BENNETT, J., & HOLTZBLATT, K. (1988). Usability engineering: Our experience and evolution. In M. Helander (Ed.), *Handbook of human-computer interaction* (pp. 791–817). Amsterdam: North-Holland.

WRIGHT, P. (1988). Issues of content and presentation in document design. In M. Helander (ed.), *Handbook of human-computer interaction* (pp. 629–652). Amsterdam: North-Holland.

14

Procedures:
Full-Scale Development
and Final User Testing

FULL-SCALE DEVELOPMENT

After several iterations of prototyping and user testing, the system will be completed in its final form. This full-scale development usually means completing additional modules of the instructional system in a manner that is consistent with the modules that have been developed for user testing. In some cases, such as design of a job aid, there may be little additional work. In other instances, such as development of a complex CAI system, the majority of the development is done at this point. The following sections provide an overview of tasks that are accomplished at this point.

Tutoring and OTJ Training

After the prototyping and user testing for tutoring and OTJ training, there will be little work left to be performed. The analyst may want to develop detailed problems or simulations to use during training, but this will probably depend on the complexity of the task and the severity of the consequences of making a performance error.

There is one final thing to note regarding on-the-job training. By working with the person who will be doing the training, you may notice that there are certain aspects of the job that are unclear or difficult even for that person.

One thing that you can do as an analyst is to determine whether there are any job aids that could help *both* trainer and trainee. As an example, I recently went to an unfamiliar library to find a book known only by the two authors' last names. I was unable to figure out how to do the required search using their computerized search system. I went to the front desk and asked the attendant for help, and she courteously attempted to perform a search with her system. Unable to perform the task, she asked a co-worker who was also unable to accomplish the search. Finally, a person obviously in a supervisory role came over and told them that this required a Boolean search. After several unsuccessful attempts, this person was able to use a very obscure command to perform the search. Life would be much easier for everyone, and reduce the need for on-the-job training, if a simple job aid were developed for everyone's use. This job aid could be developed for training new employees but also posted where everyone could use it. As it was, an on-the-job training program required the supervisor to teach each person helping at the front desk how to use the system. Given the likelihood of a high employee turnover rate, a job aid would enhance efficiency by augmenting the OTJ training.

The point is that one must be creative in developing solutions to performance problems, even in circumstances where the primary instructional design strategy is pre-specified, such as when a person is required to develop on-the-job training. By developing "training tools" that will directly support employees in their future work, one can help both trainers and trainees in a long-term fashion.

Lectures and Workshops

For the final development phase, we can first consider a relatively pure lecture type of instructional program. You will by now have a detailed outline of topics and their sequencing. You should also have at least one module that has been subjected to user testing by going through a dry run with people who are representative of the trainee population. You may decide to stop at this point. Alternatively, you can develop a very full set of lecture notes or even write a precise "speech." Which of these alternatives you choose will depend on the situation. Most occasions will probably only warrant a good set of notes. Remember to apply the guidelines provided in Chapter 6 (e.g., tie material into trainees previous knowledge, etc.).

For workshops, there will usually be more work required. This is because the most successful workshops incorporate a large amount of learner activities. User testing should have been completed with at least one example of each learning activity. For example, if the workshop consists of lecture, modeling of a variety of behaviors, and then practicing of the behaviors, the user testing should have included at least one example of each type of activity. For the remainder of the project, the analyst should develop notes, examples, and exercises that are similar in design to the prototyped activities. If possible, final user testing should be performed using the entire workshop. There are several reasons for this suggestion, but the most important is that stringing together a number of instructional activities may overload the trainee. This problem will affect both learning and motivation. There is no way to identify this problem for certain without actually trying out the entire sequence on a sample of trainees.

Linear Documents

After the prototype materials have been approved and submitted to user testing, the design team writes text and develops the graphics for the remainder of the instructional modules. Remember to stick to the style guide to ensure consistency across the various modules. After one or two revisions but before final polishing, perform additional user testing on the entire document if time and resources permit. Ask trainees to read the document and provide feedback about clarity of the text and graphics. Ask trainees to perform relevant tasks while using the text or later on, as appropriate. See Chapter 13 for guidelines on user testing. Depending on the domain, experts should also be consulted to make sure that the text is accurate.

Make any necessary revisions based on final expert comments and results of the user testing. After revisions have been made, edit the text to its final form, have graphic designers draw the figures, take photos, and compose the materials for final production. Create any indices and other types of lists that are based on the completed document.

Videotape

Develop materials. At this point you are ready to produce the videotape. In deciding whether to produce your own tape, you must consider the resources available to you. Anderson and Veljkov (1990) suggest asking yourself the following questions:

1. Do we have the necessary expertise and experience?
2. De we have the necessary equipment?
3. If the equipment is not available, can we fund the rent or purchase of the equipment?
4. Can we fund the training of in-house personnel to produce the video?

If you answer these questions and determine that you do not have in-house resources, you should contract out the video production. Develop a Request for Proposal (RFP) that specifies exactly what needs to be accomplished. It is often advisable to include samples of the script. Send the RFP out for bid to several production houses. In deciding which proposal to accept, you may want to review videotapes they have produced in the past.

If you decide to produce your own videotape (and this is based on the assumption that you have in-house production expertise), you will probably follow a sequence similar to that shown in Table 14.1. The table shows the steps normally necessary to shoot videotape material. Your job will be to work with the producer to make sure that the actors, location, shoots, etc. are capturing what you have intended in the instructional design. For example, if narration is to be added to film clips or animation, make sure that the person does not talk too quickly. Remember, this is an instructional program that presumably presents information that needs to be remembered later. In addition to shooting video clips, you will be working with graphic designers to develop the computer-based portions of the videotape.

Compose product.   In this phase, you will work with the producer and others to edit the raw footage, reducing or eliminating poor scenes and deciding on final segments. Show the segments in their correct sequence to clients, instructors, and learners. Do final user testing at this point. After user

TABLE 14.1   Procedures for Shooting Videotape Materials

1. Hire actors
    a. Pull files
    b. Interview
    c. Have actors read sample scripts
    d. Explain production requirements
    e. Choose actors
2. Research shoot locations
    a. Identify locations
    b. Tour sites
    c. Choose sites
3. Locate production equipment
4. Shoot video
5. Preview raw footage
6. Re-shoot any necessary scenes
7. Edit tape

testing data is acceptable and clients have approved of the material and sequencing, prepare the final master tape.

CAI and Nonlinear Documents

Develop materials. It is possible to begin developing final materials before the storyboarding is finished. However, performing this task too early, before client approval and user testing has been finished, may result in wasted efforts. Once the overall design, interface, system flow, and other basic design considerations have been finalized, it is time to develop all of the materials necessary for the final product. This may include shooting photos and videotapes, designing graphics on the computer, developing animation sequences, etc. Animation sequences can often be developed within an authoring environment. For more complex sequences, it is better to use a special program. If a laserdisc will be used in the application, all materials will have to be mastered onto a videotape and sent out to be pressed on a videodisc.

Computer code. Many applications will require some programming, even when they are being developed with an authoring tool. Most of today's authoring tools allow you to go outside of the application to run programs you have designed for special needs. Sometimes expert systems are embedded within a tutorial, and these will require quite a bit of time to develop. Several types of expert systems can be useful for tutorials, including traditional rule-based systems, neural networks, and case-based reasoning systems. And many of the expert system software packages allow communication with other programs so that the expert system-based tutorial is "embedded" within a simulation (e.g., Gordon et al., in press). Gordon (1991) provides a front-end analysis method for determining which classes of expert systems are most suitable for an application depending on the type of knowledge that is primarily involved in the task.

Compose product. Depending on the type of application, it may be possible to begin putting the interface screens and some of the instructional elements together before all components are finished. A good example is a hypermedia program. The basic structure, text, and graphics could be assembled with sound and video added last. If a laserdisc is to be used, the program can be finished while the disc is being mastered.

Debug. Once the computer-based system is finished, it must be thoroughly tested in several ways: (a) it should be evaluated by the design team and client or instructors to determine that it meets or exceeds the specifications; (b) it should be submitted to testing to make sure there are no "bugs" in the system; and (c) it should be submitted to full user testing. We will discuss this third item in a section below.

The goal of debugging is to try everything possible within the system to make sure that it doesn't have any errors or unintended system responses. This should include key inputs that are different from what is supposed to be done at any given screen. That is, the computer should be able to handle anything the user does without crashing. If incorrect keys are pressed, informative messages should be displayed. Check to make sure that the user can't do anything unintentional that could result in problems (such as exiting a system without saving work). Usually, the debugging is done by a member of the design team; however, it is also a good idea to have naive learners perform this task. Members of the design team are too familiar with the system and may have trouble performing unexpected input activities or combinations of activities.

Job Aids and Performance Support Systems

For simple job aids, there will probably be little work left to perform at this point. Based on user testing, the final development will mostly involve:

- Final adjustments in organization of the items in the document
- Formatting and layout of text and graphics for maximizing usability
- Final decisions on use of color to enhance visual search
- Checking for consistency and other ergonomic aspects
- Checking for typographical errors

These tasks can be performed by referring to the material in Chapter 11 and the resources listed at the end of the chapter.

For more complex computer-based performance support systems, there will be more extensive development work. This will usually include all of the tasks described in the section above on CAI. A performance support system will sometimes need introductory training just to brief users on how to use the system itself. This introduction will be developed at this point.

SUPPORTING DOCUMENTS

Supporting documents, other than primary instructional materials per se, are needed as supplements to instructional programs for several reasons:

1. The first and probably primary reason to include documents with an instructional system is to inform instructors and students how to use the program, in the broad sense of the word. For example, a videotape may have been developed with the assumption that a course instructor will follow its presentation with a discussion or hands-on workshop. This information is conveyed in the supporting document.

2. The second purpose is to provide a memory aid to the learner. It may be difficult to remember everything from a course or workshop, and it is often helpful to provide notes or other summaries of the material that was covered. An example is a workshop that gives attendees a copy of the overhead transparencies that were used during the presentation.

3. A third purpose is to provide the learner with one or more job aids to take back to the job. If the program provides instruction on complex tasks and subtasks, it may be unrealistic to expect trainees to remember the material. This is especially true if the relevant tasks are performed irregularly or infrequently. A card or small brochure with a summary of the basic concepts and especially procedures is a useful supplement to an instructional program.

For computer-based systems, supporting documentation may need to be relatively extensive. According to Landa (1984), there are three types of information that can accompany instructional software: pedagogical information, software information, and hardware information. Pedagogical information includes purpose, uses, and target audience for the instructional program. Software information includes general information about the software as well as all necessary information needed to modify the software; Price (1991) suggests including notes on the original software design, records of software changes, future enhancements, definitions, flow charts, and the name and address of the programmer. Hardware information includes a description of the platform components needed to deliver the instructional system.

All of this information will be differentially useful to instructors, learners, and programmers or others who may wish to modify the program. Thus, there are potentially three separate documents that could be developed: teacher documentation, learner documentation, and technical documentation (Price, 1991). According to Price (1991), one document can serve all three audiences, although designers may wish to develop separate documents. If one document is used, Price suggests an organization such as that shown in Table 14.2. Although the contents were developed for an academic learning program, the general approach could be adapted to industry. Price (1991) also provides detailed suggestions for contents of documents to support CAI sys-

Table 14.2 Content for CAI Documentation

Preliminaries
Title Page
Table of Contents
1. General Information
    1.1 Program Title
    1.2 Intended Audience
    1.3 Subject
    1.4 Hardware Requirements
    1.5 Program Overview
2. Information for the Teacher
    2.1 Subject Area Information
    2.2 Target Audience Grade and/or Ability Levels
    2.3 Prerequisite Skills
    2.4 List of Related Instructional Materials
    2.5 Instructional Objectives
    2.6 Suggestions for Program Use
    2.7 Teacher-Controlled Options (if any)
    2.8 Record-Keeping Functions
    2.9 Start-Up Procedures
    2.10 Description of Program Operation
    2.11 Back-Up Procedures
    2.12 List of Related Instructional Materials
    2.13 Feedback Form
3. Instructions for the Learner
    3.1 What the Program Does
    3.2 Getting Started
    3.3 Using the Program
    3.4 Record-Keeping Features
    3.5 Suggestions for Follow-Up Activities
    3.6 User Log
4. Reference Section
    4.1 Index
    4.2 Error Messages
    4.3 Glossary
    4.4 References
5. Appendices (if any)

(From Price, *Computer-aided instruction: A guide for authors,* 1991, Brooks/Cole. Reprinted with permission).

tems, as well as extensive examples. And as I noted in Chapter 11, Gottfredson and Guymon (1989) have written an extensive guide for writing software documentation.

FINAL USER TESTING

Before the program is fielded, it should receive final user testing as well as approval from all relevant parties. Final user testing should be done because results can be different from that obtained earlier in the development process. Learners are now reacting to the finished product and probably a much lengthier one.

The final user testing may include evaluation of all four factors discussed in Chapter 13; attitude toward the instructional program itself, degree of learning, ease of system use, and attitude toward the subject matter. Which of these are tested will depend on the type of instructional system and goals of the program.

There are basically two ways to perform final user testing. The first is to ask representative learners to participate in the instructional system and give feedback during the process. For example, the learner could watch a videotape with instructions to stop the tape and tell the experimenter whenever he or she has a question or didn't comprehend the information. With a CAI system, user testing would proceed as described in Chapter 13.

The second method is to ask the learner to participate in exactly the same manner as would occur when the system is fielded. After the training program has been completed, questionnaires, interviews, and task performance measures are used to assess the four factors of interest. Most of these variables were discussed in Chapter 13. Trainees should also be tested on performance after an appropriate period of time. For example, if they are asked to watch a safety training videotape, they should be tested on the material after a reasonable period of time has passed, at least 2–3 days. Otherwise, you do not obtain a realistic estimate of how much of the information will be retained. In summary, make sure that your tests get at the variables that are of ultimate importance, how trainees will do out in the field, not just immediately after the training program.

REFERENCES

ANDERSON, C. J., & VELJKOV, M. D. (1990). *Creating interactive multimedia: A practical guide.* Glenview, IL: Scott, Foresman.

GORDON, S. E. (1991). Front-end analysis for expert system design. *Proceedings of the Human Factors Society 35th Annual Meeting* (pp. 278–282). Santa Monica, CA: Human Factors Society.

GORDON, S. E., BABBITT, B. A., BELL, H. H., SORENSEN, H. B., & CRANE, P. M. (in press). Cognitive task analysis for development of an intelligent tutoring system. *Proceedings of the 1993 Annual I/ITSEC Conference,* Orlando, FL.

GOTTFREDSON, C. A., & GUYMON, R. E. (1989). *Style guide for designing and writing computer documentation.* Alpine, UT: The Gottfredson Group.

LANDA, R. K. (1984). *Creating courseware: A beginner's guide.* New York: Harper & Row.

PRICE, R. V. (1991). *Computer-aided instruction: A guide for authors.* Monterey, CA: Brooks/Cole.

15

Procedure: Program Evaluation

At this point in the project life cycle, the training program has been designed and developed with the use of *formative evaluation* and *user testing*. The last task is now to perform a final evaluation of the system, known as **summative evaluation.** Summative evaluation is an evaluation of the finished training program, usually just after it has been fielded. In this chapter, we will define summative program evaluation, provide a rationale for why it is important, give some of the major reasons for why people don't do it as often as they should, and provide some guidelines for conducting summative program evaluation.

DEFINING PROGRAM EVALUATION

In many, if not most training programs, some type of trainee evaluation is performed when the training program has just been completed. This is often a test or questionnaire at the end of the training session(s). While this type of testing is useful, we are really most interested in a question of transfer, whether our training program affects actual on-the-job performance as intended. Testing someone directly after training does not necessarily give us valid and applicable data concerning whether a person's job performance will improve or not. For that reason, we should collect data bearing on the effectiveness of the training program in changing the behavior that is *directly* relevant. That is, we must evaluate the extent to which we have had an effect on the variables that are ultimately of most interest to us. These will usually be some combination of trainee job or task performance and variables associated with the larger context (such as the number of job-related accidents in the organization).

The variables that we assess in program evaluation are often termed **criteria.** Criteria are simply those variables that represent the specific factors we are interested in changing through our training program.[1] Since a variable is some factor that can vary on a single dimension (such as number of typing errors, number of sales, etc.), a criterion is also a factor that varies on one dimension. We are usually interested in several factors, and therefore are usually interested in several criteria.

Summative evaluation determines the extent to which our training program has been successful in affecting various criteria related to trainee behavior and other organizational variables that are (presumably) affected by the trainee behavior. As Goldstein (1986) states,

Instructional analysts should be able to respond to the following questions:

1. Does an examination of the various criteria indicate that a change has occurred?
2. Can the changes be attributed to the instructional program?
3. Is it likely that similar changes will occur for new participants in the same program?
4. Is it likely that similar changes will occur for new participants in the same program in a different organization?

The questions described by Goldstein can be asked for each criterion relevant to the training program. By asking those questions, we may find that we have affected some variables but not others.

One important point to keep in mind is that we are trying to draw conclusions about ultimate job performance, not performance during our training program or immediately thereafter. This means that the summative evaluation should be conducted using the following general principles:

- **The evaluation should be conducted in an environment as similar to the ultimate performance environment as possible.**
- **The evaluation should be conducted after a realistic period of time.**
- **The evaluation should be based on tasks and task conditions representative of the ultimate job and tasks.**

[1]The word "criterion" is singular, "criteria" is plural.

While formative evaluation conducted during system development is important, it does not mean we can ignore the final testing of our program efforts. In summary, summative evaluation is the collection of criterion-related information in order to make decisions about the value of the training program, and determine modifications needed to make the training program more effective.

IMPORTANCE OF PROGRAM EVALUATION

In the design of any product or program, we can assume that the ultimate goal is to provide a system or product that is successful and accomplishes its targeted goals. Manufacturers in the U.S. have learned the hard way that they must identify what consumers want and need, and then collect data to determine whether their products are providing those things. For example, consumers have been focusing on quality of products, and manufacturers are now trying to design products that will deliver high quality. But to determine whether the products (and marketing schemes) are being successful, evaluation data must be collected after the products are fielded. For example, automobile manufacturers collect reliability data on their own and their competitor's products, and also send questionnaires out to assess owner satisfaction. In product design, it is important to determine whether the item does in fact meet the goals set by the design team.

In designing a training program, such evaluation is just as important. All of the efforts expended by the design team have been directed toward developing an effective training program. Formative evaluation has been conducted to support them in their efforts. However, the final test, summative evaluation, determines whether the efforts have been successful. Summative program evaluation is conducted for two reasons. First, to determine whether the goals listed in the system specifications document have been met. The second reason is to identify specific areas where the training program needs modification. There may be certain areas where the trainee is performing acceptably, and other areas where the trainee is still having difficulties. This information is used to go back and modify the program when possible.

There is currently a trend in business today that is related to the issue of successful product design, a trend epitomized by **Total Quality Management** (Deming, 1986; Gitlow & Gitlow, 1987). This term refers to many things, but one is a strong focus on producing high-quality products. The idea is to design products that don't just barely meet the minimum system specifications. Manufacturers are trying to design products that are of *much higher* quality than minimum standards. If we apply this concept to training, we would try to go beyond just meeting behavioral objectives. The goal would be to develop the best training program possible. In order to reach that level, you must know all the things at which you want the product to be really good. These are your measuring sticks for high quality.

Accountability

From the previous section, it's apparent that designers should feel responsible for designing effective programs. However, business today is moving beyond a simple feeling of responsibility. While it was suggested as far back as 1964 that the benefits of training should be cost accountable (Odiorne, 1964), training analysts sometimes did not take this suggestion seriously. However, analysts are now being increasingly held accountable for their work. Designers and instructors developing training programs (including those in educational institutions) are being asked to demonstrate the effectiveness of their product. In schools, this move is taking place in the form of outcomes assessment. In industry, regulatory agencies such as OSHA are being increasingly stringent about the quality of training in the workplace.

     People involved in the evaluation of training programs are also becoming more aware of the differences between evaluation that takes place during and immediately after training, versus evaluation that takes place in the job setting. For example, a recent publication by the National Research Council (1991) suggested that, too often, training programs focus only on how well learners are doing *as they go through* the program. There is an inadequate focus on performance in post-training real world settings. As Bjork and Druckman (1991) summarize:

> With respect to training, a recurring problem is that skills and knowledge acquired by trainees are not durable or flexible. At the end of training, trainees may meet rigorous performance standards, but in post-training real world settings, those same individuals may perform poorly, especially when long intervals of disuse of that skill or knowledge have intervened, or when the real world situation differs in certain respects from that present during training. . . . What makes this problem especially significant is that what trainers typically see is performance of trainees during training, not their subsequent performance in the real world (p. 14).

As an example of this problem, one common focus of training programs is participants' speed of learning. Unfortunately, research has shown that many of the things that make people learn quickly *during* training programs are things that *don't* promote good performance afterward on the job (Bjork & Druckman, 1991). These characteristics include a highly constrained task environment, frequent reinforcement, preventing learners from making errors, and massed practice. When these factors are included in the program, transfer to complex real-world environments is poor relative to when there is a more varied environment, less reinforcement, more errors by learners, and spaced practice (Bjork & Druckman, 1991).

The implications of these recent publications are that we need much more emphasis on training for later *transfer* to complex job environments. And in conjunction, we must perform more evaluation in the real world after *realistic* periods of time. That is, if we want to follow the principles embodied in total quality management, we can't just evaluate trainees during and immediately after training, but we must perform valid summative program evaluation, focusing on transfer of skills to the actual job setting.

Liability

In case the previous discussions haven't argued convincingly enough for the need to perform appropriate program evaluation, there is one additional reason. That reason is the potential professional liability associated with development of training programs. As I will discuss in greater detail in Chapter 16, training programs are increasingly being targeted in lawsuits related to accidents and injuries. People involved in the design and delivery of training programs may be found negligent and professionally liable for damages if: (a) a trainee subsequently sustains injury and/or injures another person, and (b) their training program was not developed and delivered in accord with known and published standards and principles. The need for final program evaluation has been an established principle in the instructional design community for over 20 years. Evaluation is critical for protection against future liability (see full discussion in Chapter 16).

Why Program Evaluation Is Neglected

Proper program evaluation is seldom performed, and the reasons are numerous. First, many people who develop instructional programs have moved into the field without formal training. These designers are often simply unaware that they should conduct summative evaluation. This is unfortunate because (a) it ultimately results in systems that are less effective than might be the case, and (b) it leaves their company open to heavy litigation losses.

System designers who have been formally trained are generally aware of the need to evaluate the effectiveness of a system once it is fielded. Even still, this procedure is often not performed. Why is this the case? There are a number of reasons, including:

1. Designers don't have the resources (time, money, personnel, etc.). This may be because they failed to include this item in the budget, budgets were cut, there were unforeseen expenses, time ran short, or any other number of reasons.

2. Designers (or clients) perceive that they don't have the resources. For example, designers may move onto other "critical" projects that have a higher priority than evaluation of the project just completed. They could make the time, but they perceive that relative to other needs, they don't have the time (or money).

3. There is no perceived need. It is easy for designers to believe that they have spent so much time and effort in design, usability studies, etc., that it is clear that the program works. They simply don't feel that summative evaluation needs to be done; a common and easy trap. Part of the problem is that it isn't necessarily an explicit belief of which a person is aware. It is implicit, and only manifested by a variety of excuses ... "it would just be a waste of time and money, I would only find out what I already know," and so forth.

Because it tends to be implicit and hidden, this belief that the evaluation is unneeded is difficult to counteract. Really all one can do is go into the project with the belief that the program *will* be evaluated, and then have the discipline to stick to that resolve. It can help to think to oneself, "if I follow through with it, I will be rewarded by all that data showing how effective the program really is." Another thing to keep in mind is that the evaluation data can be very useful in the future. For example, as companies become more discriminating and expect proof of quality products up front, they will ask for some type of evidence that you are a good designer and that your programs work.

4. A training program changes hands. The design team delivers the training program to a company or academic institution. The designers are done with it, and it is now in the hands of another party. It becomes a question of ownership and responsibility. Who is responsible for determining the effectiveness of the program? Often it is the client or instructor who receives the program who are implicitly responsible for doing the evaluation. But generally they don't feel that they should, they don't know how, they don't have the time, etc.

5. Designers don't want to know. Sometimes, unconsciously, designers simply don't want to know the results of a final evaluation. This is most often the case when they are consciously or unconsciously aware of the fact that they have had to make compromises; they haven't had the resources or time to do the system the way they would have liked. This reason for not performing evaluation is almost always implicit, outside of our awareness.

6. Designers aren't aware of liability issues. Finally, many designers with or without formal training are not aware of the liability and other legal issues. Liability related to training is a phenomenon that is just beginning to appear in the courts. In addition to problems of professional liability, OSHA is becoming much more strict in the quality of safety training required in industry. Companies providing inadequate training are being fined.

GOALS OF PROGRAM EVALUATION

Overview

In performing program evaluation, we ask several important questions:

1. Has a change occurred in trainee behavior?
2. Is that change in behavior a result of the training program?
3. Would the change occur with other trainees besides our sample?
4. Would the change occur in other contexts or tasks besides those we measured?

To answer these questions we collect data for the criteria we have chosen. For example, we might look at the accident rate at a paper mill before and after implementation of a safety training program. To acquire this type of information, we often rely on two different techniques. One is to collect data on criteria before and after the training program is fielded. This is known as using a **pretest/posttest** evaluation design. For each criterion which we are interested in, the same variable is measured before the training program is administered and the program is finished. Using a pretest/posttest method is an extremely common way of answering questions about whether a change has occurred in some criterion. If not done properly, this can be a very weak evaluation design. Later I will describe some methods for boosting the validity of the pretest/posttest design.

A second method commonly used for evaluating the effects of a program is to use two groups of people where one of the groups receives the training program and the other group does not. The group that receives the program is termed the **experimental** group; the group that does not receive the program is termed the **control** group. By comparing data for the two groups of people, we can infer whether the training program had an effect. The use of control groups is very common in program evaluation research.

Each of the two methods, use of pre/posttests and use of control groups, has unique advantages and disadvantages. In order to discuss these advantages and disadvantages, we must first understand each of the *potential* problems that can undermine our evaluation efforts, because each of these problems is more likely with some methods than others. These problems have been traditionally placed in one of two categories, problems of **internal validity** and problems of **external validity.**

Internal Validity

In the context of training programs, internal validity refers to the extent to which the treatment (training) program was responsible for changes in the criterion scores, and not other factors. Obviously, we would like to conclude that it was the training program, but we cannot do that if there were other factors that might be responsible for the change.

For example, assume that we measure accident rates at a paper mill for one year. We give a safety training workshop in the first week of January, and then measure accident rates for the remainder of that calendar year. If accident rates drop, can we conclude that our program was responsible? The first question to ask is whether the drop was simply a random occurrence. Statistical tests should be used to answer this question. (For information on how to conduct statistical tests, see texts such as Bordens & Abbott, 1988; Kirk, 1982.)

If the two sets of data *are* determined to be statistically different, then the next question to ask is what *caused* that difference. We might be tempted to conclude that it was our program. Unfortunately, there might be other differences between the two years *in addition* to the fact that we administered our training program. For example, there may have been changes in equipment, job procedures, employees, management style, etc. Any such changes could be responsible for the change in the criterion.

To the extent that there are other factors that could be responsible for criterion scores, in addition to our training program, we have reduced the *internal validity* of our study. We have reduced our ability to conclude that the training program was effective, because there are other factors that may have been responsible for the changes in the criterion scores. We will see below that some problems of internal validity tend to come with pretest/posttest designs, and other problems of internal validity tend to come with control group designs. All of these problems have been described in detail by Campbell and Stanley (1963) and later by Cook and Campbell (1979). The reader is encouraged to read Cook and Campbell and become familiar with these issues. The most common problems for internal validity are usually referred to as **threats to internal validity,** and are briefly summarized below:

1. History. History effects are specific events, in addition to the treatment variable, that occur between the first and second measurements of a criterion that could alternatively account for the difference in the scores.

2. Maturation. Maturation refers to any changes in the trainees themselves that could account for the criterion scores. These are changes occurring as a function of the passage of time, such as growing older, more tired, hungrier, smarter, etc.

3. Testing. Testing refers to the effects of taking some test a first time on performance scores when taking the test a second time. Usually taking a test once results in a person performing better the second time. This means that scores will improve even if the training program has no effect whatsoever. In fact, the training program could have a negative effect, and because of the positive effects of multiple tests, trainees' scores would show no change from pretest to posttest.

4. Instrumentation. The instruments used to measure various criteria may change from one time to another. This can be either physical instruments or the people who observe and make ratings. An example would be a supervisor who rates employees at one point in time, and then has personal problems causing him or her to make the ratings more severe at a later time.

5. Statistical Regression. This is problematic only when trainees are chosen on the basis of extreme scores, for example, if secretaries have particularly low typing speeds. When people are chosen on the basis of extremely low scores, they will tend to score better on a second testing, even without intervention. This phenomenon is termed **regression to the mean.** It also occurs for people chosen who have extremely high scores, but that is usually not a concern for training programs.

6. Selection. When using a treatment and control group, there may be differences between the members of the two groups that can account for the differences in criterion scores. For example, we might ask police officers to volunteer for a special training program. After the program we compare their scores on some performance variable with scores of police officers who did not participate. There may be characteristics of the volunteers (such as motivation) that cause their scores to be higher than the nonvolunteers, even without any benefits of the training program.

7. Experimental mortality. This effect refers to a differential loss of people either from the training group or from the control group.

8. Interactions. This is a category of problems where the threats listed in #1 through #6 occur in one group but not the other. For the first threat, this would mean that a history effect occurs for the treatment group but not the control group. For example, the experimental group might receive a raise whereas the control group does not. With maturation, the experimental group might become fatigued where the control group does not, etc.

Each of these threats to internal validity can be reduced through proper experimental design. Such designs will be reviewed in a later section. There are certain additional threats that occur specifically in the evaluation of programs designed to increase learning and/or human performance (Cook & Campbell, 1979). These include:

9. Diffusion or imitation of treatments. People receiving the training program may pass information on to others who are not receiving the program. The data would reveal smaller differences in the groups than would otherwise be found.

10. Compensatory equalization. This factor refers to the fact that many supervisors as well as employees or learners realize which groups are receiving the treatment. Many feel that the other people are now disadvantaged and should receive other forms of "support" to make up for that disadvantage. Unfortunately, well-meaning supervisors may eliminate differences between the groups by supporting control groups in other ways.

11. Compensatory rivalry. If participants are aware of the groups being studied, they may react to the competitiveness of the situation. Members of the control group may work harder to show that they can perform just as well. This effect has been discussed by Saretsky (1972), who terms the phenomenon a "John Henry" effect. The label refers to John Henry, a man who outdid himself to outperform a steam shovel, but died of overexertion in the process.

12. Demoralization in control groups. People who are members of control groups may realize that they are not receiving the most valuable treatment. This may lead to demoralization and an associated decrease in effort and performance.

These threats are not automatically reduced through experimental design. They are reduced through proper communication with employers and employees regarding the importance of program evaluation. If control groups are to be used, they should receive what seems to be an equally desirable alternative program. Alternatively, they can be assured that they will receive the experimental training program as soon as the evaluation is completed. In general, analysts should make sure that there will be no negative consequences for members of the control group, and communicate this to all employees involved. For more information on how to deal with these problems, readers are referred to Cook and Campbell (1979).

External Validity

External validity is the degree to which we can generalize the findings to other trainees and/or other settings (Campbell & Stanley, 1963; Cook & Campbell, 1979). To have external validity, we must first have internal validity (the assumption that our treatment was responsible for the changes in the first place). But internal validity only tells us that for this group of trainees, in this particular study, the treatment was responsible for a change. We must still try to insure that the treatment would have the same effect for others performing in other settings, external validity.

For training programs, the question of external validity is not trivial. For example, suppose you developed a training program for managers on dealing with sexual harassment charges. You implement the program in a local university and measure the effects on some set of criteria. Your data indicate that the program was highly effective. At a later time, a local business wishes to know whether the training program would work for them. How do you know whether the effect would be similar? This is a question of external validity.

There are several considerations critical for maximizing the external validity of your program evaluation efforts. The first is to use trainees, tasks, and environments that are representative of the trainees, tasks, and environments to which you wish to generalize. For example, if you are developing a training program for dealing with sexual harassment that is supposed to work for any managers in any company, then that is the population to which you hope to generalize. If that is the case, you should obtain subjects for treatment and (optionally) control groups that are fully representative of that population. For example, you should not use only lower level managers, only males, only managers in large companies, etc. If you wish to generalize to all sizes of firms, you should evaluate the program in firms of various and *representative* sizes. In summary, for program evaluation, you should adequately *sample* the people, tasks, and job environments from the populations of people, tasks, and job environments to which you wish to generalize. (See Babbie, 1989; Cook & Campbell, 1979, p. 74–80, for specific guidance on sampling methods.)

There are other factors that will affect the external validity of your evaluation study. Some of the factors suggested by Campbell and Stanley (1963) are listed below:

1. Reactive effects of testing. This refers to the phenomenon where pretesting sensitizes the trainee to the contents of the training program. This may either increase or decrease the effect of the program. For example, Bunker and Cohen (1977) found that pretesting decreased the effectiveness of a training program. Either way, it reduces the extent to which we can generalize, since under normal circumstances you would not give trainees a pretest.

2. Reactive effects of the experimental setting. In a formal study, certain factors often make participants aware that they are taking part in an experiment. These include special people and equipment that are otherwise not in place. Becoming aware that they are in a study may cause people to act or learn or perform in a manner different than they otherwise would.

Participants in organizational research frequently perform better if they know they are being studied (Aamodt, 1991; Landy, 1989). This problem is often termed the **Hawthorne effect,** because a series of studies done at a plant in Hawthorne, Illinois showed that employees tended to increase their production regardless of whether the specific changes in working conditions were positive or negative (Roethlisberger & Dickson, 1939). A variety of factors may change peoples' behavior who know they are participating in a study. These can include novelty, presence of observers, enthusiasm of the instructor, social interaction, feedback, a desire to give the experimenters what they are looking for, etc. (Goldstein, 1986; Landy, 1989). In conducting an evaluation study, it is most ideal for external validity to somehow keep participants from knowing that they are in the study.

3. Multiple treatments. If multiple treatments are administered to the same trainees, the generalizability of results will be diminished if other trainees will not have the benefit of all of the treatments. This is because the treatments seldom act independently of one another.

METHODS FOR CONDUCTING
PROGRAM EVALUATION

In this section, we will discuss methods for collecting evaluation data. Our goal is to answer the questions listed earlier (Has a change occurred? Is the change attributable to the program? Would the change occur with other trainees? Would the change occur for other tasks in other settings?), and to do so with the best internal and external validity possible. In determining how to collect the evaluation data, we must make four categories of decisions:

- What **Criteria** to Measure
- **When** to Measure the Criteria
- **Who** to Use When Measuring the Criteria
- What **Context** to Use

In the next few sections, we will review how to make these types of decisions. We will consider the four questions specifically for the evaluation of training programs; job aids will then be discussed in a separate section. Before beginning, it should be noted that evaluation is a type of research requiring a considerable amount of knowledge and skill. Readers are encouraged to obtain additional information on performing evaluation research from appropriate sources such as Babbie, 1989; Bordens and Abbott, 1988; Cook and Campbell, 1979; Ghiselli, Campbell, and Zedeck, 1981; Sommer and Sommer, 1986.

Choosing Criteria

Types of criteria to measure.   Researchers have developed a variety of schemes for classifying evaluation criteria (e.g., Kirkpatrick, 1959; Lindbom & Osterberg, 1954). For example, Kirkpatrick (1959) and Goldstein (1986) discuss four levels of criteria: (a) **reaction** to the training program; (b) **learning** of principles, facts, techniques, and attitudes; (c) **behavior** relevant to job performance; and (d) **results** of the training program related to organizational objectives.

We have previously discussed the types of variables that are important for formative user testing. Using Kirkpatrick's (1959) categories, these include reaction, learning, and behavior. If adequate formative evaluation has been performed, the instructional analyst(s) should have redesigned the program so as to attain a satisfactory level of reaction data and also a satisfactory level of learning. In summative evaluation, reaction and learning are still important, but the variables of behavior and results become much more critical.

In the discussion below, I will describe the evaluation of all four types of criteria, but with less emphasis on reaction. I will refer to Kirkpatrick's "behavior" criteria as **performance** variables, to reflect the critical "job performance" aspect. I will refer to "results" as **secondary results** to reflect the fact that these result variables are a second-order effect in the causal chain (that is, training affects learning and performance which then secondarily affects organizational factors). To see this, consider the variables illustrated in Figure 15.1. Each variable is represented as an oval in the diagram. A training program affects performance in the job place, which hopefully reduces the

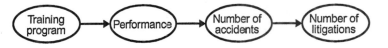

Figure 15.1   Causal chain of effects starting with training program.

number of accidents. The reduction in accidents may then result in a smaller number of litigations brought against the employer. Any such results that derive from changes in performance will be termed secondary results.

In summary, summative evaluation can potentially concern reaction to the training program, learning from the training program, job performance, and secondary results within the organization. While it is most common to measure reaction and learning (Goldstein, 1986), it is uncommon to measure all four criteria in a comprehensive fashion.

It is usually fairly clear why we should measure reaction and learning criteria. Why do we also need to measure both performance and secondary result variables? First, consider a situation where we measure only performance. Imagine that we go into the local paper mill and ask the employees to show us how they use the equipment to package rolls of paper. We find that everyone shows us the procedurally correct way to perform the task. It would be easy to be satisfied at this point that we have done our job. Unfortunately, this change in performance might have no impact whatsoever on other variables of interest (most notably accident rates). If we don't measure accident rates before and after the program, we will have no idea whether our training program has impacted the variable of most interest to the company. We cannot assume that changing behavior will change the secondary variables.

On the other hand, if we only measured secondary result variables, we would obtain data that is almost useless. Imagine that we measured accident rates at our plant before and after training. The accident rate could be higher, lower, or no different (as determined by statistical tests). What might we conclude? In each case, we don't know if the results were due to our training program or some other factors. If the accident rate went up, it doesn't mean our program had no positive effects, there might be other negative causal factors that were simply stronger. If the accident data went down, we might be tempted to assume it was due to the training program. However, the effect might have been completely caused by other factors affecting accident rate. Figure 15.2 shows an example of causal factors that could affect accident rates.

Objective and subjective measures.   Criteria can be categorized in a number of ways. One dimension is that of objective vs. subjective vari-

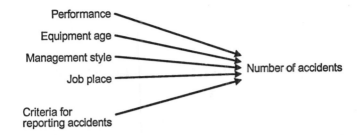

Figure 15.2 Factors affecting secondary variables in addition to performance.

ables. Objective measures are variables that do not involve any human judgment. They are things one can count such as number of products produced, days of absenteeism, number of products sold, number of accidents, number of errors, etc. If a variable is objective, various people should be able to measure that variable and come up with the same number.

The subjective measure of a variable involves human opinion, judgment, or belief. Typical subjective variables for training program evaluation include:

- Rating the training program
- Rating oneself on some dimension
- Reporting the extent of one's behavior
- Supervisor rating of an employee

Note that just because these dimensions are often "scales," such as the seven-point scale, it does not mean they are objective.

Objective and subjective measures each have problems associated with **construct validity** which means, "Are we measuring the variable we are really interested in, or something else?" Objective measures are generally considered to be more valid indicators of a variable. However, even objective measures may lose validity because of a lack of opportunity to directly observe a person and measure a relevant and completely objective variable.

Subjective measures such as rating scales are particularly prone to problems of construct validity. This means that what we are measuring is often something other than what we are trying to measure. For example, if a person is asked to report on the number of times they have behaved in some manner, they will be relying on memory, which has a number of biases (Anderson, 1990).

There are also a number of problems with asking people to rate another person's behavior. As an example, think of employee evaluations of their supervisor. They are supposed to be rating their job performance as a supervisor, but people are biased by other factors including gender, age, race, likability, etc. In general, it is best to use objective measures where possible.

Rating scales.   Ratings are a subjective form of measurement commonly used in industry (Landy & Rastegary, 1988), and in some cases may be the only feasible method for measuring performance. Unfortunately, they tend to be low in construct validity because they are "contaminated" by so many other factors, such as how well we like the person (Landy, 1989). While rating scales are prone to problems, there are certain methods to reduce those problems. These methods fall into two categories, rating scale development and rater training.

Because of the frequent use of rating scales in a variety of settings, researchers have attempted to develop rating scales that are less prone to bias. Instances include the mixed standard rating scales (Blanz & Ghiselli, 1972), behaviorally anchored rating scales (Smith & Kendall, 1963), and behavioral observation scales (Latham, Fay, & Saari, 1979; Latham & Wexley, 1977). Unfortunately, reviews of research evaluating these methods have concluded they are not significantly better than other methods (e.g., Landy, 1989; Landy & Farr, 1980, 1983; Murphy & Constans, 1987). Careful development of scales based on objective behaviors seems to be a good approach (e.g., Landy, 1989).

One successful approach developed fairly recently is that of rater training (Landy, 1989; Murphy, Garcia, Kerkar, Martin, & Balzar, 1982). For example, both McIntyre, Smith, and Hassett (1984) and Pulakos (1984) provided raters with information on the multidimensional nature of performance, the meaning of anchors on the scales, practice in rating a standard performance stimulus, and feedback on the practice exercises. They found that the accuracy of the raters was significantly improved over control group raters or raters simply instructed on avoiding errors.

Interviews and questionnaires.   Many criteria will be evaluated through the use of interviews and questionnaires. Interviews and questionnaires are a research method that require a fair degree of expertise. One very good source for interview and questionnaire methods is Sommer and Sommer's (1986) guide to behavioral research. In this section, I will review the advantages and disadvantages of interviews and questionnaires for program evaluation, and provide a short guideline based on Sommer and Sommer (1986) and Babbie (1989).

Interviews include any method of collecting data where an interviewer asks questions orally and records respondents' answers. Interviews can be conducted in person or over the phone. Advantages for performing an interview over self-administered questionnaires include (a) a higher response rate, (b) a lower number of "I don't know" answers, (c) fewer problems with confusing questions, and (d) a source of direct observation (Babbie, 1989). In addition, interviews are good at getting qualitative, uncertain, or complex information. They are advantageous because respondents are more willing to verbalize long, drawn-out information that they wouldn't take the time to write down (Sommer & Sommer, 1986). Structured interviews tend to be preferable because they are generally more valid, reliable, and easier to code later for data analysis.

Self-administered questionnaires include any questions where the respondent reads the items and responds directly on paper or on a computer. Such questionnaires have certain advantages over interviews, including economy, speed, lack of interviewer bias, and privacy (Babbie, 1989). Questionnaires may contain open-ended questions, where respondents write in their own answers, or closed questions, where respondents choose among alternatives (or both). According to Sommer and Sommer (1986), the open-ended format is preferable (1) when the researcher does not know all the possible answers to a question, (2) when the range of possible answers is so large that the question would become unwieldy in multiple-choice format, (3) when the researcher wants to avoid suggesting answers to the respondent, and (4) when the researcher wants answers in the respondent's own words (Sommer & Sommer 1986, 109). While the open-ended format has certain advantages, most experienced researchers prefer closed-ended questions (Sommer & Sommer, 1986). They tend to be more valid, reliable, and easier to score.

Questionnaires should be limited to essentially one topic. They should be short, clear, and meaningful. Terms should not be too difficult, and should be relatively neutral. That is, don't ask questions about "dangerous drugs," "noisy trucks," or "excess government spending" (Sommer & Sommer, 1986). Keep items balanced, with approximately equal numbers that will seem positive and negative to the average respondent. The questionnaire should be constructed with an introductory explanation first, followed by factual, noncontroversial questions. Finally, design the questionnaire using specific guidelines for item construction.

Criterion reliability. Reliability is the consistency of a measure; that consistency can be between two or more raters or judges, or across different points of time. If two people observe the same performance and rate it differently, there is low inter-rater reliability. We have already discussed the biases that can lead to poor accuracy of rating scores and therefore low inter-rater reliability.

However, reliability across time is also important because we often use pretest and posttest scores for the same variable. When we measure something repeatedly, the value should be the same (given that nothing has caused it to change in the meantime). As an example, if we step onto a scale and read a certain number, stepping off and then back on should result in the same number being displayed. If a different number is displayed, that scale is unreliable.

Reliability is important for the same reasons that validity is important. It allows us to make conclusions about the goodness of our criteria, and therefore about the effectiveness of our training program. If our criterion is unreliable over time, it may introduce two problems. First, any differences between a pretest and posttest might cause us to attribute the changes to our program when in reality it is simply an unreliable criterion. Second, our program might result in changes in the criterion, but those changes could be counteracted by fluctuations due to the unreliability of the measure. In either case, unreliability causes us to infer something that is actually incorrect.

Nagle (1953) has suggested that unreliability over time is critical for evaluating programs, and that certain factors often affect criterion reliability. These include:

- The size of the sample of performance (measure behavior at three stop-lights versus fifty)
- The range of ability among the participants
- The ambiguity of instructions
- Varying environmental conditions when criteria are measured
- The amount of aid provided by instruments

An additional factor for performance measures is making sure that the person has ample *opportunity* to perform the behavior, both during pretest and posttest time periods. In summary, all conditions during pretest and posttest measurement periods should be similar. Analysts should make attempts to measure reliable criteria so that valid conclusions may be drawn regarding the training program.

In summary, there are many factors to consider when determining the criteria for evaluation studies. We have barely had the opportunity to scratch the surface. Texts that address some of the issues include Landy (1989), Muchinsky (1990), Goldstein (1986) and Ghiselli et al. (1981).

Research Designs:
Giving Power to Our Conclusions

Up until this point we have seen that to draw valid conclusions, our evaluation study must be acceptable in terms of certain qualities, namely, internal validity, external validity, and construct validity. There are design decisions that will affect each of these qualities. These were stated earlier as deciding (1) what to measure, (2) who to measure, (3) when to measure, and (4) where to measure. We have just reviewed information relevant to "What to measure." We stated that there are four important types of criteria, with performance being the most critical. You should identify and measure criteria to maximize validity and reliability.

We now address the second and third questions of who to measure, and when to measure the criteria. These questions are framed in terms of what are called research designs. Research designs specify the people to measure, and the various times to measure them. In this section, I will review the most basic and common research designs. Readers are referred to Campbell and Stanley (1963) for a more thorough review of these designs plus others. All of the information below is equally applicable to **learning, performance,** and **secondary results** criteria. They are also appropriate for measures of **attitude** toward the subject matter (but not for attitude toward the training program per se).

The easiest design.   The easiest possible way to collect information about training program effectiveness is to administer the program and then test trainees on the criteria of interest. This design is termed a **one-group posttest only** design for obvious reasons. The one-group posttest only design can be symbolized as:

$$X \quad T$$

where X stands for the treatment program and T stands for testing. Each row represents one group of participants.

It can be seen that this design symbology represents one group of trainees who first receive the program, then chronologically moving to the right, receive a posttest. An extension of this design is:

$$X \quad T_1 \quad T_2$$

where T_1 represents the first posttest and T_2 represents a second later posttest. However, note that this is still a one-group posttest only design.

As an example of the one-group posttest only design, imagine that we train a group of dental patients how to correctly brush their teeth. Two days later, we go back and ask them to show us the proper procedure for brushing their teeth. We would probably be very happy if all patients showed us the correct procedure. But our satisfaction would be entirely unfounded. Because while performance might be perfect, we could not be sure that it was our training program that was responsible. Even if they performed well, it could theoretically have been caused by one of three reasons:

1. Our program worked.
2. They already had the knowledge/skills.
3. Some other variable caused them to gain the knowledge.

Unfortunately, it is impossible to tell which of these reasons is actually correct. The reason is that we have nothing with which to compare the performance. What if most of the students had done well but some had not? We would have absolutely no idea if our program had had an impact on the participants.

Technically speaking, what are the problems with this type of data collection design? If we look back to the threats to internal validity listed earlier, we find that performance could be due to four threats: history, maturation, selection, and mortality, with history and selection the most critical. This means that other events could have caused the performance, or that subjects selected could have already been able to perform the task. In summary, use of this type of design is *never justified, never valuable,* and viewed as basically worthless by the scientific design community. Needless to say, the design would provide no support whatsoever in the case of litigation.

Better but still inadequate designs.   There are certain designs that could be described as better but still not particularly good. The two designs presented in this section are better because each eliminates certain threats to internal validity. However, since neither comes close to eliminating all of them, they should be considered inadequate designs.

The first design involves adding a pretest to the design. Therefore, it is referred to as a **one-group pretest-posttest** design, and is symbolized as:

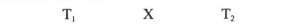

$$T_1 \qquad X \qquad T_2$$

where T_1 is a pretest and T_2 is a posttest.

What does this design do for the *validity* of our conclusions? Consider the example given earlier. If we had used this design, at least we would know if the students we chose already knew proper tooth brushing procedures. This means that we can eliminate the threat known as selection, which is a very important one. However, we still don't know if any differences between pre-test scores and posttest scores are due to some other variable such as external events (history), maturation, effects of pretesting, changes in instrumentation, or regression if we selected participants with extreme scores. In summary, we may be able to state that changes occurred, but we *cannot infer* that they were caused by our training program.

A second design that improves on the first one described is formed by adding a control group. In this design, one group of people are given the training program, and another control group is given no training program (or a less effective program, a traditional program, etc.). This design is symbolized as:

Treatment Group	X	T
Control Group		T

where the top row represents participants in the treatment condition followed by testing, and the bottom row represents the control group who only are tested.

Notice that in this design, it is not assumed that people are randomly assigned to be in one of the two groups. The groups are simply chosen in whatever method is desired. For example, participants are allowed to sign up for the treatment condition. We call this a **static group comparison** design (or non-equivalent groups design).

This design does control for certain threats. For example, any external factors (history effects) that would affect the treatment group would also affect the control group. Likewise with maturation, and so forth. The major problem with this design is that the two groups of people could begin with different characteristics. This difference could account for the differences in the measured criterion. This factor severely limits our ability to draw conclusions regarding the effectiveness of our program.

The best designs.   Restricting ourselves to relatively simple designs, there are two that are experimentally sound and allow us to draw conclusions about the effects of a training program. These will be briefly described.

The first design is a **posttest-only control group** design. This design is considered a true experimental design and is frequently used in the behavioral sciences. This design is symbolized as:

(R)	X	T
- -		
(R)		T

where R stands for random assignment of people to groups.

In this design, the two groups of participants are made equal, on the average, by taking a number of people and then randomly assigning each of them to be in one of the two (or more) groups. Because it is based on the law of averages, or probability theory, this method *only* works if you use an adequate number of participants (i.e., 20–30 people in each group). Random assignment is very powerful because it results in the groups beginning on an equal footing on all possible variables, including the criteria being measured. This eliminates the threat of selection that is so problematic for the static group comparison design.

There is only one drawback to this particular design. We can conclude that the training program worked if the scores are better for the treatment group than for the control group. However, we don't really know how much they improved from what they were before the training program. We might infer that the control scores are similar to what pretraining scores might have been, but we do not know for sure. The only way to determine how much scores have improved are to conduct pretests. This design will be discussed next. Keep in mind, however, that giving pretests opens one up to the associated threats to internal validity, and can reduce external validity by sensitizing people to the training program. In other words, each of these two designs has its particular drawbacks.

The second good design is the **pretest-posttest control group** design. It is symbolized as:

(R)	T_1	X	T_2
- -			
(R)	T_1		T_2

As in the previous study, the differential selection of participants is controlled by random assignment. However, even if random assignment can't be used, it is still a good design because the pretest (T_1) will show how much the two groups differ. In that case, any differences between the two groups for the posttest (T2) should be *larger* than the differences in the pretest. This is generally a good design because you can see the differences caused by your training program from before to after. These changes can be compared to the changes for an equivalent group of people who did not receive the program (or more realistically received a *different* program).

When to Collect Posttest Data

At this point in the planning process, you will have tentatively developed specific evaluation criteria for measuring trainee reaction, learning, performance, and any relevant secondary results. You have also identified an evaluation plan for each of the criteria. These may be the same plan, with testing to be performed at the same time, or they may be different plans depending on the type of criteria. Now we will discuss in further detail the appropriate time to conduct the posttests.

Leaving aside the issue of pretesting, we can measure criteria at various points after the training program has begun. These time points are shown in Figure 15.3. Each of the four types of criteria will be discussed separately.

1. Reaction. For the first criterion, reaction, it makes the most sense to measure the variable immediately after the training program. However, trainees may not know the extent of the benefits of the program until they are on the job for some period of time. Therefore, it is recommended that analysts ask for reactions to the program both immediately afterward, and after an appropriate period of time on the job. The specific amount of time depends on the tasks being trained. If it is something that will be performed frequently and beginning immediately, then three months would probably be adequate.

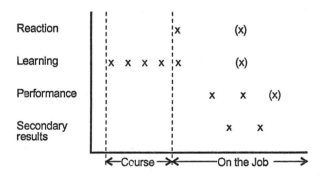
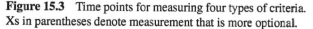

Figure 15.3 Time points for measuring four types of criteria. Xs in parentheses denote measurement that is more optional.

If it is a task that is performed infrequently, the analyst may have to wait one to two years to measure reaction again.

2. Learning. Learning is most commonly measured during and immediately after the training program. It is important to measure learning to determine where the instructional program needs modification. For this reason, learning assessment is better done at a micro-level, with question probes and problem solving, than at a general level. The methods used for assessment can come directly from any specific instructional objectives written earlier. For example, an instructional objective might have been: "State the five pieces of equipment that are necessary to set up the architecture imagery archive."

The learning assessment item would consist of asking the trainee to perform this task. The need to measure learning after some period of time will depend on the specific training domain. For example, if the training program concerns safe operation of job-related equipment, there might be some value in determining whether trainees retain important declarative knowledge, such as parts of the machine, or correct maintenance routines, after relatively long periods of time.

3. Performance. Under most circumstances, analysts should strive to evaluate performance at least twice, once fairly soon after training and once after a period of time. This will provide information related to two things: (1) how well the training program initially prepares the employee for job performance, and (2) how well the training bears up over time.

One question that must be answered by using judgment is how long to wait for the performance measures. This question will depend on the *mean frequency* of task performance. If you train a task that is performed daily, the first period of measurement can take place within the first one or two months. The second period of measurement can take place after anywhere from six months to several years. However, if the task trained is performed infrequently, or intermittently, the evaluation must be adjusted accordingly. The goal is to capture performance after realistic periods of time.

As an example, imagine that we performed a task analysis for training engineers to perform human operator reliability analyses. During the task analysis, we found that engineers will perform the task approximately twice a year. That is, the mean time between tasks will be six months. However, we also found that there may be intervals between tasks lasting up to two years. To measure performance, we could ask the engineers to perform the task within one week after training. This would provide certain information about the effectiveness of the training program. However, we should also collect performance data for the task when conducted after six months, and when conducted after two years. This would give us the most complete data on the adequacy of the program. The data should be collected in a realistic setting with no warning to trainees (so that they don't brush up on methods, reducing external validity).

4. Secondary results. Criteria that represent secondary results will not change until some period of time after performance has changed. The time of measurement for these variables will obviously depend on the specific domain, and may stretch over a relatively long time. For example, the measurement of sales, accidents, employee complaints, etc. can begin immediately after the training program but should extend for a long enough time period to acquire stable and reliable data. Other factors that could cause changes in the secondary variables should be assessed during the same time frame.

Choosing a Context

At this point, we have identified the criteria to be measured, the research design to be employed, and the time frame for posttests. All that is left is to determine the setting or context where the measurement should take place. This decision is really only critical for learning and performance variables (reaction is usually done at the training site and secondary results are measured at the organization).

The decision of where to measure learning and especially performance will strongly affect both the internal and external validity of a study. No matter where the data are collected, there are strong implications for the conclusions that can be drawn. Let us consider the two basic alternatives. The data can be collected at the job site (known as a field setting) or at another location such as the training center (sometimes known as a lab study). When we collect learning data immediately after the training program, we usually collect it at the training location. However, to assess performance data only at the training location and not on the job raises serious questions about the generalizability of the data. That is, how do you know that the trainees would perform the task in a similar manner on the job, over a period of time?

The importance of the setting for measuring performance variables depends entirely on the particular task being trained. For example, imagine that we have developed an instructional manual for assembling a new bicycle. Does it matter whether we bring a person into a laboratory and ask them to perform the task or measure task performance in their own garage? On the surface, it doesn't matter, because the task performance would probably be the same. However, there might also be important differences. For instance, a person might use many of their own tools to perform some of the required activities. Any differences between the ultimate task environment and the testing environment can be important. The extent to which there are differences will reduce the ability to generalize the evaluation data from the research context to trainees who are performing the task in the real-world environment.

Consider another example. I have developed a training videotape and workshop booklet to teach factory employees how to safely operate a lift truck. To evaluate the program, I ask an instructor to give the workshop to 20 people (randomly chosen from a sample of 40). After the workshop, we take all 40 people to a shop floor and ask each one to demonstrate safe lift truck driving procedures. All 20 workshop participants exhibit correct performance. Only 12 of the 20 in the control group exhibit correct performance. What can we conclude from this evaluation? We can conclude that our training program appears to be effective, *when* testing is conducted immediately afterward in a simulated environment and participants are not actually on the job. If you were asked to speculate on the effectiveness of the training program for actual on-the-job performance, you would be forced to make inferences. In making those inferences, you really have no sound justifications, because you don't really know the impact of the program on performance in a real job setting.

In summary, it is almost always easier to test learning than performance. And it is almost always easier to test performance in a restricted and artificial environment than in the actual job environment. However, in light of recent research and reviews (e.g., National Research Council, 1991), it would seem that we must make every attempt possible to evaluate learning and especially performance in the trainees' ultimate job environment, and after realistic periods of time.

JOB AIDS

Job aids should be evaluated using the same considerations discussed above. Potential criteria include reaction to the job aid (do people like it?), learning and comprehension, performance of tasks using the job aid, and various secondary organizational results. Because of the nature of the project being evaluated, performance is clearly the most critical variable. Performance with the job aid should be evaluated in as realistic a setting as possible.

Based on the material presented in the previous sections, it should be clear that research participants should adequately represent the population to which we are generalizing. It is not enough to bring two or three people into a laboratory and have them use the job aid for various tasks. An adequate and representative sample of the target population should be obtained, and be randomly assigned to groups.

The groups should at the least include a treatment group using the new job aid and a control group with task performance as done previously. If desired, these groups may be given a pretest; tasks performed without job aids in a natural job environment. Then the subjects randomly assigned to the job aid condition should be given the job aid. Both groups are again asked to perform the task. This is done so that any practice effects will increase performance of both groups, and we can assess the additional benefits of the job aid. It is often a good idea to allow some period of time between the initial pretest and the final posttest where half of the subjects use the job aid. For reasons described earlier, it is better if subjects are not aware of the exact treatment you are studying.

SUMMARY

In summary, we have suggested the following goals for evaluation of a training program or job aid:

1. **Collect information about four types of criteria:** reaction, learning, performance, and secondary results.
2. **Measure variables that are valid representations** of the criteria of interest, that is, variables that have good construct validity.
3. **Optimize the reliability** of criterion measures.
4. **Increase the internal validity by using appropriate research designs** and reducing demand characteristics.
5. **Use trainees that are a representative sample of the population** to which you hope to generalize.
6. **Increase the external validity of the evaluation study by using tasks, environments, and waiting times, that are representative of the ultimate job or tasks and their setting.**

REFERENCES

AAMODT, M. G. (1991). *Applied industrial/organizational psychology.* Belmont, CA: Wadsworth.

ANDERSON, J. R. (1990). *Cognitive psychology and its implications* (3rd ed.). New York: Freeman.

BABBIE, E. (1989). *The practice of social research* (5th ed.). Belmont, CA: Wadsworth.

BJORK, R., & DRUCKMAN, D. (1991). How do you improve human performance? *APS Observer, 4(6),* 13–25.

BLANZ, F., & GHISELLI, E. E. (1972). The mixed standard scale: A new rating system. *Personnel Psychology, 25,* 185–200.

BORDENS, K. S., & ABBOTT, B. B. (1988). *Research design and methods: A process approach.* Mountain View, CA: Mayfield.

BORING, E. G. (1954). The nature and the history of experimental control. *American Journal of Psychology, 67,* 573–589.

BUNKER, K. A., & COHEN, S. L. (1977). The rigors of training evaluation: A discussion of field demonstration. *Personnel Psychology, 30,* 525–541.

CAMPBELL, J. P., DUNNETTE, M. D., LAWLER, E. E., III, & WEICK, K. E., JR. (1970). *Managerial behavior, performance, and effectiveness.* New York: McGraw-Hill.

CAMPBELL, D. T., & STANLEY, J. C. (1963). *Experimental and quasi-experimental designs for research.* Chicago: Rand McNally.

COOK, T. D., & CAMPBELL, D. T. (1979). *Quasi-experimentation: Design and analysis issues for field settings.* Boston, MA: Houghton Mifflin.

DEMING, W. E. (1986). *Out of the crisis.* Cambridge, MA: Massachusetts Institute of Technology, Center for Advanced Engineering Study.

GHISELLI, E. E., CAMPBELL, J. P., & ZEDECK, S. (1981). *Measurement theory for the behavioral sciences.* San Francisco, CA: W. H. Freeman.

GITLOW, H. S., & GITLOW, S. J. (1987). *The Deming guide to quality and competitive position.* Englewood Cliffs, NJ: Prentice Hall.

GOLDSTEIN, I. L. (1986). *Training in organizations: Needs assessment, development, and evaluation* (2nd ed.). Monterey, CA: Brooks/Cole.

KIRK, R. E. (1982). *Experimental design* (2nd ed.). Monterey, CA: Brooks/Cole.

KIRKPATRICK, D. L. (1959, 1960). Techniques for evaluating training programs. *Journal of the American Society of Training Directors, 13,* 3–9, 21–26, *14,* 13–18, 28–32.

LANDY, F. J. (1989). *Psychology of work behavior* (4th ed.). Monterey, CA: Brooks/Cole.

LANDY, F. J., & FARR, J. L. (1980). Performance rating. *Psychological Bulletin, 87,* 72–107.

LANDY, F. J., & FARR, J. L. (1983). *The measurement of work performance: Methods, theory, and applications.* New York: Academic Press.

LANDY, F. J., & RASTEGARY, H. (1988). Current issues in performance evaluation. In I. Robertson & M. Smith (eds.), *Personnel evaluation of the future.* New York: John Wiley & Sons.

LATHAM, G. P., FAY, C., & SAARI, L. (1979). The development of behavioral observation scales for appraising the performance of foremen. *Personnel Psychology, 32*, 299–311.

LATHAM, G. P., & WEXLEY, K. N. (1977). Behavioral observation scales for performance appraisal purposes. *Personnel Psychology, 30*, 225–268.

LINDBOM, T. R., & OSTERBERG, W. (1954). Evaluating the results of supervisory training. *Personnel, 31*, 224–223.

MCINTYRE, R. M., SMITH, D. E., & HASSETT, C. E. (1984). Accuracy of performance ratings as affected by rater training and perceived purpose of rating. *Journal of Applied Psychology, 69(1)*, 147–156.

MUCHINSKY, P. M. (1990). *Psychology applied to work: An introduction to industrial and organizational psychology* (3rd ed.). Belmont, CA: Wadsworth.

MURPHY, K. R., & CONSTANS, J. I. (1987). Behavioral anchors as a source of bias in rating. *Journal of Applied Psychology, 72(4)*, 573–577.

MURPHY, K. R., GARCIA, M., KERKAR, S., MARTIN, C., & BALZER, W. K. (1982). Relationship between observational accuracy and accuracy in evaluating performance. *Journal of Applied Psychology, 67*, 320–325.

NAGLE, B. F. (1953). Criterion development. *Personnel Psychology, 6*, 271–288.

NATIONAL RESEARCH COUNCIL (1991). *In the mind's eye: Enhancing human performance.* Washington, DC: National Academy Press.

ODIORNE, G. S. (1964). The need for an economic approach to training. *Journal of the American Society of Training Directors, 18(3)*, 3–12.

PULAKOS, E. D. (1984). A comparison of rater training programs: Error training and accuracy training. *Journal of Applied Psychology, 69(4)*, 581–588.

ROETHLISBERGER, F. J., & DICKSON, W. J. (1939). *Management and the worker.* Cambridge: Harvard University Press.

SARETSKY, G. (1972). The OEO P.C. experiment and the John Henry effect. *Phi Delta Kappan, 53*, 579–581.

SMITH, P. C., & KENDALL, L. M. (1963). Retranslation of expectancies: An approach to the construction of unambiguous anchors for rating scales. *Journal of Applied Psychology, 47*, 149–155.

SOMMER, R., & SOMMER, B. B. (1986). *A practical guide to behavioral research: Tools and techniques* (2nd ed.). New York: Oxford University Press.

─────────── 16 ───────────

Minimizing Professional Liability

There are a variety of liability issues either directly or indirectly associated with developing and administering training programs. The types of issues that have historically been addressed in human resource development include any activities that might somehow discriminate against one type of population in selection or training processes (e.g., see Goldstein, 1986, Rothwell & Kazanas, 1992). For example, selection of certain groups of people to receive, or not to receive, training might be considered discriminatory for some reason. Other writers have addressed these issues and they will not be covered in this chapter. Instead, I am addressing a rather new issue that is becoming increasingly important as a result of certain trends in the federal and legal systems. This is the issue of who is responsible if an employee or private citizen is instructed or trained to perform a specific task, and subsequent to the training, someone is injured or killed as a consequence of the person performing the task incorrectly. The purpose of this chapter is to provide an overview of this issue and provide some basic guidelines for training program designers to protect themselves against professional design liability.

INTRODUCTION TO THE LIABILITY ISSUE

If you design and/or deliver a training program and that trainee is subsequently injured or killed in performance of the task, you could be held professionally negligent and liable for damages, or the firm for which you work could be held liable. This may sound unlikely, but it is happening with increasing frequency. Many people who develop and deliver training programs, both large and small, are not even aware that this liability issue exists or that they themselves are at risk. Professional liability can be defined as follows:

> *Professional liability is incurred when certain standards of performance are not lived up to or when services are rendered by individuals with specialized training or experience in a particular area, and the person for who service is rendered has been damaged in some way.*

Because of recent trends in the courtroom, the issue of professional liability is increasingly affecting people who develop and deliver training programs. In the next section I discuss these trends.

Recent Trends in the Courtroom

You may recall that in Chapter 1 I discussed a certain design "priority system" used by engineers when designing physical systems.[1] To repeat the important points, when an engineer is designing a system, there are several methods of eliminating or controlling hazards. In general, designers should deal with hazards by doing the following *in the order listed* (Hammer, 1989; National Research Council and Institute of Medicine Report, 1985):

1. **Design the Hazard Out:** Design the system so that either (1) the hazard is eliminated entirely, or (2) the hazard is limited to a level below which it can do no harm.
2. **Safeguard:** Where safety by design is not feasible, protective safeguards should be employed.
3. **Warn:** When designing the hazard out and safeguarding are not feasible, automatic warning devices should be employed.
4. **Train:** Instruct people concerning proper procedures for task performance and use of equipment.

[1]Readers may want to review that particular section of Chapter 1.

The designer should first strive to design a system that is inherently safe. If this cannot be done and still preserve the basic functioning of the system (e.g., a chain saw), the designer should provide adequate physical safeguards to shield the user from injury. If that solution is not feasible, then proper warnings should be used, and so forth.

According to Sanders and McCormick (1987), it is sometimes difficult to distinguish between warning and instruction. They suggest that *"warnings* [sic] inform the user of the dangers of improper use and, if possible, tell how to guard against those dangers. *Instructions* [sic] tell the user how to use the product effectively. In addition, instructions themselves may, and often do, contain warnings" (p. 547). For this reason, many guidelines tend to lump warnings and instruction together into one category.

The relevant issue for this particular chapter is the question of what happens when someone is injured or killed using a piece of equipment or a physical space designed by someone? Often there is a lawsuit. If the person was injured during the performance of his job, workmen's compensation laws in most states prohibit the employee from suing his own employer. However, he does not give up the right to sue a third party, and that is what frequently happens. Examples of third parties who might be sued are: the manufacturer of equipment involved in the injury, the driver of a vehicle involved in the accident, the architect who designed the structure (that collapsed), the engineer who designed the product, or "an inspection agency representative that certified as safe a pressure vessel that later burst" (Hammer, 1989).

Very frequently, a lawsuit is filed against the person who designed the equipment and/or the firm for which they worked at the time of the design process. This means that designers are brought to court to determine whether they properly designed the system. To understand the implications for training designers, let's look at some typical courtroom scenarios. Three of the four scenarios happen to deal with actual cases that we have seen here in the Pacific Northwest (names and details were changed). The first scenario is given to show a typical case dealing with the adequacy of the design of *a physical system.* This is followed by scenarios that show how the same *types* of cases are being extended to the adequacy of training program design.

Courtroom Scenarios

The paper mill. Joe Smith worked at the local paper mill for 14 years. He held a number of jobs at the mill, in different locations, and under different supervisors. One evening Joe was working on the line where large machines were used to wrap and stamp bales of cardboard. He noticed a problem with the cardboard being damaged and went over to inspect. As Joe reached between the machinery to look at the damaged materials, his other arm was caught between a piece of machinery and a cardboard bale. His arm was essentially destroyed.

Now Joe and his wife are suing the manufacturer that made the machinery for the paper mill 20 years earlier. Their case rests on the contention that proper care was not taken in designing the equipment. The case goes to trial with a jury. In that trial, the prosecution brings in engineering experts of various backgrounds. These experts argue that placement of one's body and body parts between the pieces of equipment was a foreseeable hazard. They argue that even 20 years earlier, engineers were aware of the need to design equipment to prevent injury from such hazards. They present data showing that there was published literature on the matter, and argue that because the designers ignored this safety issue and designed the equipment without regard for this particular hazard, they were negligent in their professional duties.

The defense must counter this argument. Based on the discussion presented above, one can see that this usually means showing that it is *not possible* to design the system so as to prevent the injuries sustained. In actuality, "not possible" means not feasible in the real world. That is, one could not design the system so that it (1) contained no hazard, (2) still retained its functions, (3) was usable under reasonable circumstances, and (4) was economically viable to produce. In summary, the defense will argue that it is not possible to completely design out the hazard and still maintain the integrity of the equipment. The experts will go on to show that the other means of addressing the hazard were followed (guarding, warning, etc.). For example, an expert for the defense might produce one of the following counter-arguments, depending on what he or she found in investigating this particular case:

- There was a metal guard on the equipment that had been removed by workers prior to the accident.
- The employee removed the guard just prior to the accident.
- There were strict instructions accompanying the equipment that the machinery was to be shut down before anyone moved into the proximal area (defined as anywhere within 15 feet).

Based on the testimony of Joe, other employees, Joe's supervisor, Joe's wife, and the experts for both sides, the jury must decide (a) whether it was realistically possible to design the system to eliminate the hazard without creating other substantial problems, and (b) whether the engineers performed their jobs in accordance with the standards and guidelines that were available at the time the system was designed. It is easy to see that the outcome of this case depends heavily on the capabilities of the experts brought in to provide *objective* information for each side as well as *subjective* opinion.

The fireplace insert. Mary Johnson was an independent person, and had lived alone on a horse ranch all of her life. One fall she decided to buy a woodburning stove. She bought an insert from a local dealer, and decided to install it herself, unpacking all of the pieces and carefully reading the instructions. She was a little unsure about some parts, but was confident enough to proceed. When the insert was in, Mary felt that she had done a good job. Four months later, Mary's house caught on fire and she lost all of her belongings plus three pets.

Mary was not the only one to encounter this problem. A professional contractor installed several hundred of the inserts into condominiums. Many caught fire and all had to be torn out and reinstalled. A class action suit was brought against the manufacturer for failing to provide adequate instructions for the installation of the units.

The defense argued that (a) all of the necessary information for installation was *literally* in the instructions in one place or another, and (b) the units were installed improperly because the people failed to follow the instructions. What they failed to point out were these points made by the plaintiff's expert witness:

- One manual was provided for multiple models, and it was difficult to determine which information and procedural steps were relevant to which model.
- There were additional pieces of material and equipment that needed to be purchased separately for proper installation. This was not noted in the instructions. For example, the instructions said to "install metal strip" but no metal strip was provided. People assumed it was not important or it would have been included with the other parts.
- There was an inconsistent use of terms and part identifications between different segments of the instructions. For example, the text might use one term and a figure might use another term for the same part.
- Users needed to refer to previous figures to understand directions, and this was never indicated in the manual.
- Finally, and most important, there was a serious *hazard* that was not indicated whatsoever. That is, there was one procedure that, if left out, would result in a fire with 100% certainty. This fact was known by the manufacturer. The hazard was not stated in the manual, in any form.

The jury in this case must determine whether the instructions were adequately designed.

The logging accident. Jim Higgins was a professor in forestry at a large university in Oregon. He knew many families with several generations of men in the logging industry together. He also knew that it was often a "Mom and Pop" type of operation with training being handled in a relatively informal on-the-job manner. Finally, Jim knew that the logging business had a very high accident rate. In 1989, Jim decided to develop a safety training program that was portable enough to take on the road to the family-run operations. He made a videotape that showed the various hazards associated with the job.

Two years after Jim began showing the videotape to loggers in the Pacific Northwest, one of the loggers to whom he had shown the videotape was killed while using a skidder. The man's family brought suit against several parties including an equipment manufacturer and Jim Higgins. Their case against the manufacturer was simply that the equipment was poorly designed. And their case against Jim rested on the assumption that his tape did not show all of the conditions under which the equipment was hazardous, and incorrectly implied that the procedure shown in the tape was appropriate under all conditions, when in fact it was not. The role of the jury was to decide how much blame to allocate to the manufacturer, to Jim Higgins, and to the logger himself.

The equipment manufacturer brought in experts who testified that it is impossible to design, guard, or even warn for all of the foreseeable hazards inherent in this type of equipment and the logging operation in general. They argued that some of the injury prevention must take place through proper training. They argued that the equipment was designed as safely as possible given the technologies at the time, and that the cause of the logger's death was improper use of the equipment. In short, the man was not trained properly. Training experts brought in by the *equipment manufacturer* described the established guidelines and procedures for development of training programs that existed in 1989. They also pointed out all of the discrepancies between what Jim did in the development of his videotape and the established guidelines and procedures. They suggested that if the jury must assign blame, clearly the vast majority of it belonged with Jim for professional negligence in development of the training program. Jim was left with little defense because he had, in fact, not known about or used any training guidelines when he developed the videotape. In addition, he had not thought it was his responsibility to determine what other training the loggers should receive, or evaluate effectiveness of the program.

The forklift accident. Tom Simpson was an electrical subcontractor who was called in to work for three days at a large warehouse. During the course of his job, he was struck and severely injured by a forklift operated by an employee of the warehouse. Tom brought suit against the company for failing to properly train the employee on operation of the forklift. The defense argues that the company followed OSHA regulations that stipulate that "Only trained and authorized operators shall be permitted to operate a powered industrial truck. Methods shall be devised to train operators in the safe operation of powered industrial trucks" (OSHA 1910.178–1989). The company provided training by giving employees written materials and showing a videotape made on site by the company.

The plaintiff successfully argued that the training provided to the employee was inadequate. An expert in the field of training and instructional design testified that, inconsistent with existing professional standards and guidelines, there was never any type of trainee evaluation showing that employees actually learned the concepts and proper procedures. In addition, there was no training whatsoever for the specific rule that was violated by the forklift driver which resulted in the injury.

Implications for Training Program Design

When viewed in the order I have described them, these scenarios suggest a direction in which things are headed in the courtroom. Recall the design list at the beginning of the chapter; **design out, guard, warn,** and **train.** It has been typical for years to put engineers on the spot for the safety of the systems they design. They must show that they developed a system in as safe and reliable a fashion as possible, whether this meant *designing a hazard out* of the system or providing adequate *safeguards.* They must also show that they did this design in accordance with existing guidelines and procedures (meaning work that was published and publicly available).

More recently, this type of professional defense has moved down the design priority list; a significant number of court cases have revolved around the adequacy of *warning labels* and *warning literature* provided with a product (DeJoy, 1989; Gill, Barbera, & Precht, 1987; Hammer, 1989; Kreifeldt & Alpert, 1985; Lehto & Miller, 1988; Sanders & McCormick, 1987; Wallace & Key, 1984; Wogalter, Fontenelle, & Laughery, 1985; Wogalter et al., 1987). These cases mostly focus on three variables:

- Whether a warning label for a particular hazard should have been provided with a product,
- Whether an existing warning label was adequately designed
- Whether product manuals and instructions were adequate

Notice in particular the second and third issues. These are *design* questions similar to those for physical equipment design. Experts are brought in to testify as to whether a warning label or owner's manual were designed in accordance with existing principles and guidelines. The person who designed the product must show that there was a design rationale, and that the rationale was consistent with established standards.

Finally, we are just beginning to see this issue of negligence in design trickle down to the lowest rung. If it is not possible to design out, guard, or warn against all hazards, people must be *trained* or instructed adequately. People design training programs just like they design physical systems and warning labels. In a court of law, a judge or jury will be asked to determine whether a person was adequately trained to perform a hazardous task. How will they decide whether the person who designed and/or implemented the training program was professionally responsible or professionally negligent?

Based on the limited cases to this date, we can say that they will do it in the same manner that they judge the other types of design. Experts will be brought in for both sides and argue whether the (training program) design was carried out properly. This means asking whether the trainer followed the principles and guidelines that were established at the time the program was designed, or whether a better job could have been done.

Where do professionals find the existing "standards" for use in designing products and procedures? OSHA and other governmental agencies publish design standards and guidelines. For example, standards for designing safety signs and labels are published by the federal government in ANSI Standard Z535.4–1991. OSHA publishes numerous guidelines for design of physical systems as well as safe procedures for a variety of tasks (e.g., OSHA 1910–1989). Unfortunately, there *are no standards yet* for the design of training programs. Some federal documents will briefly touch upon the topic of training when discussing particular hazards. For example the OSHA document dealing with ladders, OSHA 1926, states that:

> The employer shall provide a training program for each employee using ladders and stairways, as necessary. The program shall enable each employee to recognize hazards related to ladders and stairways, and shall train each employee in the procedures to be followed to minimize these hazards. (OSHA 1926.1060–1991)

The document goes on to state the types of information that must be trained (hazards, proper construction and care, maximum load, etc.), and that "retraining shall be provided for each employee as necessary so that the employee maintains the understanding and knowledge acquired through compliance with this section" (p. 349). But there are no guidelines concerning *how* to train employees, nor how to test for the knowledge and skills, or know when employees need retraining. The result is that trainers must look to the published literature in the various related fields to find the principles and guidelines.

Summary

Hopefully the previous section made a strong point. According to federal law, anyone designing a training program or performance support system has the *legal* responsibility to design that program in accordance with established principles and guidelines. If they do not do so, and a trainee or another person is subsequently injured or killed as a result of improper performance of the task, they (and/or their firm) can be found professionally liable and responsible for damages.

The only exception to this case will be when the trainer works for an injured employee's own company. This is because in almost all cases, workmen's compensation will cover on-the-job accidents to an employee; the person cannot bring a lawsuit against their own employer. Notice that even if you are training employees in your own company, you are not free from risk. For example, in the scenario above concerning the forklift accident, a subcontractor was injured, and being a third party, was able to bring a lawsuit against the company. In addition, OSHA will fine companies that fail to adequately train employees (not to mention the fact that workmen's compensation payments are higher when the company's employees have made more claims for accidents).

GUIDELINES FOR TRAINERS

Five Basic Rules

The previous material in this chapter suggests the approach that one must take to minimize risk of litigation and professional liability. The best approach a designer can use is to:

- **Design instructional systems and performance support systems in accordance with established principles, guidelines, and procedures.**

Since there are currently no specific federal guidelines and standards, designers must refer to published works in the field. The remainder of this section will describe how the designer should go about applying the design principle stated above. For discussion purposes, I have broken the general principle down into five basic design rules. The rules are based on what seems to be the best defense in any general design case. The first rule is essentially the cardinal design rule stated above, and the others are very important supplemental support mechanisms.

RULE 1. Follow published principles, guidelines, and procedures in design, development and administration of the training program, job aid, or performance support system.

It can be seen that the major criterion for determining whether a designer has done their job properly is to compare their particular process and product with established guidelines. For a training program, the analyst should use the guidelines and procedures established in the fields of instructional design, applied cognition, and human factors. One of the overriding goals of this book has been to gather the major principles and guidelines from these fields into one place. Thus, the chapters have been designed to act as a reference guide for people designing training and instructional programs. However, any designer is *technically* responsible for being familiar with *all* of the published work (a truly daunting task). In this book, each procedure has been included for a reason, and as you develop the training program, it is important to perform each step shown in Figure 1.1 very carefully.

Caution on the Use of Nonexamples. There is one guideline in this book that must be used with caution in instructional systems dealing with potentially hazardous tasks. That is the use of examples and nonexamples. Under normal circumstances, it is a good idea to make liberal use of examples and nonexamples. However, when instructing a person about proper procedures, the presentation of nonexamples can be problematic. This is because, over time, the person may forget whether a particular behavior was an example or nonexample. It simply becomes associated with the stimulus conditions. For this reason, nonexamples of safe behavior should be minimized or completely eliminated. Where it is important to show common but unsafe procedures, it should be done in a way that ensures ultimate use of the *proper* safe procedure. For example, the trainer could verbalize the unsafe procedure, but show a more salient videotape of the safe procedure. The trainees could then be drilled on safe procedures at intermittent points of time. An example is fire drills conducted at children's public schools. This reinforces the association between the stimulus conditions and the correct response.

Caution on Supporting Transfer of Training. A second issue that merits specific attention is the issue of *transfer* (see Chapter 6). Under stress, we revert back to the strongest or most dominant procedural associations between stimulus conditions and response; that is, the things that are the most well-learned. Don't assume that because people *can* correctly make the non-dominant response during training and testing that they will do so under adverse circumstances. Whatever behavior must be ultimately performed in times of stress, hazard, difficulty, or other duress, make sure that *those* are the behaviors that are practiced much the most frequently during training.

RULE 2. Adequately evaluate the program using established guidelines and procedures. Training programs often receive inadequate final evaluation. If there is any possibility of future litigation, user testing and final system evaluation cannot be overemphasized. No matter how carefully a program is developed, in accordance with all known principles, there is no guarantee it will work. User testing should be used extensively during the design and development phase. Final program evaluation is the ultimate insurance policy. Imagine a case where you have rigorously followed standard design principles and procedures. Now imagine that you have taken a large and representative sample of trainees, and using an experimentally sound research design, administered the training program and collected evaluation data. All trainees showed mastery of the material, and showed high performance levels under realistic time spans and task conditions. Armed with this information, it will be much easier to defend your program in court than if you had ONLY used appropriate guidelines and procedures for system development, but not followed up with rigorous program evaluation.

RULE 3. Prepare and keep documentation. Most designers make hundreds of minor decisions in their head and never document the rationale for those decisions. The problem is, when they are questioned about these decisions in court, they usually can't recall the rationale. In addition, rationales generated for legal purposes are often suspect. Historical documents are always superior. If you might ever be in a position to defend the adequacy of your work, you should rigorously document all of your design decisions. This means documenting the following items, preferably in detail:

- Methods that were used for performing organizational analysis, data from the organizational analysis, and conclusions.
- Methods for performing the task analysis (including what you did, people you collected data from, etc), rationale for task analysis methods, data from task analysis.
- Methods for performing trainee analysis (including data collection

methods, sampling methods, etc.), the specific people used for trainee analysis, data from trainee analysis.

- Functional specifications—for each item, give the rationale for including the specification. Where certain properties (such as practice) were NOT included, provide rationale.
- Alternatives generated for the design concept—and why they were accepted or rejected. Provide detailed rationale for final design.
- Rationale for prototyping and methods used for user testing results of each iteration of user testing and how the design was modified (with rationale).
- During or after prototyping and user testing, write a system specification document with the details of the program and the rationale for each decision.
- Any design decisions and guidelines used to develop the final product.
- Limitations or constraints on the final product.
- Methods for final user testing and program evaluation, including how the sample of trainees was selected, rationale for research design.
- Results of final user testing and program evaluation.

RULE 4. Provide appropriate guidelines (and disclaimers) to anyone who will be administering a program that you have developed. There may be times when you are constrained in program development and cannot provide the entire spectrum of training that is dictated by your task analysis. For example, you may perform the task analysis for forklift operation and realize that much of the job involves perceptual skills. You realize that proper training would involve hands-on practice under a variety of conditions. If you are asked to make a videotape on safe forklift operating procedures, you will realize that by definition the training program could be considered inadequate because a videotape would not provide the perceptual hands-on training. You have two choices. The first is to refuse to develop a training program that is not what it should be.

The other choice is to develop the program, but include written and other types of documentation stating that the program is not meant to be a complete training system for this particular task. For example, a videotape should include a written disclaimer as well as a verbal statement at the beginning of the program that the skill requires other types of training such as practice with actual machinery; the videotape should not be used without training of other types. The document should include a list of appropriate types of training for the instructor or whoever will be administering the training tape.

RULE 5. Insure that trainees are properly evaluated at regular intervals for both conceptual knowledge and knowledge of proper rules and procedures. If trainees are tested at proper time delays, in accordance with the principles stated in Chapters 13 and 15, you will know whether employees are in need of additional training. In addition, in the case of injury or death, you will also have solid evidence that the employee did in fact possess the required knowledge and skills. This will put the burden of responsibility on the employee for failing to perform proper procedures (which he or she had previously demonstrated) rather than on your training program.

While there is no guarantee against findings of professional liability, following these rules will go a long way toward protecting a designer against them.

Design of Warnings

Many instructional systems that deal with potentially harmful activities contain warnings of some type. In particular, manuals dealing with the installation, use, and maintenance of physical systems will more often than not contain warnings. There is an entire literature on the design of warnings and warning labels (e.g., see Lehto & Miller, 1988), and the majority of that material is beyond the scope of this book. Readers are referred to appropriate sources (e.g., American National Standards Institute, 1987; Lehto & Miller, 1986) for guidelines on how to properly design warnings. In general, a warning should always include the following four elements (Sanders & McCormick, 1987; Wogalter et al., 1987):

- **Signal Word.** The signal word conveys the gravity of the risk. Signal words conventionally used are *danger* (immediate hazard resulting in severe personal injury or death), *warning* (hazards that could result in severe personal injury or death), and *caution* (for hazards or unsafe practices that could result in minor personal injury or property damage).
- **Description of the hazard** stating the specific nature of each one.
- **Statement of the consequences associated with the hazard.**
- **Statement of behavior needed to avoid the hazard.**

An example of a warning with all of these elements is (Strawbridge, 1986):

DANGER:

Contains Acid

To avoid severe burns, shake well before opening

Some warnings also rely on icons such as the one shown in Figure 16.1. Even warnings that have provided the critical elements have sometimes been found inadequate in legal cases. Kreifeldt and Alpert (1985) cite a case where the warning shown in Figure 16.2 was inadequate because it failed to "specify the danger of using the product in the vicinity of closed and concealed pilot lights" (Kreifeldt & Alpert, 1985).

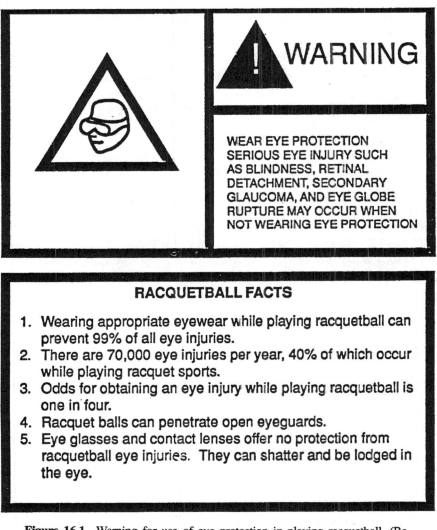

Figure 16.1 Warning for use of eye protection in playing racquetball. (Reprinted with permission from *Proceedings of the Human Factor Society, 35th Annual Meeting,* 1991. Copyright by the Human Factors and Ergonomics Society. All rights reserved.)

CAUTION: FLAMMABLE MIXTURE. DO NOT USE NEAR FILE OR FLAME.

CAUTION! WARNING! EXTREMELY FLAMMABLE! TOXIC! CONTAINS

NAPHTHA, ACETONE, AND METHYL ETHYL KETONE. ALTHOUGH THIS

ADHESIVE IS NO MORE HAZARDOUS THAN DRYCLEAING FLUIDS OR

GASOLINE, PRECAUTIONS MUST BE TAKEN. USE WITH ADEQUATE

VENTILATION. KEEP AWAY FROM HEAT, SPARKS, AND OPEN FLAME. AVOID

PROLONGED CONTACT WITH SKIN AND BREATHING OF VAPORS. KEEP

CONTAINER TIGHTLY CLOSED WHEN NOT IN USE

Figure 16.2 Warning found to be inadequate in a court case. (From Kreifeldt & Alpert, "Use, misuse, warnings: A guide for design and the law," 1985, *Interface 85: Fourth Symposium on Human Factors and Industrial Design in Consumer Products* (pp. 77–82).

While there has been some agreement as to what constitutes the minimal warning label, there has not been particularly good consensus regarding what constitutes a *good* or *adequate* warning label. An additional problem is the fact that people often ignore warning labels even when they are provided (e.g., Dorris & Purswell, 1977; Gill, Barbera, & Precht, 1987). Nevertheless, any instructional system that contains warnings of any type should comply with the general guidelines currently available including use of the components described above.

REFERENCES

ANSI ACCREDITED STANDARDS COMMITTEE. (1987). *Safety signs and colors.* Gaithersburg, MD: National Bureau of Standards.

DEJOY, D. M. (1989). Consumer products warnings: Review and analysis of effectiveness research. *Proceedings of the 33rd Meeting of the Human Factors Society* (pp. 936–940). Santa Monica, CA: Human Factors Society.

DINGUS, T. A., HATHAWAY, J. A., & HUNN, B. P. (1991). A most critical warning variable: Two demonstrations of the powerful effects of cost on warning compliance. *Proceedings of the Human Factors Society 35th Annual Meeting* (pp. 1034–1038). Santa Monica, CA: Human Factors Society.

DORRIS, A., & PURSWELL, J. (1977). Warnings and human behavior: Implications for the design of product warnings. *Journal of Products Liability, 1,* 255–264.

GILL, R., BARBERA, C., & PRECHT, T. (1987). A comparative evaluation of warning label designs. *Proceedings of the Human Factors Society 31st Annual Meeting* (pp. 476–478). Santa Monica, CA: Human Factors Society.

GOLDSTEIN, I. L. (1986). *Training in organizations: Needs assessment, development, and evaluation* (2nd ed). Monterey, CA: Brooks/Cole.

HAMMER, W. (1989). *Occupational safety management and engineering* (4th ed.). Englewood Cliffs, NJ: Prentice Hall.

KREIFELDT, J., & ALPERT, M. (1985). Use, misuse, warnings: A guide for design and the law. In T. Kvalseth (ed.), *Interface 85: Fourth Symposium on Human Factors and Industrial Design in Consumer Products* (pp. 77–82). Santa Monica, CA: Consumer Products Technical Group of the Human Factors Society.

LEHTO, M. R., & MILLER, J. M. (1986). *Warnings: Volume 1: Fundamentals, design, and evaluation methodologies.* Ann Arbor, MI: Fuller Technical Publications.

LEHTO, M. R., & MILLER, J. M. (1988). The effectiveness of warning labels. *Journal of Products Liability,* 11, 225–270.

NATIONAL RESEARCH COUNCIL and INSTITUTE OF MEDICINE REPORT. (1985). *Injury in America.* Washington, DC: National Academy Press.

OSHA 1926 (1991). Safety and health regulations for construction. In OSHA 29 CFR Code of federal regulations.

ROTHWELL, W. J., & KAZANAS, H. C. (1992). *Mastering the instructional design process: A systematic approach* San Francisco, CA: Jossey-Bass.

SANDERS, M. S., & MCCORMICK, E. J. (1987). *Human factors in engineering and design* (6th ed.). New York: McGraw-Hill.

STRAWBRIDGE, J. (1986). The influence of position, highlighting, and imbedding on warning effectiveness. *Proceedings of the Human Factors Society 30th Annual Meeting* (pp. 716–720). Santa Monica, CA: Human Factors Society.

WALLACE, W., & KEY, J. (1984, December). Human factors experts: Their use and admissibility in modern litigation. *For the Defense,* 16–24.

WOGALTER, M., FONTENELLE, G., & LAUGHERY, K. (1985). Behavioral effectiveness of warnings. *Proceedings of the Human Factors Society 19th Annual Meeting* (pp. 679–683). Santa Monica, CA: Human Factors Society.

WOGALTER, M. S., GODFREY, S. S., FONTENELLE, G. A., DESAULNIERS, D. R., ROTHSTEIN, P. R., & LAUGHERY K. R. (1987). Effectiveness of warnings. *Human Factors, 29(5),* 599–612.

Index

A

Accumulation model of instruction, 217
Accountability, 356
Activity sampling, 106–8
Adaptive aiding, 5
Advance organizers, 145, 155, 268
Animation, 240, 301
   for prototyping, 301
ARCS model of motivation, 147–49
ARCS motivational strategies, 151–53
Associative memory retrieval, 134
Associative scaling algorithms, 105–6,
   107
Associative stage of expertise, 55
Attention, 131–32, 147, 151, 154
   resources, 131–32
Attitude change, 49–50
Attitudes, 49–50
Attribution theory, 149–51
Audiovisual techniques, 178–79,
   182–84

Authoring tools, CAI, 233–40
Automated rules, 39, 44, 146
Automatic processing, 47–48, 146
Automation of subtasks, 146, 157
Autonomous stage of expertise, 55

B

Benchmark tasks, 340
Behavioral objectives, 171
Behavioral modeling, 209
Belief systems, 49–50
Brainstorming, 247–48
Business games, 208–9

C

CAI (*See* Computer-assisted instruction)
Causal attribution, 149–51
CBT (*See* Computer-based training)

CD-I, 221
CD-ROM, 221
Cognitive processing capacity, 131–35
Cognitive resources, 145, 158, 340
Cognitive stage of expertise, 55
Cognitive workload, 145–46
  reducing, 158
Cognitive task analysis, 67–69
  Means and Gott, 98–99
Component display theory, 307–8
Computer-assisted instruction, 178–79,
      194–96, 197–201, 216–45,
      280, 298–99, 301, 318–22,
      338–40, 347–48
  advantages, 199
  animation, 240
  basic programming languages,
      234–35
  benefits, 197–98
  debugging, 347–48
  design concept, 251–52
  development, 318–22, 347–48
  disadvantages, 200–201
  drill and practice, 223–24
  effectiveness, 197–98
  final user testing, 351
  first generation, 217
  high-end programming languages,
      238–41
  history, 216–19
  hybrid designs, 231–33
  HyperCard, 235–38
  hypertext, 198, 225–29
  hypermedia, 198, 225–29, 232
  in classrooms, 216–17
  information access, 223, 226–27, 319
  information generation, 223, 227–29
  instructional games, 223, 230–31
  interface design, 280–88
  mid-level programming languages,
      235–38
  multimedia, 218–23, 257–58, 280
  research results, 197–98
  scripting language, 236–38

  simulations, 223, 224, 229–30, *See
      also* Simulations
  software, 222, 233–41
  supporting documents, 348–51
  ToolBook, 235–38
  tutorials, 223, 224–25
  types, 198
  user testing, 338–40
Computer-based semantic networks, 85
Computer-based training, 198 (*See also*
      Computer-assisted instruction)
Computer interface design, 173,
      280–88
  color, 284–85
  design goals, 281
  interaction, 285–86
  specifying requirements, 173
  static screen design, 282–84
Computerized Readability Editing
      System, 272
Computer-managed instruction, 198
Concept mapping, 84
Concept networks, 40–41, 42–43,
      84–87, 99–101, 303
Conceptual graph analysis, 99–105
Conceptual graph structures, 42–43,
      84–86, 99–101, 102–3, 303–5
Construct validity, 367
Controls and displays analysis, 93
Controlled memory search, 134
Controlled cognitive processing,
      47–48, 134
Criteria, performance, 165–66
  types of, 365–66
Criterion reliability, 369–70
Critical decision method, 96
Critical incident technique, 96–97

D

Decision matrix, 248
Decision support system, 4, 5
Declarative knowledge, 39–47, 66–67

acquiring, 129, 136–39, 154–56
conceptual, 39, 44, 55, 137
instruction of, 129, 136–37, 224–29
organization of, 137, 138
rule, 39, 44, 55, 137
types, 42–43
Delphi method, 91–92
Design and development, 14–16
Design concept (*See also* Generating
    design concept)
preliminary, 293
Design constraints, 9–10, 11, 14, 18,
    128, 167–68, 173–75, 177,
    246–49, 259
Design cycle, 166
Design documentation, 392–93
Designing instructional materials (*See*
    Instructional materials)
Designing out hazards, 8, 383, 383–89
Design model, 10, 11–19
traditional, 10–11, 295–97
Design specifications, 246, 256 (*See
    also* Functional design specifi-
    cations)
Design team members, 9
Developing the design concept (*See*
    Generating design concept)
Development, program, 309–23
combining development proce-
    dures, 323–25
computer-assisted instruction,
    318–22
documents, 312–13
job aids, hardcopy, 313–14
lecture, 311–12
on-the-job training, 310–11
scripts, 315, 322
simulations, 323
storyboards, 316, 322
tutoring, 310–11
workshops, 311–12
videotape, 314–17
Digital cameras, 221
Disclaimers, 393

Discovery learning, 186–88
Discrimination, in learning, 140–141
Displays, predictive, 126
Document analysis, 70–72, 101
Document and equipment analysis,
    70–72
Document design, 261–80, 312–13
color, 266, 271, 277
content design, 266–67
content organization, 268–69
design process, 263–64, 312–13
evaluation tools, 272
final development, and, 345
format design, 266
illustrations, 271–72
principles, 264–72
procedures, 269–70
user biases, 263
Document evaluation tools, 272
Document user testing, 336–37
Domain-general strategies, 39, 45, 48,
    187
Domain-specific strategies, 39, 45, 48,
    187
Domain mental models, 54
DVI (Digital Video Interactive), 221

E

Elaboration theory, 144
Elaborative rehearsal, 134
Electronic performance support
    systems, 4
example, 122–24
Electronic text, 180, 318–22
Elementary component model, 146
Engineering hazard design priorities, 8
Ergonomic system design, 10, 119–20,
    294, 297
Error-tolerant systems, 126
Evaluation:
criteria, 171–72
formative, 15–16, 353

Evaluation (*cont.*)
  system, 16–17, 157–58
Examples and nonexamples, 140–41,
    155, 156, 268
Experiential learning, 186–88
Expert performance, 55–57
Extended Task Analysis Procedure
    (ETAP), 93–95
Evaluation tools, document, 272
External hard drives, 220
External validity, for program evalua-
    tion, 363–64
Extrinsic rewards, 148–49, 153

F

Final user testing, 16, 343, 348, 351
Feedback loop, 166
Feedback, performance, 157
Flow charts, 87–89, 270, 303
  for CAI design, 320–21
Focus groups, 92
Formative evaluation, 15–16, 296,
    330–31, 353
Front-end analysis, 11–14, 25–36, 37,
    164, 212
Functional Analysis System Tech-
    nique (FAST), 97–98
Full-scale development, 16, 343
  lecture, 344–45
  workshops, 344–45
Functional design specifications, 11–
    14, 36, 129, 164–76, 177, 246,
    248, 249, 256, 259, 296, 300
  list of contents, 168
  posttraining goals and objectives,
    167, 170–72
  program goal, 167, 168–69
  system constraints, 173–74
  system performance requirements,
    167, 172–73
  writing, 13, 129, 164–76
Functional flow diagram, 108

G

Generalization, in learning, 139–40
Generating design concept, 14, 246–56,
    259
  product development plan, 253
  project approval, gaining, 253–54,
    259
Generative instructional presentation,
    201
Generative learning, 136, 155
Goals, instructional (*See* Instructional
    goals)
Goals, training program, 167–69 (*See
    also* Functional design specifi-
    cations)
GOMS, 86, 95–96, 101
Grapevine project, 226
Graphical user interface (GUI), 299
Group interviews, 75, 92

H

Hawthorne effect, 364
Hierarchical networks, 85–87 (*See
    also* Concept networks)
Hierarchical task analysis, 93, 94
High-fidelity simulations, 203, 205
Human factors (*See also* Ergonomic
    system design)
  task analysis for, 69
Human relations training, 210
HyperCard, 235–38
Hypermedia, 198, 223, 225–29,
    319
Hypertext, 198, 223, 225–29, 319
  design, 319

I

ICAI (*See* Intelligent computer-
    assisted instruction)

Identical elements theory of transfer, 142

Identical productions theory of transfer, 142

Inert knowledge, 138

Information access, computer-based, 223, 226–27, 319

Information-processing model, 130–35 (*See also* Memory)

Integrated information systems, 4, 5

Interface prototypes, 320–22

Initial system development, 15

Inquiry learning, 178–79, 186–87

Instructional content, 303, 306–9

Instructional design model (*See* Design model)
    traditional, 10–11, 295–97

Instructional design (*See also* Instructional materials and Design model)
    content, 303
    feedback loop, 166
    list of guidelines, 153–58
    principles, 153–58
    strategies, 153–58
    techniques, 177–212
    text philosophy, 19

Instructional goals, 164, 165–66

Instructional materials:
    advance organizers, 145, 155, 268
    content, 303, 306–9, 318
    designing, 302–9
    sequencing, 144–45, 303–6, 318
    sequencing for video, 314–15
    zoom metaphor, 144

Instructional objectives, 164, 165–66, 170–72, 182

Instructional sequence, 144–45, 303–6

Instructional strategies, 129
    advance organizers, 145, 155, 268
    list of guidelines, 153–58
    motivational strategies, 151–53
    part task vs. whole task training, 145–46

    sequencing materials, 144–45
    writing as specifications, 172–73

Instructional system design, ergonomic, 10, 297

Instructional techniques, 177–212 (*See also* Instructional strategies)
    audiovisual techniques, 178–79, 182–84
    combining methods, 207
    computer-assisted instruction, 178–79, 194–96, 197–201
    inquiry learning, 178–79, 186–87
    intelligent computer-assisted instruction, 201–3
    interpersonal skills training, 203, 208–11
    leadership training, 203, 208–10
    lecture, 178–79, 184–86, 311–12
    on-the-job training, 3, 178–79, 205–7
    programmed instruction, 178–79, 189–97
    simulations, 203–5
    team training, 211–12
    text, 178–79, 180–82
    tutoring, 178–79, 188–89
    videotape, 178—79, 182–84

Intelligent computer-assisted instruction, 3, 198, 201–3

Intelligent tutoring system, 198 (*See also* Intelligent computer-assisted instruction)

ITS (*See* Intelligent tutoring system)

Intelligent support systems, 5

Interactive multimedia, 3, 225 (*See also* Multimedia)

Interface design, 7 (*See also* Computer interface design)

Interface redesign, 8

Internal validity:
    for program evaluation 360–63
    threats to, 361–63

Interpersonal skills training, 203, 208–11

Interviews, 32–33, 73–76, 101–05,
    368–70
  group, 75–76, 91–92
  for program evaluation, 368–70
  structured, 74–75, 101–05
  unstructured, 73–74, 101
Intrinsic rewards, 148–49, 153
Iterative design processes, 18–19

J

Job aids, 4, 5, 8, 120–25, 181, 262–
    80, 313–14, 348
  benchmark tasks, 340
  content, 277–80
  design, 262–72, 272–80, 313–14
  example, 123–25
  final development, 348
  final user testing, 351
  format, 273–74
  indications for, 121–22
  organization, 274–77
  program evaluation, 378–79
  user testing, 340
Job redesign, 8, 119–20

K

Knowledge (*See also* Declarative
    knowledge and Procedural/
    skill knowledge)
  acquiring, 135–41
  application, 137–39
  conceptual, 44
  declarative, 39–44
  definition, 37
  explicit, 40
  explicit procedural, 44
  inert, 138
  measuring, 62–110
  motor skills, 39
  procedural/skill, 39–46

  rule, 44
  types, 38–51, 166
Knowledge-based processing, 46
Knowledge, skills, abilities, 50,
    66–67

L

Laserdisc, 221
Learned capabilities, Gagne, 48–49
Learner control, of computer, 197,
    200, 318
Learning, 37
  definition, 37
  generative learning activities, 136
  assessing, 354, 360, 365–68, 371,
    376
Learning organization, 2
Learning styles, 111
Lecture, 178–79, 184–86, 250–51,
    311–12, 344–45
  user testing, 335–36, 344–45
Levels of processing, Rasmussen, 46–
    47
Liability, professional (*See* Profes-
    sional liability)
Long-term memory, 130, 134–35 (*See
    also* Memory)

M

Matrices, for task analysis, 82
Memory:
  associative memory, 134–35
  capacity, 131–135
  components, 130
  decay, 133
  enhancing, 136–37, 154–56
  information-processing model of,
    130–35
  long-term memory, 130, 134–35
  measuring, 332

patterns of activation, 134
rehearsal, 133, 134
retrieval, 134–35, 138, 154–56
search processes, 134–35
sensory memory, 130, 131–32
short-term memory, 130
storage systems, 130
strengthening, 134–35
selection, 132
working memory, 130, 132–33
Mental models, 51–54, 126, 274, 319
application to training, 54
conceptual model, 53
definition, 51–52
domain mental models, 54
training mental models, 126, 319
types of knowledge and, 53
user's mental model, 53, 319
Methods, instructional, 177–212
(*See also* Instructional
techniques)
Microworlds, 186, 232
Minnesota Educational Computing
Consortium (MECC), 217
Mock-up, 294, 323
Modeling student learning, 202
Motivation, 146–53, 158
ARCS model, 147–49
ARCS strategies, 151–53
attribution theory, 149–51
causal attribution, 149–51
effects of training methods, 183–84,
185, 187, 188, 196, 199, 204,
209
extrinsic rewards, 148–49, 153
Motivation strategies, ARCS, 151–53
writing as specifications, 172–73
Multimedia, 3, 19, 218–23, 257–58
budget form, 254
CD-ROM, 221, 225
effectiveness, 225
for training, 222–23
input devices, 221
software, 222

storage devices, 220–21
technology, 219–22
intrinsic rewards, 148–49, 153

N

Negative transfer of training, 143
Need analysis, training, 13, 118–26
Needs assessment, 11–13, 17–18, 25
Nonexamples, 140–41, 155, 156, 268,
391

O

Objectives, posttraining (*See* Posttraining goals and objectives)
Object-oriented programming, 234
Observation, performance, 34, 80–81
One-group pretest-posttest design,
372–73
One-group posttest only design,
371–72
On-line encyclopedia, 180
On-the-job training, 3, 178–79, 205–
7, 252–53, 310–11
user testing, 335
Operational sequence diagram, 87
Optical storage, 221
Organizational analysis, 11, 13, 25–36
acquiring information, 32
evaluating ability, 35
information to acquire, 27–31
initiating circumstances, 26
interviews, 32
new job, 29–31
observation, 34
questionnaires, 33
rationale, 25
Organizational goals, 27
OSHA guidelines, 389–90
OSHA fines, 390
OTJ training (*See* On-the-job training)

P

Part-task training, 145–46
Pattern recognition, perceptual, 141
Perceptual pattern recognition, 141
Perceptual skills, 39, 45
Performance, 3, 6, 25–36, 55–57,
    118–19
  expert, 55–57
  measuring, 376–77
  novice vs. expert, 55–57, 137
Performance criteria, 165–66
Performance enhancement, 7–9, 119,
    125–26
  priorities for, 8
Performance requirements, system,
    172–73
Performance support, 8, 120–25,
    125–26
Performance support systems, 1, 3–5,
    8–9, 11–14, 16–17, 20, 28, 30,
    31, 38, 63, 118–26, 248, 259,
    261, 295, 298, 313, 391
  adaptive aiding, 5
  benchmark tasks, 340
  decision support system, 4, 5
  definition, 3
  electronic, 4
  example, 122–24
  final development, 348
  final user testing, 351
  intelligent support system, 5
  real-time, 126
  user testing, 340
Perseus project, 226
Person analysis, 11
Photo-CD, 221
PLATO, 216–17
Position Analysis Questionnaire, 91
Positive transfer of training, 143
Posttraining goals and objectives, 167,
    170–72
Posttest only control group design,
    373–75

Predictive displays, 126
Pretest-posttest control group design,
    374–75
Pretest/posttest evaluation design, 359
Problem space, analysis, 98
Proceduralization, 137
Procedure/skill knowledge:
  acquiring, 129, 139–41, 156–58
  definition, 41
  discrimination, 140–41
  generalization, 139–40
  instruction of, 129, 223–24, 224–25
  effects of practice, 141, 223–24
  proceduralization, 137
  production rules, 42, 137–38
  types of, 39, 44–46
  transfer, 141–43
Product development plan, 253, 259
Production rules, 42, 137–38, 139–41
    (*See also* Procedure/skill
    knowledge)
Professional liability, 8, 20–21, 166,
    175, 357, 382–96
  guidelines to protect against, 390–94
Program evaluation, 353–79
  choosing the context, 377–78
  choosing criteria, 365
  construct validity, 367
  control group, 359
  criteria, 354
  definition, 353–55
  goals, 359–64
  importance, 355–57, 392
  internal validity, 360–63
  for job aids, 378–79
  objective measures, 366–68
  research designs, 371–75
  secondary results, 365
  subjective measures, 366–68
  types of criteria, 365–66
  variables measured 354
Program goal, 167, 168–69
Programmed instruction, 178–79,
    189–97, 224

branching, 192–97
  examples, 191, 193, 195
  linear, 190–92
  traditional, 189–90
Project approval, gaining, 253–54,
    259, 300
Protocol analysis (*See* Verbal protocol
    analysis)
Prototyping, 15, 293–326
  advantages, 302
  intelligent computer-assisted in-
    struction, 301
  methods, 301
  origins, 293–95
  rapid prototyping, 295
  tools, 299
  Wizard of Oz, 301
Psychomotor skills, 39, 46

Q

Questionnaires, 33–34, 77–78, 332,
    336–37, 367–70
Question probes, 74, 81, 101–5, 340

R

Rating scales, 368
Reaction, measuring, 375–76
Reactive effects, in program evalua-
    tion, 364
Redesign, system, 119–20, 123
  combined with training, 125
Rehearsal:
  memory, 133, 134
  elaborative rehearsal, 134
  rote rehearsal, 133
Requirements, system performance,
    172–73
Resource analysis, training, 13, 118,
    126–28
Resources, cognitive, 145
Retrieval, memory, 134–35

Role-playing, 208, 209–10
Rote rehearsal, 133
Rule-based processing, 47

S

Schemas, 53
Script, 315, 322, 324
Scripting language, 236–38
Secondary results, 365–66, 371, 377
Semantic networks, 85–87 (*See also*
    Concept networks)
  computer-based, 85
Sensory memory, 130, 131–32
Sequencing instructional materials,
    144–45, 303–6
Signal word, 394
Similarity ratings, 77
Simulations, 3, 178–79, 186, 198,
    219, 203–5, 208–12, 232, 241,
    251–52, 301, 306, 310, 318–
    20, 323
  computer-based, 203
  design concept for, 251
  high-fidelity, 203, 205
  interpersonal skills, 203
  leadership, 203
  prototyping, 301, 323
  strengths, 204
  transfusion medicine tutor, 232
  weaknesses, 204–5
Skill-based processing, 47
Skill knowledge (*See* Procedural/skill
    knowledge)
Sorting, for task analysis, 76–77
Specs, 164
Specifications, functional (*See* Func-
    tional design specifications)
Static group comparison design, 373
Stage model of expertise, 55
Storyboard, 301, 316, 322, 324
Strageties, instructional (*See* Instruc-
    tional strategies)

Strength of memory associations, 134–35, 136

Structural networks, 83, 85 (*See also* Concept networks)

Subjective norm, 49

Summative evaluation, 295–96, 353–79

Supporting documents, 348–51

System design, 9–10, 119–20
    engineering design, 9
    ergonomic, 10, 119–20

System constraints, 173–74 (*See also* Design constraints)

System performance requirements, 167, 172–73

System redesign, 119–20
    example, 120

T

Task analysis:
    activity sampling, 106–8
    associative scaling algorithms, 105–6, 107
    choice of methods, 108–9
    cognitive task analysis, 67–69
    conceptual graph analysis, 99–105
    controls and displays analysis, 93
    critical incident technique, 96–97
    data collection methods, 70–81
    data representation methods, 81–90
    Delphi method, 91–92
    example for training, 257
    Extended Task Analysis Procedure (ETAP), 93–95
    flow charts, 87–89
    focus groups, 92
    Functional Analysis System Technique (FAST), 97–98
    functional flow diagram, 108
    GOMS, 86, 95–96
    hierarchical task analysis, 93, 94
    human factors task analysis, 69
    interviews, 73–76
    list of methods, 71
    matrices, 82
    Means and Gott cognitive task analysis, 98–99
    networks, 83–84
    observation, 80–81
    Position Analysis Questionnaire, 91
    protocol analysis, 78–80
    questionnaires, 77–78
    similarity ratings, 77
    sorting, 76–77
    specific methods, 90–108
    structural network, 83
    task simulation, 81
    timeline charts, 89–90
    traditional task analysis, 63–67
    types of knowledge, and, 67
    verbal protocol analysis, 78–80

Task interface redesign, 8, 119–20

Task hierarchies, 63–66, 270, 303

Task lists, 63–66

Task simulation, 81

Task specification (*See* task analysis)

Task timeline, 255

Taskwork skills, 212

Taxonomy of instructional techniques, 178

Team performance, measurement, 212

Team training, 211–12

Teamwork skills, 212

Techniques, instructional, 177–212

Text, 178–79, 180–82, 250
    design, 262–80
    electronic text, 180

Text organization, 20, 269–69

Think-aloud protocols (*See* Verbal protocol analysis)

Three-stage model of expertise, 55–56

Timeline charts, 89–90

Total quality management, 10, 355

Trainee analysis, 11–13, 18, 62, 110–11, 174, 257, 273

Training (*See also* Instructional strate-

gies and Instructional techniques)
evaluation of, 157–58
methods, 177–212
part-task, 145–56
transfer of, 141–43
whole-task, 145–46
Training design priorities, 8
Training need analysis, 13, 118–26
Training program:
delivery needs, 126–28
development needs, 126–28
Training resource analysis, 13, 118, 126–28
Transfer of training, 141–43, 258
identical elements theory of transfer, 142
identical productions theory of transfer, 142
negative, 143
positive, 143
under stress, 391–92
Transfusion Medicine Tutor (TMT), 232, 251–52
Transparency, computer, 285
Troubleshooting, task analysis for, 98–99
Tutoring, 123, 178–79, 181, 184, 186–89, 194, 197–200, 222, 232–33, 251–52, 298, 310–11
user testing, 335
Typographical design principles, 266

U

Usability, 332–33, 337
assessing, 333, 337

User testing, 15–16, 166, 293–95, 297–302, 309–14, 320–26, 330–42
collecting data, 334
appropriate subjects, 334
variables measured, 332–34
User mental models, training, 54

V

Verbal protocol analysis, 78–80, 340
Videotape:
development, 314–17, 345–47
instructional method, 182–84
production script, 315
storyboard, 316
user testing, 337–38
Virtual reality, 3, 219

W

Warning labels, 388–89, 394–96
Warnings, 8, 383–84, 388–89, 394–96
Whole-task training, 145–46
Wizard of Oz, 301
Working memory, 130, 132–33
Workshops, 250–51, 301, 311–12, 332, 344–45
user testing, 335–36, 344–45
Writers' Workbench, 272
Writing functional specifications, 13, 129, 164–76 (See also Functional design specifications)
Writing script, 315, 322, 324